PERGAMON INTERNATIONAL LIBRA
of Science, Technology, Engineering and Social Studies
The 1000-volume original paperback library in aid of education, industrial training and the enjoyment of leisure
Publisher: Robert Maxwell, M.C.

Advanced Engineering Thermodynamics

SECOND EDITION

THERMODYNAMICS AND FLUID MECHANICS SERIES
GENERAL EDITOR: W. A. WOODS

Other Titles of Interest in the Pergamon International Library

Bradshaw	Experimental Fluid Mechanics, 2nd Edition
Bradshaw	An Introduction to Turbulence and its Measurement
Buckingham	The Laws and Applications of Thermodynamics
Daneshyar	One-Dimensional Compressible Flow
Dixon	Fluid Mechanics, Thermodynamics of Turbomachinery, 2nd Edition
Dixon	Worked Examples in Turbomachinery (Fluid Mechanics and Thermodynamics)
Haywood	Analysis of Engineering Cycles, 2nd Edition
Morrill	An Introduction to Equilibrium Thermodynamics
Peerless	Basic Fluid Mechanics

Advanced Engineering Thermodynamics

SECOND EDITION

BY

ROWLAND S. BENSON

Professor of Mechanical Engineering, University of Manchester Institute of Science and Technology

PERGAMON PRESS

OXFORD · NEW YORK · TORONTO · SYDNEY · PARIS · FRANKFURT

U.K.	Pergamon Press Ltd., Headington Hill Hall, Oxford OX3 0BW, England
U.S.A.	Pergamon Press Inc., Maxwell House, Fairview Park, Elmsford, New York 10523, U.S.A.
CANADA	Pergamon of Canada Ltd., 75 The East Mall, Toronto, Ontario, Canada
AUSTRALIA	Pergamon Press (Aust.) Pty. Ltd., 19a Boundary Street, Rushcutters Bay, N.S.W. 2011, Australia
FRANCE	Pergamon Press SARL, 24 rue des Ecoles, 75240 Paris, Cedex 05, France
WEST GERMANY	Pergamon Press GmbH, 6242 Kronberg-Taunus, Pferdstrasse 1, Frankfurt-am-Main, West Germany

First edition 1967

Second edition 1977

Library of Congress Cataloging in Publication Data

Benson, Rowland S
Advanced engineering thermodynamics. 2nd Edition
(Thermodynamics and fluid mechanics series)
Includes indexes.
1. Thermodynamics. I. Title.
TJ265.B49 1977 621.4'021 76-50049
ISBN 0-08-020719-7
ISBN 0-08-020718-9 pbk.

In order to make this volume available as economically and rapidly as possible the author's typescript has been reproduced in its original form. This method unfortunately has its typographical limitations but it is hoped that they in no way distract the reader.

Printed in Great Britain by A. Wheaton & Co., Exeter

CONTENTS

PREFACE TO SECOND EDITION

The basic thermodynamics covered in a final year honour course has not changed since the first edition so that the text is essentially the same. In view of the almost general use of the S.I. system in thermodynamics the tables in the text have been presented in these units as well as a number of worked examples and exercises. Since the publication of the first edition there has been an increase in interest in combustion-generated pollution. The thermodynamics of these processes can be handled by the methods outlined in Chapter 4; to illustrate the technique this chapter has been extended with an example of the rate controlled nitric oxide reactions.

A number of readers have used the data in Table A1 to generate their own thermodynamic properties, in particular in cycle and combustion calculations in computer programs. To assist new readers these polynomial coefficients have been presented in a form suitable for computing and a few simple algorithms are presented in the text. Because these coefficients have been referred to in published papers with their original symbols these have been retained in the headings of Table A1 in addition to the new symbols.

The opportunity has been taken to correct some minor printing errors in the first edition. I wish to thank all those who notified me of them. Finally, I wish to thank my former secretary, Mrs. J.A. Munro, who typed the master script upon which the printed copy is based.

MANCHESTER, April 1976.

PREFACE TO FIRST EDITION

It behoves the author of a new textbook on thermodynamics to state his case to the reader. The term "Advanced" in the title is relative and refers primarily to the grade of the text, which is directed to the undergraduate final year honours course in engineering. The changes in the first-year thermodynamic syllabus as a consequence of the introduction of the so-called "Keenan" approach has been followed by similar changes in the Applied Thermodynamics or Heat Engines course in the final year of study. Current practice is to offer separate honours courses in classical thermodynamics, heat transfer, internal-combustion engines, turbomachinery, power plants (cycles), gas dynamics and so forth. Classical thermodynamics forms the basis for the applied thermodynamics topics covered in engineering courses. The provision of a separate lecture course for this subject has many advantages over the old system in which, if the topic was discussed at all, it was included as a minor part in a series of lectures in applied thermodynamics.

The science of thermodynamics was founded by engineers in the nineteenth century; unfortunately they did not actively pursue its development and the major advances were made by chemists, physicists and mathematicians. The non-participation of engineers in the development of thermodynamics was reflected in engineering courses where, for nearly half a century, there was almost a standstill in the content and quality. The advances in engineering science during and since the war have called attention to the inadequacy of these courses and, in the United States, the chemists' and physicists' approach to thermodynamics was introduced into engineering degrees. In the last ten years or so this has been followed in the United Kingdom. The first year course has been adequately covered by textbooks using the new approach, but the final year has been dependent on textbooks written primarily for chemists or physicists. There is a need for a short textbook covering those aspects of classical thermodynamics appropriate to a final-year engineering course and which would form a basis for the applied subjects. It was considered that, whilst such a textbook might whet the appetite of some students to a broader study of the subject in the standard texts by Keenan, Roberts and Miller, Zemansky, Denbigh and others, at the same time the book should be complete by itself for those students whose bias was directed to the more practical application of thermodynamics.

No claim is made to the originality of the presentation of the material, although there is perhaps a different emphasis in the development of certain topics from that in the standard texts referred to above. In particular, the formulation of the thermodynamic data for gas mixtures has been devised in such a manner that the various expressions can be used directly in computer programming. This should be of value for cycle analysis. The order of presentation is, in general, different from current texts and courses. It is considered that a discussion on equilibrium is necessary before the general development of the subject - this topic is scarcely discussed in engineering courses; Chapter 1 is therefore concerned solely with equilibrium. The material in Chapter 2 on the general thermodynamic relations is a happy hunting ground for students with mathematical facility and is fairly straightforward. In Chapter 3 the approach follows elementary chemical thermodynamics with a brief discussion of the quantum theory; more advanced chemical thermodynamics is described in Chapter 4 with the introduction of the chemical potential. Certain new material is added at the end of Chapter 4, including a proof of Lighthill's ideal dissociating gas state equation as well as a discussion of "frozen" flow. The exercises in Chapters 3 and 4 have been devised to include a number of practical applications of classical thermodynamics to engineering subjects. The interest of engineers in direct conversion of heat to electricity and cryogenics has prompted the inclusion of some of this material in Chapter 5 - this has been covered by chemists and physicists for some years. The chapter ends with some irreversible thermodynamics, a new topic of importance to engineers in thermoelectricity and similar fields.

The range of units used by engineers is extensive and, whilst it would have been desirable to keep to one set of units, it was difficult to decide which was the best. The básic relations have, therefore, been developed without units. For numerical examples the units are given both in the text and in the exercises. The tables at the end of the text have been prepared using the latest thermodynamic data kindly provided by Dr. A. Russo of Cornell Aeronautical Laboratory, Buffalo, N.Y. The programmes for these tables were prepared under my direction by Mr. W.G. Cartwright and Mr. R. Dale of my department; the calculations were carried out on the Manchester University Atlas Computer. I would like to thank all concerned.

The manuscript was read by Mr. W.G. Cartwright to whom I wish to express my grateful thanks for his comments and suggestions. I would also like to thank the editor of the series, Professor J.H. Horlock, for his helpful comments. The body of the material has been included in courses in mechanical engineering in the Universities of Liverpool and Manchester, whilst for the past three years the original draft of this text has been used by third-year honours students in the Department of Mechanical Engineering, Faculty of Technology, University of Manchester. I would be grateful for any comments and criticisms.

Acknowledgement is made to the Senates of the Universities

of Manchester and Liverpool for permission to reproduce examination questions but responsibility for the solutions given is entirely the author's.

I would like to thank Mrs. M.S. Ehren and Mrs. J.A. Munro for the preparation of the typed manuscript and Mr. E. Clough for the drawings.

Finally, my grateful thanks are given to my wife and family for their patience and forbearance whilst I spent many evenings and week-ends preparing this book.

MANCHESTER, June 1966.

EDITORIAL INTRODUCTION

The books in the Thermodynamics and Fluid Mechanics Division of the Commonwealth Library have been planned as a series. They cover those subjects in thermodynamics and fluid mechanics that are normally taught to mechanical engineering students in a three-year undergraduate course.

Although there will be some cross-reference to other books in the division, each volume will be self-contained. Lecturers will therefore be able to recommend to their students a volume covering the particular course which they are teaching. A student will be able to purchase a short, low-price, soft-cover book containing material which is relevant to his immediate needs, rather than a large volume in which most of the contents are outside his current field of study.

The book meets the immediate requirements of the mechanical engineering student in his undergraduate course, and of other engineering students taking courses in thermodynamics and fluid mechanics.

CHAPTER 1

EQUILIBRIUM OF THERMODYNAMIC SYSTEMS

Introduction to concept of equilibrium, maximum work of thermodynamic systems, development of Gibbs and Helmholtz functions. Equilibrium of thermodynamic systems, conditions for stability and spontaneous change.

Notation

a	constant in Van der Waals' equation
b	constant in Van der Waals' equation
E	internal energy
F	Helmholtz free energy function
g	specific Gibbs free energy function
G	Gibbs free energy function
H	enthalpy
m	mass
m_l	mass of liquid
m_v	mass of vapour
M	number of mols
p	pressure
Q	heat transfer to system
R_{mol}	universal gas constant
S	entropy
t	time
T	temperature absolute
U	internal energy in the absence of motion, gravity, etc.
V	volume
W	work from system
W'	shaft work

Concept of Equilibrium, Spontaneous Change and Criterion of Stability

In the study of engineering situations the question of stability of systems is of major importance. We may be concerned with spontaneous changes in the state of a system or at the other extreme with the states of equilibrium. In thermodynamics we are interested in similar situations. For example, a vessel may contain two gases separated by a partition. If the partition is removed we may wish to establish whether the gases will remain separated or whether they will mix or if they mix under what conditions there will be no further change in state. Alternatively, the gases may react with one another and we may require to know whether a reaction is possible or not and if so what is the final composition of the gases. Or we may require to know whether some of the initial reactants are still present when the system settles to a steady state.

Before we start our study we must define our terms. We say that a system is in a state of equilibrium† if no active unbalanced tendency towards a change in state exists. The equilibrium may be stable, unstable, neutral or metastable. We will be primarily concerned with criteria for stable equilibrium. By stable equilibrium we mean that if we slightly perturb the state of the system it will return to its original state. A system is not in equilibrium if there is a spontaneous change in the state. It is important how we define the system. For example, let us consider two bars of metal, one at a uniform temperature T_1 and the other at a uniform temperature T_2. If we enclose both bars in heat insulators then experience tells us that they will remain at the same temperature. Hence, the separate bars, or the two systems enclosing the bars, are in thermal equilibrium. If we now bring the two bars together and place them in contact, at the same time removing the insulating walls on the contact faces only, we know from experience that the temperature of one bar will drop and the other increase until they are both once again in thermal equilibrium, this time with each other. We will observe that on contact there was a spontaneous change in the state of both bars. It will be noticed, however, that in order to produce this change in state the boundaries defining the system were altered. Thus in defining the conditions for equilibrium we must also state at the same time the constraints on the system.

In the simple example given above we can measure directly

†J.H. Keenan, Thermodynamics, J. Wiley & Sons, New York, 1940. S.R. Montgomery, The Second Law of Thermodynamics, Pergamon Press, 1965.

the condition for equilibrium by means of a thermometer. Of course it is important to remember that the measurement of temperature is dependent on the concept of equilibrium (in this case thermal equilibrium) and in this experiment we have, strictly speaking, only three known temperatures, the two temperatures corresponding to the initial temperatures of each bar and the third corresponding to the final temperature of the two bars. In many problems the equilibrium criterion is not so readily obtained. We must therefore look for some general criterion.

This criterion we can obtain from the second law of thermodynamics. In the sixth corollary† to the second law it is stated that the entropy of an isolated system either increases or in the limit remains constant. Now an isolated system is a system of constant internal energy. If such a system eventually reaches the maximum entropy level then, since the entropy cannot decrease, any disturbance to the system will only cause the state to return to the maximum entropy. Let us examine this by means of an example.

Any system plus surroundings can be considered to be an isolated system (Fig. 1.1). Let us consider a system B at uniform temperature T_2 immersed in the surroundings A also at a uniform temperature T_1 (as shown in Fig. 1.1). We will

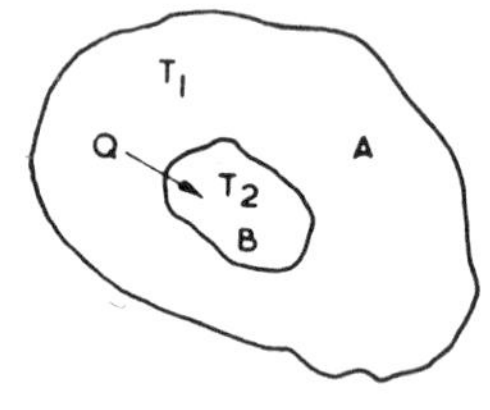

FIG. 1.1.

stipulate the following constraints, namely that the system and surroundings have fixed volume invariant with temperature. If the temperature T_1 is greater than T_2, there will be a spontaneous heat interaction at the boundary of the system and surroundings. Let dQ be the quantity of heat transferred in a small interval of time dt. If both system and surroundings are maintained at the uniform temperature T_2 and T_1 respectively, then the entropy changes are:

†J.H. Keenan, Thermodynamics, J. Wiley & Sons, New York, 1940. S.R. Montgomery, The Second Law of Thermodynamics, Pergamon Press, 1965.

For surroundings A, $$dS_A = -\frac{dQ}{T_1}. \qquad (1.1)$$

For system B, $$dS_B = +\frac{dQ}{T_2}. \qquad (1.2)$$

Since there are no changes in volume in A or B, the first law of thermodynamics states that the heat transferred equals the change in internal energy, that is $dQ = dE$. For the isolated system, comprising system B and surroundings A, the change in internal energy dE is zero or

$$dE = dE_A + dE_B = 0$$

and the total change in entropy will be

$$dS = dS_A + dS_B \qquad (1.3)$$

Substituting (1.1) and (1.2) into (1.3)

$$dS = dQ\left(\frac{1}{T_2} - \frac{1}{T_1}\right) \qquad (1.4)$$

which after rearrangement becomes

$$dS = \frac{dQ}{T_1 T_2}(T_1 - T_2). \qquad (1.5)$$

Since $T_1 > T_2$, then $dS > 0$.

If the entropy change dS takes place in time dt then equation (1.5) may be written in the form

$$\dot{S} = \frac{\dot{Q}}{T_1 T_2}(T_1 - T_2) \qquad (1.6)$$

and $$\dot{S} > 0 \qquad (1.7)$$

(using Newton's notation for time derivatives).

Hence the isolated system is producing entropy, or we can say the entropy is increasing. If the process is such that we maintain uniform, but different, temperatures throughout A and B then as T_2 approaches T_1 the rate of entropy production is reduced.

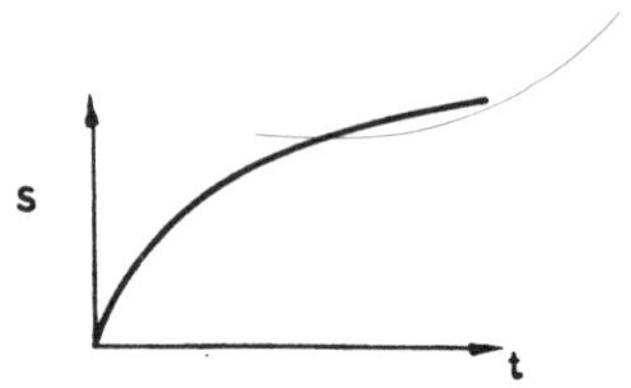

FIG. 1.2.

We might expect a curve of the type shown in Fig. 1.2. In the limit when $T_1 = T_2 = T$ there will be no further increase in entropy.

Let us assume that when the temperatures of system B and surroundings A are the same (= T) we can cause the temperature of A to rise dT_1 with a heat flow dQ from B at the same time producing a drop dT_2 in the temperature of B. To simplify the analysis we will assume that the heat capacities of the system and surroundings are equal, then $dT_1 = dT_2 = dT$. The change in entropies will be

$$dS_A = \frac{dQ}{T+dT} \tag{1.8}$$

$$dS_B = \frac{-dQ}{T-dT} \tag{1.9}$$

the <u>total</u> change in entropy for the <u>isolated</u> system

$$dS = dS_A + dS_B = dQ\left\{\frac{1}{T+dT} - \frac{1}{T-dT}\right\} \tag{1.10}$$

$$dS = \left\{\frac{dQ}{T^2-dT^2}\right\}(-2dT) = -2dQ\,\frac{dT}{T^2} \tag{1.11}$$

to the first-order small quantities.

In this case there is a <u>decrease</u> in entropy. The same results would have been obtained if we had raised the temperature of the surroundings. The point of maximum entropy has clearly been obtained when both system and surroundings are at the same temperature and there is no change in volume.†

One other important point can be observed in this example. Since from the second law the entropy of an isolated system can only increase, it is clear that the variation in temperature dT discussed above cannot take place; in this case the volume is fixed (one of the initial constraints) and no other variations in state are possible in the system. The isolated system must therefore be in stable equilibrium. This follows since our perturbation of the state of the system (in this case the temperature) produces an absurd situation (decrease in entropy) and the system must immediately revert to its original state, which is the stable equilibrium state. We may observe the following three points from the systems we have examined but which apply quite generally:

(1) If the properties of an isolated system change spontaneously there is an increase in entropy of the system.

(2) When the entropy of an isolated system is at a maximum the system is in equilibrium.

(3) If for all the possible variations in state of the isolated system there is a negative change in entropy then the system is in stable equilibrium.

In the last case we know that the possible variation cannot take place, but this we discover after we have made the test for the stability of the system. We note of course the constraints which we cannot vary (in this case the volume).

We may generalize and summarize therefore for an isolated system:††

(i) $dS)_E > 0$ Spontaneous change. (1.12)

(ii) $dS)_E = 0$ Equilibrium. (1.13)

(iii) $\Delta S)_E < 0$ Criterion of stability. (1.14)

---

†This method of examining a function is frequently used in mathematics to test for a maximum. In this case dS is negative both sides of the maximum value of S.

††Where we use Keenan's notation $\Delta S)_E$ meaning variation in entropy at constant E. The E outside the half bracket indicates constant E. The Δ indicates an increment in S to any order of small quantities.

Equilibrium of System

In practice it is more convenient to formulate the above criteria for the system B only. Other state functions are used for this purpose.

Let us consider a system immersed in surroundings at constant temperature T. We impose a constraint on the system that the heat interactions can only occur at temperature T. Initially the system is at this temperature. Some process then occurs within the system in which heat and work interactions may take place. Finally the system is in equilibrium at temperature T and the entropy of the system and surroundings is at a maximum.

From the second law of thermodynamics we have for system plus surroundings (isolated system)

$$(dS)_{system} + (dS)_{surroundings} \geqslant 0 \qquad (1.15)$$

and for the surroundings†

$$(dS)_{surroundings} = -\frac{dQ}{T} \qquad (1.16)$$

hence

$$(dS)_{system} - \frac{dQ}{T} \geqslant 0 \qquad (1.17)$$

Integrating between the initial and final states

$$T(S_2-S_1) \geqslant Q \quad \text{for the system} \qquad (1.18)$$

where Q is the heat transferred to the system.

Now from the First Law of Thermodynamics in the absence of motion, gravity, electricity, magnetism and capillarity

$$dQ - dW = dU \qquad (1.19)$$

hence

$$Q - W = U_2-U_1$$

for the system (1.12) and

$$W = Q - (U_2-U_1). \qquad (1.20)$$

---

†The negative sign is used since we are examining the system, heat flow to the system is +dQ, heat flow from the surroundings is -dQ.

Substituting for Q from (1.18)

$$W \leq T(S_2-S_1)-(U_2-U_1). \qquad (1.21)$$

Let

$$F = U - TS$$

then

$$W \leq (F_1-F_2) \qquad (1.22)$$

The function F = U-TS is called the Helmholtz Potential or Helmholtz Free Energy Function or simply Helmholtz Function. Since U, T, S are all properties, F is a property and has the same units as U.

The maximum work obtained in the above process is equal to the decrease in the Helmholtz potential.

Now since the decrease in Helmholtz potential of the system is associated with an increase in entropy of the system and surroundings then the minimum value of the potential corresponds to the equilibrium condition. Thus a given system of prescribed volume and temperature moves to the equilibrium state when it produces work and the criterion therefore for spontaneous change for a closed system† is $dF < 0$.

A decrease in entropy will correspond to an increase in F; hence the criterion for stability is $\Delta F)_T > 0$. It is interesting to note that the latter criterion corresponds to work being done on the system.

For a system of constant volume in which W = 0 then

$$F_1 \geq F_2. \qquad (1.23)$$

For reversible processes $F_1 = F_2$; for all other processes there is a decrease in Helmholtz potential and as before, equilibrium is reached at the minimum value of F.

If we change one of the constraints to the system another useful criterion can be obtained. Let us assume that the system is of variable volume but at constant pressure p and temperature T. There will be work done by the system to maintain the constant pressure. If we let W' be the shaft work then

---

†We use the term closed system since we use the first law in the form shown in (1.19).

$$W = W' + p(V_2 - V_1) \tag{1.24}$$

substitution in equation (1.20)

$$W' + p(V_2 - V_1) = Q - (U_2 - U_1) \tag{1.25}$$

or

$$W' + p(V_2 - V_1) \leq T(S_2 - S_1) - (U_2 - U_1) \tag{1.26}$$

hence

$$W' \leq T(S_2 - S_1) - (U_2 - U_1) - p(V_2 - V_1) \tag{1.27}$$

or

$$W' \leq (G_1 - G_2) \tag{1.28}$$

where

$$G = U + pV - TS = H - TS. \tag{1.29}$$

The function G is called the Gibbs Potential or Gibbs Free Energy Function or simply Gibbs Function and is an extensive property of the system.

As before we can summarize the equilibrium condition for the constraints of constant pressure and temperature:

(1) $dG)_{p,T} < 0$ Spontaneous change. (1.30)

(2) $dG)_{p,T} = 0$ Equilibrium. (1.31)

(3) $G)_{p,T} > 0$ Criterion of stability. (1.32)

The maximum useful work which can be performed by the system is equal to the decrease in the Gibbs function. The minimum value of the Gibbs function corresponds to the equilibrium condition. We should notice that for spontaneous change and equilibrium the differentials dS, dF or dG strictly apply to the rates of change of S, F or G with respect to some variable x which may be the concentration or time or any other quantity.

Equilibrium of System Obeying Van der Waals' Equation of State

To illustrate the above procedures we will examine the equilibrium criterion for a perfect substance.

All gases can exist in the liquid phase and some in the solid phase, indeed the "ideal" gas law

$$pV = MR_{mol}T \qquad (1.33)$$

where M is the number of mols and R_{mol} the universal gas constant has only limited application mainly at low pressures and moderate temperatures. The deviations from the law are of practical importance in liquefaction processes, these will be discussed later. The "ideal" gas equation can be replaced by the so-called Van der Waals' equation, which although not corresponding to any real substances is a useful approximation and enables us to distinguish between the liquid and gas phases.

One form of Van der Waals' equation is

$$\left(p + \frac{a}{V^2}\right)(V-b) = MR_{mol}T \qquad (1.34)$$

where a and b are constants. Typical isothermals (constant temperature) are shown in Fig. 1.3.

The slope $(\partial p/\partial V)_T$ of the curves (I) and (II) is always negative; that is, if we compress the gas the pressure will increase and if we expand the gas the pressure will decrease. The slope of curves (IV) and (V) may be positive, negative or zero. For the positive slope a <u>decrease</u> in volume produces a <u>decrease</u> in pressure and an <u>increase</u> in volume an <u>increase</u> in <u>pressure</u>. This situation is obviously unstable; our physical reasoning will tell us so, but we can also test this by thermodynamic analysis in the following manner.

The procedure will be to determine the variation in the Gibbs function along an isotherm. The reason we select the Gibbs function will be seen from inspection of the curves (IV) and (V) in Fig. 1.3. For certain pressures there are three possible volumes. Thus the stability criterion we select corresponds to constant T and p, and we determine the magnitude of G at these parts of the curve.

From the second law of thermodynamics we can write the following expression for a pure substance in the absence of motion, gravity, electricity, magnetism and capillarity:

$$T\,dS = dU + p\,dV, \qquad (1.35)$$

capital letters refer to extensive properties.† Equation (1.29) for the Gibbs function is

†Except T the absolute temperature which is an intensive property.

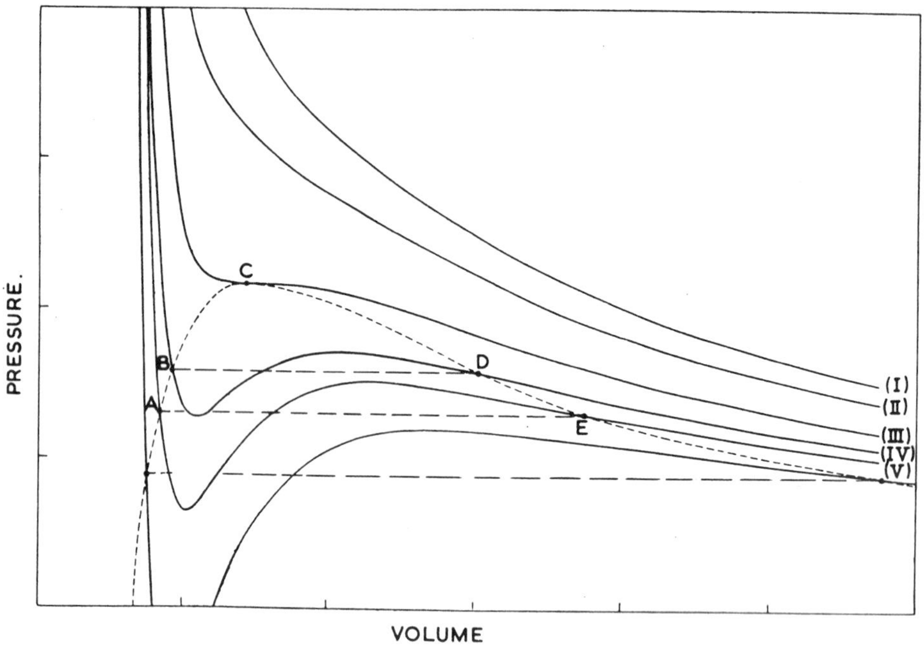

FIG. 1.3.

$$G = U + pV - TS \tag{1.29}$$

which in differential form becomes

$$dG = dU + p\,dV + V\,dp - T\,dS - S\,dT. \tag{1.36}$$

Inserting equation (1.35) into (1.36) and simplifying we obtain

$$dG = V\,dp - S\,dT. \tag{1.37}$$

Along an isotherm (dT = 0) we have

$$\left(dG\right)_T = V\,dp \tag{1.38}$$

or

$$\left(\frac{\partial G}{\partial p}\right)_T = V. \tag{1.39}$$

In Fig. 1.4 (a) (upper graph), the volume is plotted against the pressure for a curve of type (IV) for a Van der Waals gas. The change in Gibbs function G along the isotherm is the area under this curve. We can draw therefore a G-p graph by examining the V-p curve using equation (1.38). We should note that since the volume of the substance is always positive the slope of the G-p graph will be positive.

Integrating equation (1.38) along an isotherm

$$G = \int_A V\,dp + G_a. \tag{1.40}$$

Referring now to Fig. 1.4 (a) the point corresponding to A in the V-p diagram is a in the G-p diagram. Along the V-p curve from A to B the Gibbs function increases to G_b. At B the V-p curve changes direction, and there is a decrease in the net area under the curve until the point C is reached. The slope of the G-p curve is always positive (and equal to the volume at any point from (1.39)); the point b is therefore a discontinuity in this curve. Since at any pressure p the volume at a point on bc is less than the volume at a point on ab, corresponding to the same pressure, the slope of bc will be less than the slope of ab at this pressure, hence the curve bc will lie above ab and the general shape of abc will have the form shown in Fig. 1.4 (b). At C the V-p curve once again changes direction; there is now an increase in net area under the graph and consequently an increase in G. The point C is therefore a discontinuity in the G-p curve. From the same argument for the location of bc the curve cd can be shown to lie below bc.

If we examine the G-p diagram it is seen that the curves ab and cd cross at e. The point e corresponds to three volumes on the V-p diagram and this we will examine shortly. In the region between p_c and p_b the Gibbs function may have two or three values for each pressure. One of these values will correspond to the equilibrium state of the substance. The criterion for equilibrium along a line of constant pressure for an isotherm is, from equation (1.31), the minimum value of the Gibbs function. Thus the equilibrium state corresponds to the curve aed (since this curve is below ebc). Along the curves ce and eb the state is not thermodynamically stable, indeed it is metastable (i.e. a slight perturbation will change the state to the equilibrium state). The states corresponding to the curve bc are wholly unstable, this corresponds to BC in the V-p diagram, i.e. the condition of positive $(\partial V/\partial p)_T$ or positive $(\partial p/\partial V)_T$.

We now return to the point e, referring to the Fig. 1.4(b). The point e in the G-p diagram lies along ab and cd, and hence

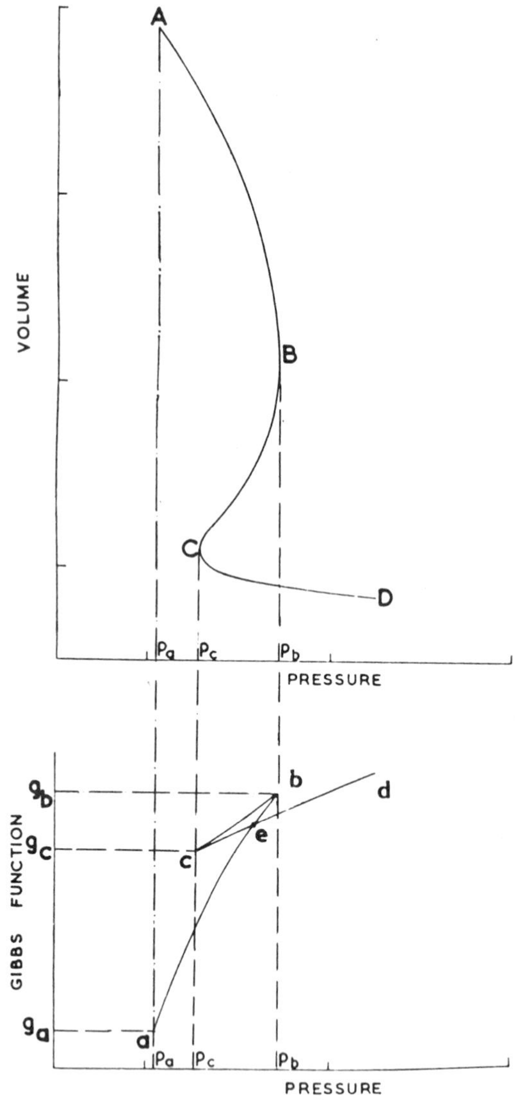
A
B
C
D
VOLUME
p_a
p_c
p_b
PRESSURE
g_b
g_c
g_a
b
d
e
c
a
GIBBS FUNCTION
p_a
p_c
p_b
PRESSURE

FIG. 1.4 (a).

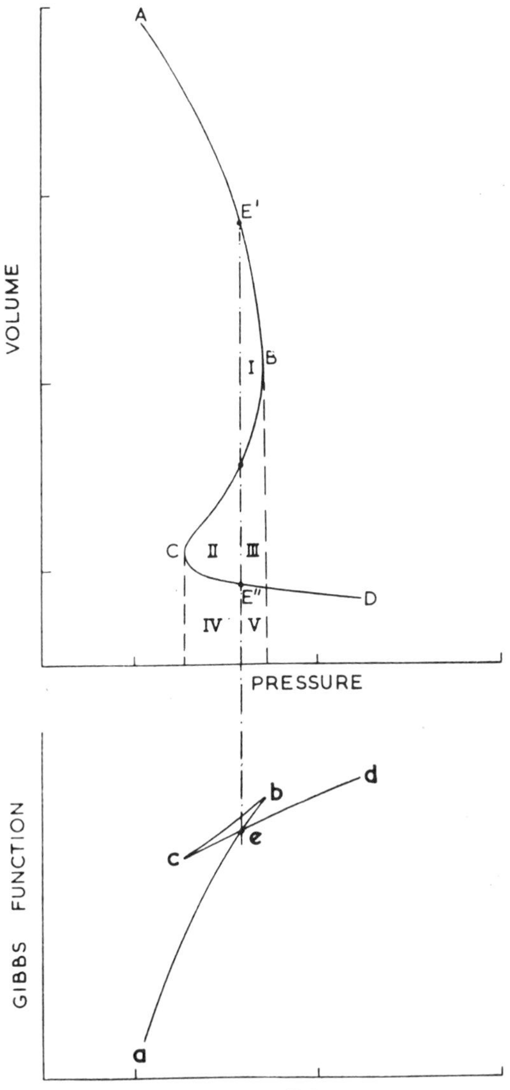
A
E'
VOLUME
I
B
C
II
III
E''
D
IV
V
PRESSURE
b
d
e
c
GIBBS FUNCTION
a
PRESSURE

FIG. 1.4 (b).

the corresponding points in the V-p diagram will be E' and E". These two points correspond to two phases of the substance. The state point E' corresponds to the vapour (or gaseous) phase and the state point E" corresponds to the liquid phase. During transition from one state to another both phases are present, coexisting in equilibrium. This can be demonstrated as follows.

Consider a system consisting of a fixed mass m containing liquid (mass m_l) and vapour (mass m_v) of the same pure substance. The p-V, G-p property diagrams for unit mass of the substance may be represented by Fig. 1.4 (b). If g_l is the specific Gibbs function for the liquid, and g_v the specific Gibbs function for the vapour, then, from Fig. 1.4 (b), it will be seen that at e

$$g_l = g_v = g.$$

The Gibbs function for the mixture of the phases is†

$$G = m_l g_l + m_v g_v = (m_l + m_v) g$$

$$G = mg.$$

Since

$$m = m_l + m_v = \text{constant}.$$

Let a small quantity of heat be exchanged between the system and the surroundings, the system being maintained at constant pressure and temperature. There will be no change in the specific Gibbs function for each phase. The change in Gibbs function for the system is therefore

$$dG)_{p,T} = dmg$$

but the mass m is constant hence dm = 0 and

$$dG)_{p,T} = 0.$$

The system is therefore in equilibrium. Thus a mixture of two phases at the same pressure and temperature can coexist in equilibrium. The quantity of each phase in the system will depend on the volume (or energy) of the mixed phase. The minimum volume corresponds to the wholly liquid state, the

†This follows from the Gibbs-Dalton laws (Chapter 3, p.83).

maximum volume corresponds to the wholly vapour state. The mixed phase lies between the two limits of volume E' and E". Since the Gibbs function is constant for the mixed phase the locus of the isotherm for this region will be the straight line joining E' and E". The true V-p diagram for the Van der Waals' gas is therefore AE'E"D, which corresponds to aed in the G-p diagram, the point e being multivalued in the V-p diagram.

To locate the points E" E' we notice that the Gibbs function at E" equals the Gibbs function at E'. From equation (1.40), therefore,

$$G_{E''} - G_{E'} = \int_{p_{E'}}^{p_{E''}} V\,dp = 0 \tag{1.41}$$

in terms of areas under the V-p curve

$$G_{E''} - G_{E'} = (I + III + V) - (II + III + IV + V) + (IV)$$

$$= I - II$$

or Area I = Area II. Hence the points E'E" are located by making the areas I and II equal. This is called Maxwell's construction.

Reverting to Fig. 1.3 it will be seen that the curves (IV) and (V) have a horizontal section. If we joint the points ABCD the locus of the pure phases is obtained. The line ABC corresponds to the liquid line and the line CDE to the vapour line. The region to the left of ABC is the liquid phase and within the broken curve the two-phase region.

If a substance in the vapour phase state point E' (Fig. 1.4 (b)) is rapidly compressed at constant temperature then, in the absence of nuclei such as dust particles, liquid droplets or ions, measurement of the temperature and pressure will show that the state of the substance lies on the curve E'B in the V-p diagram and eb in the G-p diagram. The Gibbs function is therefore above the minimum value. Such a state is metastable and the vapour is supersaturated. A small disturbance such as, say, a dust particle will cause a rapid change in state to e. Now since the compression has reduced the volume the new state will correspond to the mixed phase and will lie on the straight line E'E". The dust particle causes nucleation, that is the formation of liquid particles. Equilibrium is reached when the proportions of the liquid phase and vapour phase satisfy the minimum Gibbs function for the pressure, temperature and volume of the system.

Exercises

1. Determine the criteria for equilibrium for a thermally isolated system:

(a) at constant volume,
(b) at constant pressure.

2. Determine the criteria for isothermal equilibrium of a system:

(a) at constant volume,
(b) at constant pressure.

3. Show that if a liquid is in equilibrium with its own vapour and an inert gas in a closed vessel, then

$$\frac{dp_v}{dp} = \frac{\rho_v}{\rho_1}$$

where p_v is the partial pressure of the vapour, p is the total pressure, ρ_v is the density of the vapour, ρ_1 is the density of the liquid.

4. An incompressible liquid of specific volume v_1, is in equilibrium with its own vapour and an inert gas in a closed vessel. The vapour obeys the law

$$p(v-b) = R_{mol}T.$$

Show that

$$L_n \frac{p_v}{p_o} = \frac{1}{R_{mol}T} \quad (p-p_o)\ v_1-(p_v-p_o)b$$

where p_o is the vapour pressure when no inert gas is present, p is the total pressure.

5. The Van der Waals' equation for water is given by

$$p = \frac{0.73T}{v-0.485} - \frac{1.397 \times 10^3}{v^2}$$

where p = pressure atm abs, v = ft^3/mol, T = ^{o}R.

Draw a p-v diagram for the following isotherms: 941°R, 976°R, 1030°R, 1080°R, 1164°R, 1790°R.

Discuss your results and calculate the specific volumes v for the liquid and vapour phases for those isotherms which correspond to two phases. (Use Maxwell's construction).

Compare the computed specific volumes with Steam Table values and explain the differences in terms of the value of $p_c v_c / R_{mol} T_c$.

CHAPTER 2

THERMODYNAMIC PROPERTIES OF SYSTEMS OF CONSTANT CHEMICAL COMPOSITION

General thermodynamic relations for systems of constant chemical composition, development of Maxwell relations, derivatives of specific heats, coefficients of h, p, T, Clausius-Clapeyron equations, and Joule-Thomson effect, application to liquefaction system with Van der Waals' gas-inversion curves.

<u>Notation</u>

a	constant in Van der Waals' equation
b	constant in Van der Waals' equation
c	velocity
C_p	specific heat at constant pressure
C_v	specific heat at constant volume
E	internal energy
f	specific Helmholtz free energy function
F	Helmholtz free energy function
g	specific Gibbs free energy function
G	Gibbs free energy function
h	specific enthalpy
H	enthalpy
k	isothermal compressibility
k_s	adiabatic compressibility
K	isothermal bulk modulus
l	latent heat
m	mass
p	pressure
p_c	critical pressure
p_i	inversion pressure
p_R	reduced pressure
q	heat transfer per unit mass
Q	heat transfer to system
R_{mol}	universal gas constant
s	specific entropy
S	entropy
T	temperature absolute
T_c	critical temperature absolute
T_i	inversion temperature absolute
T_R	reduced temperature
u	specific internal energy in the absence of motion, gravity, etc.
v	specific volume
v_c	critical specific volume
v_R	reduced specific volume
α	linear coefficient of expansion
β	volume coefficient of expansion
γ	ratio of specific heats
μ	Joule-Thomson coefficient
ρ	density

THERMODYNAMIC PROPERTIES

Whether one is in the bustling field of industry or the cloistered calm of a university precinct the study of thermodynamic systems requires at some stage the insertion of numerical values for the thermodynamic properties. In industry these values are required to compute the operating conditions for such diverse plants as steam turbogenerator power stations (using conventional or nuclear fuel), oil refineries, all types of chemical plants, laundries using steam plant or even central heating installations. The production of pure industrial gases from the air using liquefaction processes (for example oxygen or nitrogen) or the liquefaction of helium for superconductors require a knowledge of the properties of these substances at extremely low temperatures. At the other end of the temperature scale the design of rocket motors requires thermodynamic data at the highest temperatures. The experimental determination of these properties is the province of the physicist, although today the engineer is actively interested in this field. Some thermodynamic properties such as pressure, temperature and some specific heats can be determined experimentally. Others are difficult to determine in the laboratory, whilst some such as entropy, the free energy function, are derived quantities. In this chapter we will develop relationships between those properties which can be measured experimentally and those which cannot. The systems we will examine will be of pure substances of constant chemical composition in the absence of motion, gravity, electricity, magnetism, capillarity; in Chapter 5 the same ideas will be extended to more complex thermodynamic systems.

We know from experience that any thermodynamic property may be represented by a function of one or more other thermodynamic properties. It will be recalled† that the definition of a property is any quantity whose change is fixed by the end states, i.e. independent of the process. This is equivalent to the mathematical definition of a continuous function of one or more variables. We can therefore use the methods of partial differential calculus to determine the functional relationships between thermodynamic properties.

There are at least eight thermodynamic properties of pure

†J.H. Keenan, Thermodynamics, published by J. Wiley & Sons, 1940. S.R. Montgomery, The Second Law of Thermodynamics, published by Pergamon Press, 1965.

substances in systems in the absence of motion, gravity, magnetism, electricity, capillarity. It is therefore not possible, even if it were desirable, to examine all the relationships which can be developed in the space available. The interested reader is referred to Bridgman† or to any of the texts given in the references.

The properties we are concerned with are of two types, called the intensive and the extensive properties. The intensive properties are properties whose magnitude is independent of the mass of a system, such as pressure p, temperature T, density ρ (later in the fourth chapter we will introduce the chemical potential). The extensive properties are properties proportional to the mass of the system. Examples of these are the volume V, the internal energy U (or E), the enthalpy H, the Gibbs free energy function G, the Helmholtz free energy function F and the entropy S.†† In addition to the intensive and extensive properties we shall use the specific value of the extensive property. This is the extensive property per unit mass (or mol or molecule): thus we have the specific volume v, the specific internal energy u, the specific enthalpy h, the specific Gibbs function g, the specific Helmholtz function f and the specific entropy s. To this group we will add the specific heats (C_p and C_v) and the other derived properties of this type. For the unit of mass we shall use the kilogram molecule††† or mol. Finally we should remember that we can only express a thermodynamic property as a function of other properties. Thus work W and heat transfer Q are not properties and the first law

$$dQ - dW = dU$$

is not a functional relationship between Q, W and U.†††† On the other hand, the second law in the form

$$T\,dS = dU + p\,dV$$

is a functional relationship between S, U and V.

† P.W. Bridgman, A Condensed Collection of Thermodynamic Formulas, Harvard University Press.

†† We differentiate between the extensive and the specific values by using capital letters for the former and small letters for the latter.

††† See Chapter 3, p. 66 for definition

†††† In some texts dQ and dW are written in the format dQ and dW because these are inexact differentials whilst dU is an exact differential, this procedure will be used in Chapter 5.

Equation of State

Whilst the zeroth, first and second laws of thermodynamics refer to the observed inter-relationship between heat and work interactions they cannot alone or together describe completely the behaviour of particular systems. In order to do so one further relationship is required, the state equation. Although it is usual to define the state equation in terms of those quantities which can be readily measured by experiment (for example, p, V and T), there is no reason why the relationship cannot be defined between any other properties such as G, H, V, S, etc. Indeed later we will give extensive examples in the use of the latter properties. To begin with we will use the p-V-T relationship.

Let us consider a system in which we can measure the pressure, volume and temperature. Let us carry out two sets of experiments.† In the first set we will hold the temperature constant and we will decrease the system volume by increasing the pressure. We note that for each pressure there is a fixed volume. We notice that the pressure and the volume of the system can be varied independently of the temperature. This situation can be expressed mathematically as

$$V = f(p)_{T,m}{}^{\dagger\dagger} \qquad (2.1a)$$

or in terms of specific volume

$$v = f(p)_T. \qquad (2.1b)$$

In the second set of experiments we will hold the pressure constant and we will decrease the system volume by decreasing the temperature. We note for each temperature there is a fixed volume. As before we observe that the temperature and volume can be varied independently of the pressure, hence

$$V = f(T)_{p,m} \qquad (2.2a)$$

or

$$v = f(T)_p. \qquad (2.2b)$$

In the two sets of experiments we have stipulated that one or other of the two variables is held constant and hence the volume must depend on both variables (p,T). Thus we can write for a given system of constant mass

---

†The reader will recognize that if the system were a gas these experiments are similar to Boyle's and Charles' experiments.

††The terms outside the bracket -T,m-are constant.

$$V = V(p,T). \tag{2.3}$$

This is the equation of state.

The algebraic relationship or the form of the function represented by (2.3) must be established by experiment.† This may be carried out directly by performing a large number of experiments, in which p and T are varied and V is measured. The results may be plotted or we can use the results of the previous two sets of experiments together with the methods of partial differential calculus.

Since V = V(p,T) is a continuous function we can express this in differential form†† as

$$dV = \left(\frac{\partial V}{\partial p}\right)_T dp + \left(\frac{\partial V}{\partial T}\right)_p dT. \tag{2.4}$$

The first term in the bracket $(\partial V/\partial p)_T$ is the slope of a graph of the results of the first set of experiments, plotted as a volume-pressure diagram, and the second term in the bracket $(\partial V/\partial T)_p$ is the slope of a graph of the results of the second set of experiments, plotted as a volume-temperature diagram.

For most substances we can carry out constant temperature and constant pressure experiments of some form or other. The two differential coefficients are usually associated with certain physical coefficients of the system. For a system maintained at constant temperature the ratio of the change in pressure to the change in volume per unit volume is called the isothermal bulk modulus K†††

$$K = -\left(\frac{dp}{\frac{dV}{V}}\right)_T \tag{2.5a}$$

$$K = -V\left(\frac{\partial p}{\partial V}\right)_T. \tag{2.5b}$$

---

†Using the kinetic theory the ideal gas relationship can be developed if we directly associate the temperature with the kinetic energy of the molecules.

††As we will be concerned with many variables it is usual to indicate the constant terms outside the brackets as shown.

†††The reader will recognize this definition from his studies in applied mechanics and fluid mechanics.

The minus sign is introduced in order that K should be a positive number. It is more usual in thermodynamics to use the reciprocal of the isothermal bulk modulus, called the <u>isothermal compressibility k</u>

$$k = -\frac{1}{V}\left(\frac{\partial V}{\partial p}\right)_T = -\frac{1}{v}\left(\frac{\partial v}{\partial p}\right)_T . \qquad (2.6)$$

For a system maintained at constant pressure the ratio of the change in volume per unit volume per degree rise in temperature is called the coefficient of expansion† β

$$\beta = \left(\frac{\frac{dV}{V}}{dT}\right)_p \qquad (2.7a)$$

or

$$= \frac{1}{V}\left(\frac{\partial V}{\partial T}\right)_p = \frac{1}{v}\left(\frac{\partial v}{\partial T}\right)_p . \qquad (2.7b)$$

Substituting the two coefficients k and β into (2.4) we obtain

$$\frac{dV}{V} = -k\, dp + \beta\, dT. \qquad (2.8)$$

Integration of equation (2.8) will lead to the state equation in algebraic form. As an example let us make the following assumptions:

Let

$$k = \frac{1}{p} \text{ and } \beta = \frac{1}{T} . \quad \text{Then}$$

$$\frac{dV}{V} = -\frac{dp}{p} + \frac{dT}{T} \qquad (2.9)$$

which on integration gives

$$p\frac{V}{T} = \text{constant}. \qquad (2.10)$$

---

† In physics the linear coefficient of expansion (α) is $\left(\frac{\left(\frac{dl}{l}\right)}{dT}\right)_p$. For an isotropic material $\beta = 3\alpha$.

This is the state equation for an ideal gas.† Looking a little deeper into our assumptions we see that the assumption $k = 1/p$ leads to Boyle's Law and the assumption $\beta = 1/T$ leads to Charles' Law. Hence an ideal gas obeys Boyle's and Charles' Laws. Except for this simple case it is not possible to obtain the state equation directly by the methods outlined above and empirical formulae based on experiment or theoretical molecular models are used. It is only proposed to discuss one well-known empirical equation already referred to in the previous chapter, namely the Van der Waals' equation.

$$\left(p + \frac{a}{v^2}\right)(v-b) = R_{mol}T \tag{2.11a}$$

where a and b are constants and v is the specific volume. We will refer to a substance obeying the Van der Waals' equation as a "Van der Waals' gas".††

In the previous chapter it was shown that the above expression does not hold over the whole range of p, V and T. In Fig. 2.1 the pressure-volume graph is shown for a "Van der Waals' gas", for the stable equilibrium states. We will examine this gas in the light of our previous discussion.

Rewriting equation (2.11a) in the form

$$p = \frac{R_{mol}T}{v-b} - \frac{a}{v^2}. \tag{2.11b}$$

It can be shown, after algebraic manipulation, that the isothermal compressibility k and the coefficient of expansion β are

$$k = -\frac{1}{v}\left(\frac{\partial v}{\partial p}\right)_T = -\frac{1}{v}\left(\frac{2a}{v^3} - \frac{R_{mol}T}{(v-b)^2}\right)^{-1} \tag{2.12}$$

$$\beta = \frac{1}{v}\left(\frac{\partial v}{\partial T}\right)_p = -\frac{R_{mol}}{v(v-b)}\left(\frac{2a}{v^3} - \frac{R_{mol}T}{(v-b)^2}\right)^{-1} \tag{2.13}$$

---

†In some texts the term <u>perfect</u> gas is used for the ideal gas. It will be shown later than the expression for the ideal gas implies the specific heats are a function of temperature. Keenan defines a <u>perfect</u> gas as a gas obeying equation (2.10) with constant specific heats (see p.41). J.H. Keenan, <u>Thermo dynamics</u>, Wiley, 1940, S.R. Montgomery, <u>The Second Law of Thermodynamics</u>, Pergamon Press, 1965.

††The constants a and b have been associated with the behaviour of a certain molecular model.

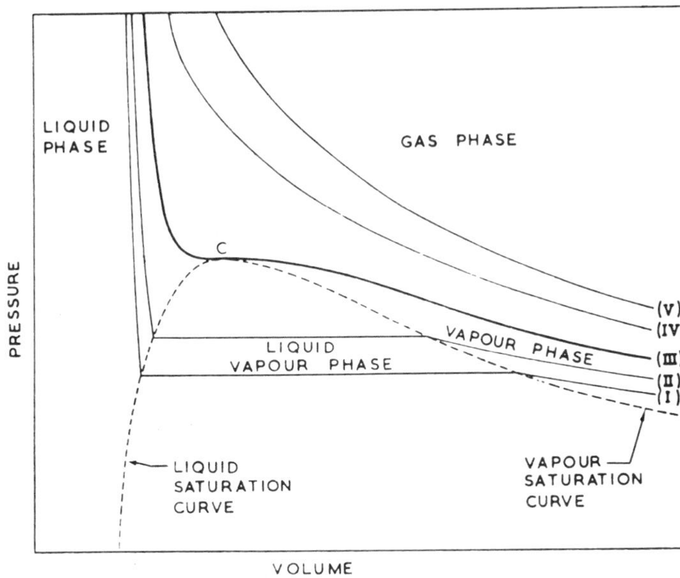

FIG. 2.1.

By substituting for v in (2.12) and (2.13) from (2.11b) it would be possible to obtain functions

$$k = k(p,T)$$

$$\beta = \beta(p,T).$$

In general the isothermal compressibility (k) for all substances is dependent on both pressure and temperature but for low pressures the temperature effects are not very significant. Similarly the coefficient of expansion (β) is dependent on the pressure and temperature, but for low pressures the effect of pressure is not significant. Whilst the isothermal compressibility (k) is always a positive number, this is not always the case for the coefficient of expansion (β). For ice and water, for example, at low temperatures 0 to 70°K and 273 to 277°K the coefficient (β) is negative. At the phase change temperature, 273.1°K, β is positive for ice and negative for water.†

†See Heat and Thermodynamics by M.W. Zemansky for further details.

Law of Corresponding States

The properties of a "Van der Waals' gas" can be further examined with the aid of Fig. 2.1. On the saturation curves there is a discontinuity in the slope of the p-v isotherm at all points except C. At this point curve (III) touches the saturation curve tangentially and it will be seen that:

$$\left(\frac{\partial p}{\partial V}\right)_T = 0 \quad \text{and} \quad \left(\frac{\partial^2 p}{\partial V^2}\right)_T = 0. \tag{2.14}$$

The liquid and vapour phases cannot be distinguished at this temperature and pressure. Both the liquid and the vapour have the same density and a change in phase from liquid to vapour takes place without the addition of latent heat. The conditions at C are called the critical conditions for a substance or the critical point. The pressure, temperature and volume are called the critical pressure p_c, critical temperature T_c and the critical volume v_c. Since all pure substances exhibit this phenomenon, as well as the general form of p-v-T relations shown in Fig. 2.1, it has been suggested that the critical state may be related to any other state through a special empirical law which would hold for all substances. This is called the law of corresponding states.†

If we replace the pressure, temperature and volume by the ratio of the pressure, temperature and volume to the critical pressure, critical temperature and critical volume respectively, i.e.

$$p_R = \frac{p}{p_c}, \quad T_R = \frac{T}{T_c}, \quad v_R = \frac{v}{v_c},$$

it is suggested that the general law would have the form

$$v_R = v_R(p_R, T_R).$$

The quantities with suffix R are called the reduced pressure, reduced temperature and reduced volume.

Let us consider the "Van der Waals' gas"

$$p = \frac{R_{mol}T}{v-b} - \frac{a}{v^2}. \tag{2.15}$$

†The purpose of this law would be to calculate the p-v-T data for any substance if limited information only were available.

At the critical point

$$\left(\frac{\partial p}{\partial v}\right)_T = -\frac{R_{mol}T_c}{(v_c-b)^2} + \frac{2a}{v_c^3} = 0 \tag{2.16}$$

$$\left(\frac{\partial^2 p}{\partial v^2}\right)_T = \frac{2R_{mol}T_c}{(v_c-b)^3} - \frac{6a}{v_c^4} = 0 \tag{2.17}$$

and

$$v_c = 3b,\ T_c = \frac{8a}{27bR_{mol}},\ p_c = \frac{a}{27b^2}. \tag{2.18}$$

If we substitute the reduced quantities into equation (2.11) we obtain the law of corresponding states.

From equations (2.18) we have

$$\left.\begin{aligned} a &= 3p_c v_c^2 \\ b &= \tfrac{1}{3} v_c \\ R_{mol} &= \tfrac{8}{3}\frac{p_c v_c}{T_c} \end{aligned}\right\} \tag{2.19}$$

and

$$\left(p_R + \frac{3}{v_R^2}\right)(3v_R-1) = 8T_R. \tag{2.20}$$

The algebraic form of the law of corresponding states, as given by (2.20), does not satisfy many substances and has limited value in computing the properties. The use of reduced values to calculate thermodynamic properties gives results of the order of 10 per cent accuracy over certain pressure ranges.†

The deviation of real substances from the Van der Waals' gas can be deduced by examining the conditions at the critical

---

†See Keenan, p.358, for a full discussion of the law of corresponding states and the generalized presentation of p-v-T data.

point. If all substances obeyed the law of corresponding states then the numerical value of $p_c v_c / R_{mol} T_c$ should be independent of the substance.† From Van der Waals' equation this constant is

$$\frac{p_c v_c}{R_{mol} T_c} = 0.375 \tag{2.21}$$

Values between 0.2 and 0.3 are obtained in practice.

The isotherm (III) (Fig. 2.1) is called the critical isotherm. Above the temperature corresponding to this isotherm it is impossible to liquefy a gas. We can define the regions for the different phases of a substance on a pressure-volume diagram such as Fig. 2.1. The liquid phase lies to the left of the liquid saturation curve and the critical isotherm, the vapour phase lies between the vapour saturation curve and the critical isotherm, the gas phase lies above the critical isotherm and the two-phase (liquid-vapour) region is in the area enclosed by the saturation curve. As pointed out in our discussion on equilibrium states the phase change for a Van der Waals' gas takes place at constant pressure and temperature.

No substance obeys the Van der Waals' equation quantitatively but the p-v-T curves generally follow the same pattern. Hence trends indicated by this expression elucidate many

---

†Experiments show that at low pressures all gases obey the law

$$\frac{pv}{R_{mol} T} = 1.0$$

Now from the law of corresponding states

$$v_R = f(p_R, T_R)$$

and it follows that

$$\frac{p_R v_R}{T_R} = f(p_R, T_R).$$

At low pressures

$$\frac{pv}{R_{mol} T} = \frac{p_R v_R}{T_R} \frac{p_c v_c}{R_{mol} T_c} = 1.0.$$

If the law of corresponding states holds for all gases, then the first term on the right-hand side of the above equation is the same for all gases, therefore, $\dfrac{p_c v_c}{R_{mol} T_c}$ must be the same for all gases.

important facts about real substances. For a real substance a solid phase also exists, but this is not given by the Van der Waals' equation. Finally, we may note that at low pressures and large volumes the Van der Waals' curves tend to hyperbolae, and a and b are small compared with p and v. This region corresponds to the ideal gas.

Thermodynamic Relations for Pure Substance

The p-v-T data can also be used to compute the specific heats of substances either directly or indirectly. Alternatively if one specific heat can be measured the other can be evaluated from the p-v-T data. From the specific heat data we can calculate other properties such as U, H, G, F and S. For these methods it is not always convenient to express the state equation in the form v = v(p,T) and alternative forms such as U = U(S,V) or H = H(S,p), etc., may be used. For this purpose it is convenient to use the state equation in differential form, as illustrated in equation (2.4), together with one or other of the laws of thermodynamics. The procedure will be to use the methods of partial differential calculus. Since one or two mathematical expressions will be used throughout the analysis it is proposed for convenience to discuss these here.

Let $z = z(x,y)$ be a continuous function, then

$$dz = \left(\frac{\partial z}{\partial x}\right)_y dx + \left(\frac{\partial z}{\partial y}\right)_x dy \qquad (2.22)$$

or if we let

$$M = \left(\frac{\partial z}{\partial x}\right)_y, \quad N = \left(\frac{\partial z}{\partial y}\right)_x$$

$$dz = M\,dx + N\,dy, \qquad (2.23)$$

This is analogous to the differential form of the state equation (see equation (2.4)).

For continuous functions

$$\left(\frac{\partial M}{\partial y}\right)_x = \left(\frac{\partial N}{\partial x}\right)_y ; \qquad (2.24)$$

this equation will be used in developing the Maxwell relations.

In addition to the functional relationship $z = z(x,y)$, x and y can be expressed as $x = x(u,v)$, $y = y(u,v)$ where u and v are variables and the functions are continuous.

It can be readily shown that†

$$\left(\frac{\partial z}{\partial u}\right)_v = \left(\frac{\partial z}{\partial x}\right)_y \left(\frac{\partial x}{\partial u}\right)_v + \left(\frac{\partial z}{\partial y}\right)_x \left(\frac{\partial y}{\partial u}\right)_p . \qquad (2.25)$$

If we now let $v = z$, $u = x$ then $x = x(z)$ and we have

$$\left(\frac{\partial z}{\partial u}\right)_v = 0, \quad \left(\frac{\partial x}{\partial u}\right)_v = 1 \qquad (2.26)$$

hence equation (2.25) becomes

$$0 = \left(\frac{\partial z}{\partial x}\right)_y + \left(\frac{\partial z}{\partial y}\right)_x \left(\frac{\partial y}{\partial x}\right)_z$$

which after rearrangement becomes the well-known cyclic equation

$$\left(\frac{\partial z}{\partial y}\right)_x \left(\frac{\partial x}{\partial z}\right)_y \left(\frac{\partial y}{\partial x}\right)_z = -1 \qquad (2.27)$$

Equations (2.22) to (2.27) will be used at various times in the subsequent analysis.

Let us consider a system comprising a pure substance in the absence of motion, gravity, electricity, magnetism, chemical reaction. For reversible processes the first law is

$$dQ - dU = p\,dV \qquad (2.28)$$

and the second law with the first law

$$dQ = T\,dS = dU + p\,dV. \qquad (2.29a)$$

Using specific properties (2.29a) becomes

$$T\,dS = du + p\,dv. \qquad (2.29b)$$

The enthalpy is by definition

$$h = u + pv. \qquad (2.30)$$

†See R.P. Gillespie, Partial Differentiation (Oliver and Boyd), p.19.

The Gibbs free energy function

$$g = h - Ts \tag{2.31}$$

and the Helmholtz free energy function

$$f = u - Ts. \tag{2.32}$$

Equations (2.30) to (2.32) in differential form are

$$dh = du + p\,dv + v\,dp \tag{2.33}$$

$$dg = dh - T\,ds - s\,dT \tag{2.34}$$

$$df = du - T\,ds - s\,dT. \tag{2.35}$$

Rearranging (2.29b) we have

$$du = T\,ds - p\,dv. \tag{2.36}$$

Substituting (2.36) into (2.33)

$$dh = T\,ds - p\,dv + p\,dv + v\,dp$$

and

$$dh = T\,ds + v\,dp. \tag{2.37}$$

Substituting (2.37) into (2.34) and (2.36) into (2.35)

$$dg = v\,dp - s\,dT \tag{2.38}$$

$$df = -\,p\,dV - s\,dT. \tag{2.39}$$

Grouping (2.36) to (2.39) together we have four <u>state</u> equations in differential form

$$du = T\,ds - p\,dv \quad \text{or} \quad u = u(s,v) \qquad \text{I(a)}$$

$$dh = T\,ds + v\,dp \quad \text{or} \quad h = h(s,p), \qquad \text{II(a)}$$

$$df = -\,p\,dv - s\,dT \quad \text{or} \quad f = f(v,T), \qquad \text{III(a)}$$

$$dg = v\,dp - s\,dT \quad \text{or} \quad g = g(p,T). \qquad \text{IV(a)}$$

Using identities (2.23) and (2.24) we obtain the following additional relations:

$$T = \left(\frac{\partial u}{\partial s}\right)_v, \quad -p = \left(\frac{\partial u}{\partial v}\right)_s, \quad \left(\frac{\partial T}{\partial v}\right)_s = -\left(\frac{\partial p}{\partial s}\right)_v, \qquad \text{I(b)}$$

$$T = \left(\frac{\partial h}{\partial s}\right)_p, \quad v = \left(\frac{\partial h}{\partial p}\right)_s, \quad \left(\frac{\partial T}{\partial p}\right)_s = \left(\frac{\partial v}{\partial s}\right)_p, \qquad \text{II(b)}$$

$$-p = \left(\frac{\partial f}{\partial v}\right)_T, \quad -s = \left(\frac{\partial f}{\partial T}\right)_v, \quad \left(\frac{\partial p}{\partial T}\right)_v = \left(\frac{\partial s}{\partial v}\right)_T, \qquad \text{III(b)}$$

$$v = \left(\frac{\partial g}{\partial p}\right)_T, \quad -s = \left(\frac{\partial g}{\partial T}\right)_p, \quad \left(\frac{\partial v}{\partial T}\right)_p = -\left(\frac{\partial s}{\partial p}\right)_T; \qquad \text{IV(b)}$$

equating the pairs for T, v, p, s we have

$$\left(\frac{\partial u}{\partial s}\right)_v = \left(\frac{\partial h}{\partial s}\right)_p, \qquad (2.40)$$

$$\left(\frac{\partial g}{\partial p}\right)_T = \left(\frac{\partial h}{\partial p}\right)_s, \qquad (2.41)$$

$$\left(\frac{\partial u}{\partial v}\right)_s = \left(\frac{\partial f}{\partial v}\right)_T, \qquad (2.42)$$

$$\left(\frac{\partial g}{\partial T}\right)_p = \left(\frac{\partial f}{\partial T}\right)_v. \qquad (2.43)$$

The group of equations called the Maxwell relations are:

$$\left(\frac{\partial T}{\partial v}\right)_s = -\left(\frac{\partial p}{\partial s}\right)_v, \qquad (2.44)$$

$$\left(\frac{\partial T}{\partial p}\right)_s = \left(\frac{\partial v}{\partial s}\right)_p, \qquad (2.45)$$

$$\left(\frac{\partial p}{\partial T}\right)_v = \left(\frac{\partial s}{\partial v}\right)_T, \qquad (2.46)$$

$$\left(\frac{\partial v}{\partial T}\right)_p = -\left(\frac{\partial s}{\partial p}\right)_T. \qquad (2.47)$$

In some texts equations (2.44) to (2.47) are called the Maxwell first, second, third and fourth equations respectively.

Applications of Thermodynamic Relations

The specific heats at constant pressure and constant volume can be measured for some substances fairly readily, but for others either one or the other may be difficult to evaluate experimentally. When only limited experimental specific heat data are available we may use the above relations to compute the specific heats over a range of pressures and temperatures if other information such as p-v-T data are available. If the specific heat data are known we can also determine the internal energy and enthalpy. We will discuss here a number of methods using the above relations, for this purpose illustrating each method by reference to the ideal gas and the "Van der Waals' gas".

By definition the specific heat at constant volume is

$$C_v = \left(\frac{\partial u}{\partial T}\right)_v \tag{2.48}$$

and

$$\left(\frac{\partial u}{\partial T}\right)_v = \left(\frac{\partial u}{\partial s}\right)_v \left(\frac{\partial s}{\partial T}\right)_v . \tag{2.49}$$

Now from I(b)

$$T = \left(\frac{\partial u}{\partial s}\right)_v \tag{2.50}$$

hence

$$C_v = T\left(\frac{\partial s}{\partial T}\right)_v . \tag{2.51}$$

It can be shown by the same method that

$$C_p = T\left(\frac{\partial s}{\partial T}\right)_p . \tag{2.52}$$

Now

$$\left(\frac{\partial C_v}{\partial v}\right)_T = \left(\frac{\partial}{\partial v}\, T\left(\frac{\partial s}{\partial T}\right)\right)$$

$$= T\,\frac{\partial^2 s}{\partial v \partial T} \qquad (2.53)^{\dagger}$$

also from (2.46)

†We are differentiating equation (2.51) with respect to v holding T constant, hence we obtain a cross derivative and the constant notation is dropped.

$$\left(\frac{\partial s}{\partial v}\right)_T = \left(\frac{\partial p}{\partial T}\right)_v .$$

Therefore

$$\frac{\partial}{\partial T}\left(\frac{\partial s}{\partial v}\right)_T = \frac{\partial}{\partial T}\left(\frac{\partial p}{\partial T}\right)_v$$

or

$$\frac{\partial^2 s}{\partial T \partial v} = \left(\frac{\partial^2 p}{\partial T^2}\right)_v . \tag{2.54}$$

For continuous functions the cross derivatives are equal (equation (2.24)) and hence, equating (2.53) to (2.54),

$$\left(\frac{\partial C_v}{\partial v}\right)_T = T\left(\frac{\partial^2 p}{\partial T^2}\right)_v . \tag{2.55}$$

The right-hand side of this equation is in terms of the p-v-T state relations. Empirical experimental results may therefore be used to evaluate the specific heat at constant volume.† As an example let us consider (a) the ideal gas and (b) the "Van der Waals' gas".

(a) Ideal Gas

$$pv = R_{mol}T$$

$$v\left(\frac{\partial p}{\partial T}\right)_v = R_{mol}$$

$$v\left(\frac{\partial^2 p}{\partial T^2}\right)_v = 0$$

Since $v \neq 0$ then

$$\left(\frac{\partial^2 p}{\partial T^2}\right)_v = 0$$

†It is, of course, necessary to know at least one value of specific heat at each temperature, this mathematically being the constant of integration along an isochoric (constant volume) line. Macroscopic thermodynamic analysis does not lead us to predict this value, which must be measured by experiment. In statistical thermodynamics, by selecting a suitable model for the molecule of a substance, the specific heats can be computed from the internal energy of the molecule.

hence, from equation (2.55),

$$\left(\frac{\partial C_v}{\partial v}\right)_T = 0.$$

The specific heat at constant volume is therefore independent of volume.

(b) "Van der Waals' Gas"

Using the state equation in the form

$$p = \frac{R_{mol}T}{v-b} - \frac{a}{v^2}$$

then

$$\left(\frac{\partial p}{\partial T}\right)_v = \frac{R_{mol}}{v-b}$$

$$\left(\frac{\partial^2 p}{\partial T^2}\right)_v = 0$$

hence

$$\left(\frac{\partial C_v}{\partial v}\right)_T = 0.$$

As before the specific heat at constant volume is independent of the volume. Although the specific heats at constant volume are independent of volume they may well be dependent on the temperature. The above expression gives us no information on this point.†

Expressions similar to (2.55) can be obtained for the specific heat at constant pressure in terms of p, v and T.

The variation of u (or h) with p, v, T is an important relationship required in studies of thermodynamic systems. To obtain this relationship we will introduce four new equations derived through the Maxwell relations. We will develop two of these equations here and leave the remaining two for the reader to derive.

---

†See footnote to p. 41

We will use the following form of state equation for the first expression

$$s = s(T,v)$$

hence

$$ds = \left(\frac{\partial s}{\partial T}\right)_v dT + \left(\frac{\partial s}{\partial v}\right)_T dv. \tag{2.56a}$$

Now from (2.51)

$$T\left(\frac{\partial s}{\partial T}\right)_v = C_v$$

and from the Maxwell relation (2.46)

$$\left(\frac{\partial s}{\partial v}\right)_T = \left(\frac{\partial p}{\partial T}\right)_v.$$

Substituting this relation into (2.56a) and rearranging we obtain

$$T\,ds = C_v\,dT + T\left(\frac{\partial p}{\partial T}\right)_v dv. \tag{2.56b}$$

This is called the first T ds equation.

The second and third T ds equations are:

Second T ds equation

$$T\,ds = C_p\,dT - T\left(\frac{\partial v}{\partial T}\right)_p dp. \tag{2.57}$$

Third T ds equation

$$T\,ds = C_v\left(\frac{\partial T}{\partial p}\right)_v dp + C_p\left(\frac{\partial T}{\partial v}\right)_p dv. \tag{2.58}$$

The fourth equation is called the energy equation.

The state equation we shall use is

$$u = u(T,v)$$

hence

$$du = \left(\frac{\partial u}{\partial T}\right)_v dT + \left(\frac{\partial u}{\partial v}\right)_T dv. \tag{2.59a}$$

Now

$$C_v = \left(\frac{\partial u}{\partial T}\right)_v$$

from the second law

$$du = T\,ds - p\,dv$$

and the first T ds equation is

$$T\,ds = C_v dT + T\left(\frac{\partial p}{\partial T}\right)_v dv.$$

Substitution into equation (2.59a) leads to

$$C_v dT + T\left(\frac{\partial p}{\partial T}\right)_v dv - p\,dv = C_v dT + \left(\frac{\partial u}{\partial v}\right)_T dv$$

or

$$\left(\frac{\partial u}{\partial v}\right)_T = \left[T\left(\frac{\partial p}{\partial T}\right)_v - p\right]. \qquad (2.59b)$$

From the energy equation (2.59b) we can obtain the internal energy u in terms of experimentally determined p-v-T relations, if limited data on the specific heats (and hence internal energy) are available. The procedure will be illustrated once again using the ideal gas and the "Van der Waals' gas" as examples.

(a) Ideal Gas

$$pv = R_{mol}T$$

hence

$$v\left(\frac{\partial p}{\partial T}\right)_v = R_{mol}.$$

Substitution into (2.59) gives

$$\left(\frac{\partial u}{\partial v}\right)_T = \left[\frac{R_{mol}T}{v} - p\right] = 0.$$

Thus the internal energy of an ideal gas is independent of its specific volume.

Using the state equation

$$u = u(T,v)$$

$$du = \left(\frac{\partial u}{\partial T}\right)_v dT + \left(\frac{\partial u}{\partial v}\right)_T dv$$

$$du = \left(\frac{\partial u}{\partial T}\right)_v dT = C_v dT$$

hence $$u = u(T) \text{ for an ideal gas.}$$

If we consider the identity

$$\left(\frac{\partial u}{\partial v}\right)_T = \left(\frac{\partial u}{\partial p}\right)_T \left(\frac{\partial p}{\partial v}\right)_T = 0.$$

The term $(\partial p/\partial v)_T$ for an ideal gas is equal to $-p/v$ and since $p \neq 0$

$$\left(\frac{\partial u}{\partial p}\right)_T = 0.$$

Thus the internal energy is independent of the pressure or as before

$$u = u(T) \text{ for an ideal gas.}^{\dagger}$$

---

†In some texts the formal definition of an ideal gas is

$$pv = R_{mol}T \quad \text{(a)}$$

and

$$\left(\frac{\partial u}{\partial p}\right)_T = 0. \quad \text{(b)}$$

In other texts the second equation is replaced by

$$u = u(T) \quad \text{(c)}$$

Sometimes $u = u(T)$ is defined as a linear function. In these cases the gas is called a perfect gas and the corresponding equations are

$$pv = R_{mol}T \quad \text{(d)}$$

and

$$u = C_v T \quad \text{(e)}$$

where the specific heat at constant volume is a constant. A semi-perfect gas is then defined as a gas which obeys (d) but the specific heats are functions of temperature.

$$C_v = f(T). \quad \text{(f)}$$

Both the perfect and semi-perfect gases obey (a), (b) and (c) and are therefore ideal gases.

(b) "Van der Waals' Gas"

$$p = \frac{R_{mol}T}{v-b} - \frac{a}{v^2}$$

and

$$\left(\frac{\partial p}{\partial T}\right)_v = \frac{R_{mol}}{v-b} .$$

Hence from (2.59b)

$$\left(\frac{\partial u}{\partial v}\right)_T = \left(\frac{R_{mol}T}{v-b} - \frac{R_{mol}T}{v-b} + \frac{a}{v^2}\right) = \frac{a}{v^2}.$$

As before

$$du = \left(\frac{\partial u}{\partial T}\right)_v dT + \left(\frac{\partial u}{\partial v}\right)_T dv$$

$$du = C_v dT + \frac{a}{v^2} dv \qquad (2.60)$$

and

$$u = u(T,v).$$

For a "Van der Waals' gas" the internal energy is, therefore, a function of the temperature and the volume. Since C_v is either constant or a function of T, equation (2.60) can be directly integrated to determine the algebraic relationship between u, T and v.

Relationships Between the Specific Heats at Constant Volume and the Specific Heats at Constant Pressure

By the methods outlined above we can calculate the specific heats over a range of pressure, temperature and volume if limited data on specific heats are available. For some substances either one or other of the specific heats can be determined experimentally but the other is somewhat difficult to determine. It is convenient to have some relationship between the specific heats whereby the experimental evaluation of one can be used to derive the other. The following procedure can be used for this purpose.

We start with the equation of state

$$s = s(T,v)$$

and

$$ds = \left(\frac{\partial s}{\partial T}\right)_v dT + \left(\frac{\partial s}{\partial v}\right)_T dv.$$

By using the identity (2.25) we have

$$\left(\frac{\partial s}{\partial T}\right)_p = \left(\frac{\partial s}{\partial T}\right)_v + \left(\frac{\partial s}{\partial v}\right)_T \left(\frac{\partial v}{\partial T}\right)_p \tag{2.61}$$

and from (2.51) and (2.52)

$$\left(\frac{\partial s}{\partial T}\right)_p = \frac{C_p}{T} \quad \text{and} \quad \left(\frac{\partial s}{\partial T}\right)_v = \frac{C_v}{T}.$$

Substitution in (2.61) gives

$$\frac{C_p - C_v}{T} = \left(\frac{\partial s}{\partial v}\right)_T \left(\frac{\partial v}{\partial T}\right)_p.$$

Using the Maxwell relation (2.46)

$$\left(\frac{\partial s}{\partial v}\right)_T = \left(\frac{\partial p}{\partial T}\right)_v$$

we obtain

$$\frac{C_p - C_v}{T} = \left(\frac{\partial p}{\partial T}\right)_v \left(\frac{\partial v}{\partial T}\right)_p. \tag{2.62}$$

Once again we have the specific heats in terms of the p-v-T. A more useful form of (2.62) is obtained if we used the identity (2.27) and substitute p, s and T for x, y and z. We have after simplification

$$\left(\frac{\partial p}{\partial T}\right)_v = -\left(\frac{\partial p}{\partial v}\right)_T \left(\frac{\partial v}{\partial T}\right)_p$$

hence

$$C_p - C_v = -T\left(\frac{\partial v}{\partial T}\right)_p^2 \left(\frac{\partial p}{\partial v}\right)_T. \tag{2.63}$$

Since T and $(\partial v/\partial T)_p^2$ are either zero or positive, we see that the difference in specific heats is dependent on the sign of $(\partial p/\partial v)_T$ or the slope of an isothermal in the p-v state diagram. Now all known substances have either zero or negative slopes in the p-v diagram; hence the specific heat at constant pressure is therefore either equal to or greater than the specific heat at constant volume. It is interesting to note that the specific heats are equal at the absolute zero and when $(\partial p/\partial T)_v$ is zero (for water this corresponds to 4°C).

For an ideal gas

$$pv = R_{mol}T$$

$$\left(\frac{\partial p}{\partial v}\right)_T = -\frac{p}{v} \quad \text{and} \quad \left(\frac{\partial v}{\partial T}\right)_p = \frac{R_{mol}}{p}$$

therefore

$$C_p - C_v = R_{mol}. \tag{2.64}$$

Since for an ideal gas $C_v = f(T)$ then

$$C_p = C_v + R_{mol} = f(T).$$

The difference in specific heats is a <u>constant</u> corresponding to the universal gas constant.

Equation (2.63) can be written directly in terms of the isothermal compressibility k and the coefficient of expansion β.

$$C_p - C_v = \frac{Tv\beta^2}{k}.$$

Before leaving the various functional relations we have obtained for the specific heats we can look at one important indirect use of these properties. In the calculations of the speed of sound we require to know the <u>adiabatic</u> bulk modulus or <u>adiabatic</u> compressibility k_s. If we know the specific heats of the substance and the isothermal compressibility k we can calculate k_s from the expression

$$\frac{C_p}{C_v} = \frac{k}{k_s}. \tag{2.65}$$

This expression can be proved by using three state equations $s = s(T,v)$, $s = s(T,p)$, $T = T(p,v)$, the identity (2.27) and the Maxwell relations.

Using the identity (2.27) we have

$$\left(\frac{\partial s}{\partial T}\right)_v \left(\frac{\partial T}{\partial v}\right)_s \left(\frac{\partial v}{\partial s}\right)_T = \left(\frac{\partial s}{\partial T}\right)_p \left(\frac{\partial T}{\partial p}\right)_s \left(\frac{\partial p}{\partial s}\right)_T \tag{2.66}$$

and the Maxwell relations

$$\left(\frac{\partial s}{\partial v}\right)_T = \left(\frac{\partial p}{\partial T}\right)_v, \quad \left(\frac{\partial s}{\partial p}\right)_T = -\left(\frac{\partial v}{\partial T}\right)_p \tag{2.67}$$

the specific heats

$$\frac{C_v}{T} = \left(\frac{\partial s}{\partial T}\right)_v, \quad \frac{C_p}{T} = \left(\frac{\partial s}{\partial T}\right)_p. \tag{2.68}$$

Substituting (2.67) and (2.68) into (2.66) we have after simplification

$$\frac{C_p}{C_v} = -\left(\frac{\partial p}{\partial v}\right)_s \left(\frac{\partial T}{\partial p}\right)_v \left(\frac{\partial v}{\partial T}\right)_p \tag{2.69}$$

Now from the state equation $T = T(p,v)$

$$\left(\frac{\partial v}{\partial p}\right)_T = -\left(\frac{\partial T}{\partial p}\right)_v \left(\frac{\partial v}{\partial T}\right)_p,$$

hence

$$\frac{C_p}{C_v} = \frac{\left(\frac{\partial p}{\partial v}\right)_s}{\left(\frac{\partial p}{\partial v}\right)_T} \tag{2.70}$$

We define the adiabatic compressibility k_s as

$$-\frac{1}{v}\left(\frac{\partial v}{\partial p}\right)_s$$

hence

$$\frac{C_p}{C_v} = \frac{k}{k_s} = \gamma. \tag{2.65}$$

Referring to equation (2.70) it will be seen that the slope of isotropes in pressure-volume diagrams are steeper or equal to the slopes of isothermals depending on the values of C_p and C_v. Thus more displacement work is done in an isentropic compression process than in an isothermal compression process and vice versa for an expansion process.

The Clausius-Clapeyron Equation

For substances in which two phases can exist in equilibrium some of the properties are functions of one property only in the two-phase state. For example, in the water-steam mixed phase the pressure is a function of temperature. We discussed earlier the phase change for a "Van der Waals' gas" and we noted that the phase change takes place at constant pressure and temperature with the Gibbs function being the same for both phases. This applies to all substances.† The change in internal energy plus the work done by the substance due to the change in volume is called the latent heat (1). The relationship between the pressure of the two phases in equilibrium, the latent heat and the change in volume can be determined in a number of ways.

If suffix 1 refers to one phase and suffix 2 refers to another phase, then at the same pressure and temperature for a change in phase

$$g_1 = g_2 \tag{2.71a}$$

and for a phase change at p + dp, T + dT

$$g_1 + dg_1 = g_2 + dg_2 \tag{2.71b}$$

hence

$$dg_1 = dg_2.$$

Now

$$dg = v\,dp - s\,dT \text{ from IV(a) (p.35)}$$

and

$$v_1 dp - s_1 dT = v_2 dp - s_2 dT$$

---

†Proof. Consider a system in which a phase change takes place at constant p and T. If m_1 is mass in phase 1, m_2 mass in phase 2 and g_1, g_2 the corresponding Gibbs functions, then

$$m_1 + m_2 = m \text{ constant} \tag{a}$$

and

$$m_1 g_1 + m_2 g_2 = G. \tag{b}$$

For equilibrium between phases equation (1.31) states

$$dG)_{p,T} = 0. \tag{c}$$

Now from (a)

$$dm_1 + dm_2 = 0 \tag{d}$$

and from (b) and (c)

$$dG = dm_1 g_1 + dm_2 g_2 = 0. \tag{e}$$

Comparing (d) and (e) we have

$$g_1 = g_2.$$

Gibbs function is the same in each phase.

or

$$\frac{dp}{dT} = \frac{s_2 - s_1}{v_2 - v_1}.$$

The phase change is a reversible process.†

Hence, for a change from phase 1 to phase 2 at constant p and T

$$q = l = \int_1^2 du + \int_1^2 p\,dv = \int_1^2 T\,ds = T(s_2 - s_1)$$

Therefore,

$$\frac{dp}{dT} = \frac{s_2 - s_1}{v_2 - v_1} = \frac{l}{T(v_2 - v_1)}$$

or

$$T\frac{dp}{dT} = \frac{l}{v_2 - v_1}. \qquad (2.72)$$

This is called the Clausius-Clapeyron equation and is used to calculate either the pressure p, the specific volume v or the latent heat l from experimental data.

Liquefaction of Gases

There is a large number of industrial processes using extremely low temperatures. Liquefaction of gases to separate oxygen and nitrogen from the atmosphere in the Linde process is well known. The expansion of a gas against a piston or in a turbine is one method of producing a low temperature. The disadvantage of this method is that the temperature drop is a function of the pressure drop and initial temperature. As the initial temperature falls the temperature drop decreases unless the pressure drop is increased and the law of diminishing returns raises practical limitations to the decrease in temperature one can usefully obtain. On the practical side lubrication problems may become important as the temperature is lowered. This problem is particularly important if impurities in the substance being cooled are to be avoided. With the introduction of gas bearings in high-speed expander turbines problems of contamination have been reduced.

---

† Equation (1.28) becomes, for a process in which there is no useful work (i.e. shaft work),

$$0 \leqslant (G_1 - G_2).$$

The equality sign holds for reversible processes. In this case $g_1 = g_2$, hence $g_1 - g_2 = 0$ and the process is reversible.

An alternative method to the expansion engine is flash evaporation in which there is a reduction in temperature of the substance as it evaporates. This process can be used in conjunction with the expansion engine or turbine.

For very low temperatures a property of the substance itself may be used to achieve liquefaction. This is due to the Joule-Thomson effect. When a gas is forced through a porous plug it may be observed that, below certain initial entry temperatures, there is a temperature drop after passing through the plug. The lower the initial temperature the greater the temperature drop for a given pressure drop. This phenomenon is the Joule-Thomson effect. The expansion engine or turbine and the throttling processes described above are used in series in modern gas liquefying plants.

The properties of substances which control the Joule-Thomson effect can be evaluated from the p-v-T and specific heat data. Before we examine the necessary relations to evaluate the criteria, a description of the Joule-Thomson experiment will be given. Figure 2.2 shows a schematic arrangement of the apparatus. Gas under pressure is forced through a porous plug. A constant high pressure is maintained on one side of the plug and a constant low pressure on the other. The apparatus is thermally insulated from the surroundings. Within the plug an irreversible non-equilibrium process is taking place and it is not possible to evaluate the properties of the gas. Upstream and downstream of the plug equilibrium conditions prevail and we can measure the pressure and temperature. In the Joule-Thomson experiment the upstream pressure and temperature are held constant; the downstream pressure is varied and the temperature measured. The results of such an experiment are shown in Fig. 2.3 in which the downstream temperature is plotted against the downstream pressure.

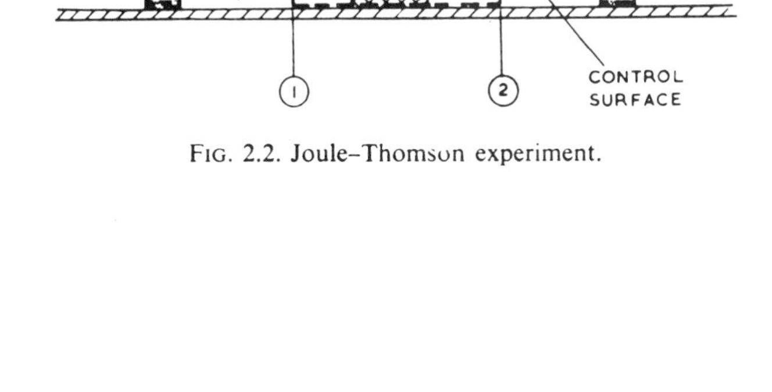

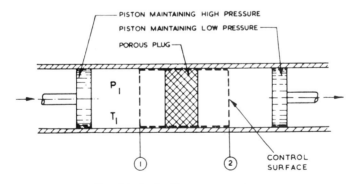

FIG. 2.2. Joule–Thomson experiment.

On the same curve the fixed upstream pressure p_1 and the temperature T_1 are shown. The experiments show that for a fixed initial pressure p_1 and temperature T_1 there is first an increase in downstream temperature as the pressure downstream of the plug is lowered. A maximum temperature is attained, then further decrease in pressure reduces the downstream temperature until eventually there is a temperature decrease across the plug.

The process may be analysed using the steady flow equation.† Applying the equation to the control volume (Fig. 2.2) for a horizontal system

$$\dot{Q}-\dot{W}' = \dot{m}\left(h_2-h_1 + \frac{c_2^2}{2} - \frac{c_1^2}{2}\right) \qquad (2.73)$$

where $\dot{Q}$ is the rate of heat transfer, $\dot{W}'$ is the rate of shaft work, $\dot{m}$ is the mass flow rate, c is the velocity, 1 and 2 are stations upstream and downstream of the plug.

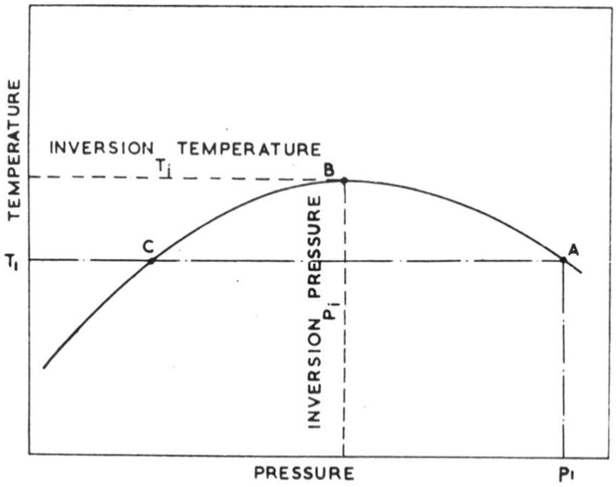

FIG. 2.3.

For this experiment

$$\dot{Q} = \dot{W}' = 0, \quad c_2 \doteqdot 0, \quad c_1 \doteqdot 0$$

hence

$$h_2 = h_1. \qquad (2.74)$$

†See Spalding and Cole, Engineering Thermodynamics, p.120.

Thus the enthalpy of the substance upstream of the plug equals the enthalpy of the substance downstream of the plug, and the curve in Fig. 2.3 is a constant enthalpy line (called an isenthalpic line). It should be emphasized that the steady flow energy equation gives *no* information on the conditions within the plug. The isenthalpic line is *not the path* of the change in state of the substance as it passes through the plug.

Referring to the results of this experiment we could adjust the upstream pressure p_1 and temperature T_1 to lie anywhere on the curve in Fig. 2.3; the downstream temperature, depending on the pressure drop, would then lie on the same curve to the left of the initial point. If the upstream pressure is to the right of the point B, say at A, there will either be an increase, a decrease, or no change in temperature across the plug depending on the pressure drop. On the other hand, if the upstream pressure is to the left of B there will always be a drop in temperature across the plug. The maximum temperature drop will occur if the upstream pressure is at B. The point B is called the *inversion point* and the temperature the *inversion temperature*.

The slope of the isenthalpic line indicates whether the Joule-Thomson effect will be "heating" or "cooling". (These terms have been placed in inverted commas because it should be clear to the reader that there are *no* heat interactions in the experiment and that a temperature drop does *not* indicate a heat loss *or* a temperature rise a heat gain. The terms used here are really associated with the notion of temperature being a measure of "hotness" - not the rigorous thermodynamic definition. By tradition these terms are used in cryogenics and we say the Joule-Thomson effect "cools" the gas.) The slope of the isenthalpic line $(\partial T/\partial p)_h$ is called the Joule-Thomson coefficient μ. A negative μ indicates either "heating" or "cooling" depending on the pressure drop (if the pressure p = p_A in Fig. 2.3 and p_2 is greater than p_C there will be "heating"; if p_2 is less than p_C there will be "cooling") a positive value of always indicates "cooling".

The relationship between μ, v, T, p and C_p can be established using the methods outlined earlier.

From II(a)

$$dh = T\,ds + v\,dp$$

hence for a constant enthalpy process†

†The word process is used loosely here; as far as the equilibrium states are concerned the process is at constant enthalpy.

$$0 = T\left(\frac{\partial s}{\partial p}\right)_h + v. \qquad (2.75)$$

The state equation is

$$s = s(T,p).$$

Along an isenthalpic line

$$\left(\frac{\partial s}{\partial p}\right)_h = \left(\frac{\partial s}{\partial T}\right)_p \left(\frac{\partial T}{\partial p}\right)_h + \left(\frac{\partial s}{\partial p}\right)_T.$$

Substituting for $(\partial s/\partial p)_h$ in (2.75) we have

$$0 = \left\{\left(\frac{\partial s}{\partial T}\right)_p \left(\frac{\partial T}{\partial p}\right)_h + \left(\frac{\partial s}{\partial p}\right)_T\right\} T + v. \qquad (2.76)$$

Now

$$C_p = T\left(\frac{\partial s}{\partial T}\right)_p \quad \text{from equation (2.52)}$$

and

$$\left(\frac{\partial s}{\partial p}\right)_T = -\left(\frac{\partial v}{\partial T}\right)_p \quad \text{from equation (2.47).}$$

Substitution into (2.76) and rearranging gives the Joule-Thomson coefficient

$$\mu = \left(\frac{\partial T}{\partial p}\right)_h = \frac{1}{C_p}\left\{T\left(\frac{\partial v}{\partial T}\right)_p - v\right\}. \qquad (2.77)$$

The coefficient of expansion for the gas is

$$\beta = \frac{1}{v}\left(\frac{\partial v}{\partial T}\right)_p.$$

Hence

$$\mu = \frac{v}{C_p}(\beta T - 1) \qquad (2.78)$$

The inversion temperature, T, corresponds to the condition $\mu = 0$ and from equation (2.78) is directly related to the coefficient of expansion, β.

$$T_i = \frac{1}{\beta}. \qquad (2.79)$$

In Fig. 2.4 a series of isenthalpic curves is shown. The locus of the inversion points is called the inversion curve. Within the region to the left of the inversion curve Joule-Thomson "cooling" will take place. The maximum "cooling" will be obtained for any given upstream temperature and downstream pressure if the initial pressure lies on the inversion curve.

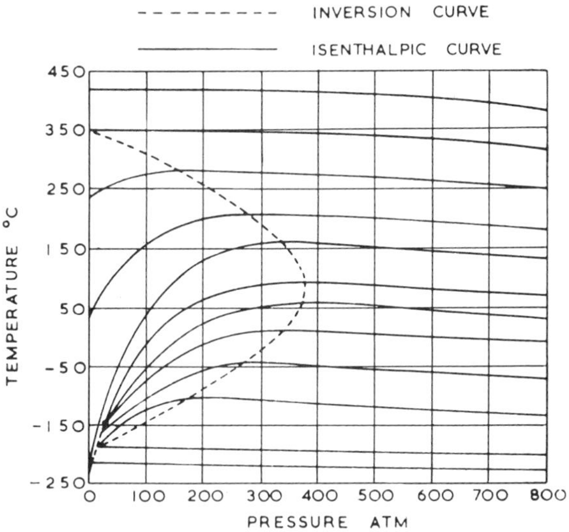

FIG. 2.4. Isenthalpic curves and inversion curves.

For a "Van der Waals' gas" the equation of the inversion curve is

$$p_i = \frac{a}{b^2}\left(1 - \sqrt{\frac{bR_{mol}T_i}{2a}}\right)\left(3\sqrt{\frac{bR_{mol}T_i}{2a}} - 1\right) \qquad (2.80)$$

where p_i is the inversion pressure corresponding to the inversion temperature T_i. There are two inversion temperatures for each pressure.

The Joule-Thomson "cooling" effect cannot take place above the maximum inversion temperature nor below the minimum inversion temperature. These values are $2a/bR_{mol}$ and $2a/9bR_{mol}$ respectively for a "Van der Waals' gas".

It will be recalled that the critical temperature is $8a/27bR_{mol}$. The ratios of the inversion temperatures to the critical temperature are

$$\left(\frac{T_i}{T_c}\right)_{max} = \frac{\text{Maximum Inversion Temperature}}{\text{Critical Temperature}} = 6.75$$

$$\left(\frac{T_i}{T_c}\right)_{min} = \frac{\text{Minimum Inversion Temperature}}{\text{Critical Temperature}} = 0.75.$$

Typical values for real substances are given in Table 2.1. The ratio $(T_i/T_c)_{max}$ varies from 4.85 to 6.2.

TABLE 2.1

	Maximum inversion temperature °K T_i	Critical temperature °K T_c	$\left(\frac{T_i}{T_c}\right)_{max}$
Carbon dioxide	∿1500	304	∿4.95
Argon	723	134	5.4
Nitrogen	621	126	4.85
Air	603	117	5.15
Hydrogen	202	33	6.2
Helium	25	4.7	∿5.32

For nearly all substances the maximum inversion temperature is above the normal ambient temperature and hence the Joule-Thomson effect can be obtained. In the case of hydrogen and helium it is necessary to pre-cool the gas below the maximum inversion temperature. For liquefaction processes it is also necessary to cool the gas below the critical temperature since the liquid phase does not occur above this value. A gas becomes liquefied in a Joule-Thomson flow when the isenthalpic line either touches or crosses the liquid-vapour phase line. A diagrammatic arrangement of a typical liquefaction plant is shown in Fig.2.5.

The Joule-Thomson cooling effect is generally associated with gases. However, the same phenomenon is present in all phases of real substances and there are many practical applications of the phenomenon. All vapour compression refrigerators depend on this effect when the refrigerant passes through the throttling valve. The Joule-Thomson heating effect can be observed in the throttling calorimeter used for the measurement of the dryness faction of steam.

Expression (2.78) leads to an indirect method of determining the specific heat of a substance. It has also been used to evaluate the absolute temperature scale and hence the absolute zero.

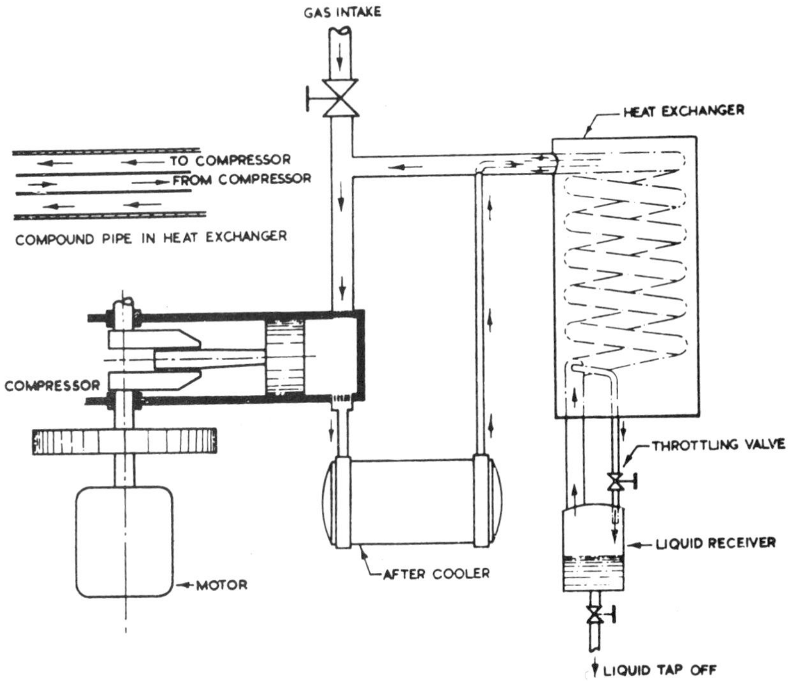

FIG. 2.5. Diagrammatic arrangement of gas liquefaction plant.

Exercises

1. The Dieterici formula for a pure substance is given by

$$p = \frac{R_{mol}T}{v-b} e^{-a/R_{mol}Tv}$$

Determine (1) The constants a and b in terms of the critical pressure and temperature.

(2) The compressibility factor at the critical condition.

(3) The law of corresponding states.

2. Derive expressions for $\left(\frac{\partial C_v}{\partial v}\right)_T$, for substances obeying the following laws:

(1) $$p = \frac{R_{mol}T}{v-b} e^{-a/R_{mol}Tv}$$

(2) $$p = \frac{R_{mol}T}{v-b} - \frac{a}{Tv^2}$$

(3) $$p = \frac{R_{mol}T}{v-b} - \frac{a}{v(v-b)} + \frac{c}{v^3}.$$

Discuss the physical implication of your results.

3. For an ideal gas $C_p - C_v = R_{mol}$, examine a "Van der Waals' gas" and a "Dieterici Gas" and comment on your results for the difference in specific heat for these gases compared with the ideal gas.

4. Derive an expression for the law of corresponding states for a gas represented by the following expression:

$$p = \frac{R_{mol}T}{v-b} - \frac{a}{Tv^2}.$$

5. Show that

(1) $$h-u = T^2\left(\left(\frac{\partial(f/T)}{\partial T}\right)_v - \left(\frac{\partial(g/T)}{\partial T}\right)_p\right)$$

(2) $$\frac{C_p}{C_v} = \left(\frac{\partial^2 g}{\partial T^2}\right)_p \Bigg/ \left(\frac{\partial^2 f}{\partial T^2}\right)_v.$$

6. Show

(a) $$T\,ds = C_p\,dT - T\left(\frac{\partial v}{\partial T}\right)_p dp$$

(b) $$T\,ds = C_v\left(\frac{\partial T}{\partial p}\right)_v dp + C_p\left(\frac{\partial T}{\partial v}\right)_p dv$$

(c) $$\left(\frac{\partial h}{\partial p}\right)_T = v - T\left(\frac{\partial v}{\partial T}\right)_p.$$

7. Show that for a gas obeying the state equation

$$pv = (1 + \alpha)\,R_{mol}T$$

where α is a function of temperature only the specific heat at constant pressure is given by

$$C_p = -R_{mol}T \frac{d^2(\alpha T)}{dT^2} \ln p + C_{po}$$

where the pressure p is in atmospheres and C_{po} is the specific heat at one atmosphere absolute.

8. Show that if the ratio of the specific heats is 1.4 then

$$\left(\frac{\partial p}{\partial T}\right)_s = \frac{7}{2}\left(\frac{\partial p}{\partial T}\right)_v .$$

9. Show that the Joule-Thomson coefficient (μ) is given by

$$\mu = \frac{1}{C_p}\left[T\left(\frac{\partial v}{\partial T}\right)_p - v\right].$$

Hence or otherwise show that the inversion temperature (T_i) is

$$T_i = \left(\frac{\partial T}{\partial v}\right)_p v.$$

The equation of state for air may be represented by

$$p = \frac{0.73T}{v - 0.585} - \frac{343.8}{v^2}$$

where p = pressure in atmospheres absolute, T = temperature in degrees Rankine, v = specific volume in ft^3/lb mol, and R_{mol} = 0.73 atm ft^3/lb mol °Rankine.

Determine the maximum and minimum inversion temperatures and the maximum inversion pressure for air.

(Univ. Manch.)

10. In the figure the last stage of a liquefaction plant is shown in diagrammatic form. Derive the relationship between p_1 and T_1 for the maximum yield of liquid at conditions p_L, T_L, h_L for a gas obeying the state equation

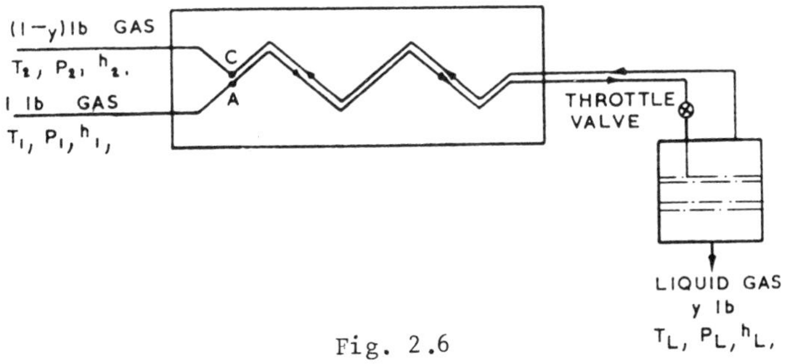

Fig. 2.6

$$p + \frac{346}{V^2} (V - 0.618) = R_{mol}T$$

where p is the pressure in atmospheres, V is the volume in ft^3, R_{mol} is 0.73 atm $ft^3/mol\ ^oR$.

If T_1 is 200°R, calculate p_1 for the maximum yield.

(Univ. Manch.)

11. An empirical relationship between the saturation pressure and temperature for a certain fluid is

$$T = 100\ p^{0.2} + 383^oR$$

where p is in p.s.i.a. The latent heat at 555°R is 124 Btu/lb and the density of the liquid is 40 lb/ft^3. Calculate the density of the vapour and the change in entropy due to the phase change at 555°R.

12. The change in melting points at 1000 atm for ice, acetic acid and tin are given below. Calculate the change in specific volume and entropy during the phase change. State any assumptions made in the calculation.

	Melting point °K	Latent heat of fusion cal/g	Change in melting point °K
Ice	273.2	79.8	- 7.5
Acetic acid	289.8	44.7	+ 25
Tin	505	14	+ 3.4

13. Show that

$$\left(\frac{\partial u}{\partial v}\right)_T = T \left(\frac{\partial p}{\partial T}\right)_v - p.$$

The energy of the radiation which is in equilibrium with an enclosure depends only on the volume v and the wall temperature T. If the energy density is given by $u = aT^4$ and the radiation pressure is $p = ku$, calculate k.

Show that: (1) For an isentropic change in volume vT^3 = constant.

(2) For an isothermal change in volume, the entropy increase per unit volume is $s = \frac{4}{3} aT^3$.

14. Show that for a Rayleigh process (FG)

(a)
$$\frac{1}{V}\left(\frac{\partial T}{\partial V}\right)_{FG} = \frac{\rho}{N^2}\,(1-N^2)\left(\frac{\partial T}{\partial p}\right)_{\rho}$$

where $F = p + \rho V^2$ = impulse function, $G = \rho V$ = mass flux, V = velocity, N = Isothermal Mach number.

(b) For a perfect gas

$$\left(\frac{dT}{d\rho}\right)_{FG} = \frac{T}{pc^2}\,(N - 1)$$

c = Isothermal speed of sound.

15. 50 lb of copper is increased in pressure reversibly and isothermally from zero to 500 atm at 540 R. It is then reduced to zero pressure reversibly and adiabatically. Making realistic assumptions, derive the work done during the whole process and the change in intrinsic energy.

At 540 R the

specific volume v = 0.00180 ft /lb,
volume coefficient of expansion β = 27.4×10^{-6} per R,
isothermal compressibility K = 3.74×10^{-10} ft^2/lbf,
specific heat at constant pressure C_p = 0.092 Btu/lb°R.

(Univ. Liv.)

16. The virial equation of state is

$$pv = R_{mol}T\left(B_1 + \frac{B_2}{v} + \frac{B_3}{v^2} + \ldots\right)$$

Compare this equation with Van der Waals' equation of state and determine the first two virial coefficients, as functions of temperature and the Van der Waals' constants.

Determine the critical temperature and volume (T_c, v_c) for the Van der Waals' gas, and show that

$$B_2 = \frac{v_c}{3}\left(1 - \frac{27T_c}{8T}\right).$$

(Univ. Liv.)

17. The equation of state for a certain gas is

$$\frac{pv}{R_{mol}T} = 1 + p\,e^{-AT}$$

where A is constant.

Show that if data are available on the specific heat at constant pressure at some datum pressure p_o, then the value of C_p at the state (T,p) may be obtained from the value C_{po} at the state (T,p_o) using the expression:

$$C_p - C_{po} = R_{mol}T\,A\,e^{-AT}(2-AT)(p-p_o)$$

(Univ. Manch.)

18. The equation of state for a certain gas is

$$v = \frac{R_{mol}T}{p} + \frac{k}{R_{mol}T}$$

where k is a constant. Show that in any constant enthalpy process from an initial state 1 to a final state 2 the variation of temperature with pressure is given by

$$T_1^2 - T_2^2 = -\frac{4k}{C_p R_{mol}}(p_1 - p_2)$$

If the initial and final pressures are 50 bars and 2 bars respectively and the initial temperature is 300°K, calculate

(a) the value of the Joule-Thomson coefficient at the initial state, and

(b) the final temperature of the gas, given that

$$k = -11.0 \text{ kJm}^3/(\text{kg mol})^2$$

$$C_p = 29.0 \text{ kJ/kg mol deg.C)}$$

(Univ. Manch.)

19. Say how the equation of state in the form of a relationship between pressure, volume and temperature may be used to extend limited data on the entropy of a substance.

A certain gas A has the equation of state

$$pv = R_{mol}T\,(1 + \alpha p)$$

where α is a function of temperature only. Show that

$$\left(\frac{\delta s}{\delta p}\right)_T = - R_{mol}\left(\frac{1}{p} + \alpha + T\frac{d\alpha}{dT}\right)$$

Another gas B behaves as an ideal gas. If the entropy per mol of gas A is equal to that of gas B when both are at pressure p_0 and at the same temperature T, show that at pressure p and temperature T the entropy per mole of gas B exceeds that of gas A by an amount

$$R_{mol}(p - p_0)\left(\alpha + T\frac{d\alpha}{dT}\right)$$

(Univ. Manch.)

20. A gas has the equation of state

$$\frac{pv}{R_{mol}T} = a - bT$$

where a and b are constants. If the gas is compressed reversibly and isothermally at the temperature T' show that the compression will also be adiabatic if

$$T' = \frac{a}{2b}$$

(Univ. Manch.)

21. A gas has the equation of state

$$\frac{pv}{R_{mol}T} = 1 + Ap(T^3 - 9.75\ T_cT^2 + 9T_c^2T) + Bp^2T$$

where A and B are positive constants and T_c is the critical temperature. Determine the maximum and minimum inversion temperatures, expressed as a multiple of T_c.

(Univ. Manch.)

22. Show that, for a pure substance,

$$\left(\frac{\partial s}{\partial T}\right)_p = \left(\frac{\partial s}{\partial T}\right)_v + \left(\frac{\partial s}{\partial v}\right)_T\left(\frac{\partial v}{\partial T}\right)_p$$

Hence show that

$$C_p - C_v = T\left(\frac{\partial p}{\partial T}\right)_v\left(\frac{\partial v}{\partial T}\right)_p$$

Maxwell relations, together with expressions for the specific heats C_p and C_v in terms of T and s may be used without proof. The following table gives values of the specific volume of water, in units of cubic centimetres per gram. A quantity of water is initially at 30 C, 20 bar and occupies a volume of 0.2 m^3. It is heated at constant volume to 50 C and then cooled at constant pressure to 30 C. Calculate the net heat transfer to the water.

Temperature \ Pressure	20 bar	200 bar
30 C	1.0034	0.9956
50 C	1.0112	1.0034

Specific volume of water, cm^3/g

(Univ. Manch.)

23. A gas has the equation of state

$$\frac{pv}{R_{mol}T} = 1 + Np + Mp^2$$

where N and M are functions of temperature. Show that the equation of the inversion curve is

$$p = -\frac{dN}{dT}\Big/\frac{dM}{dT}$$

If the inversion curve is parabolic and of the form

$$(T-T_o)^2 = 4a\,(p_o-p)$$

where T_o, p_o and a are constants, and if the maximum inversion temperature is five times the minimum inversion temperature, show that $a = \frac{T_o^2}{9p_o}$ and give possible expressions for N and M.

CHAPTER 3

THERMODYNAMIC PROPERTIES OF IDEAL GASES AND IDEAL GAS MIXTURES OF CONSTANT COMPOSITION

Thermodynamics of gas mixtures: revision of Avogadro's hypothesis, the mol, general gas law, internal energy, enthalpy and entropy of perfect gases, internal energy, enthalpy, Gibbs and Helmholtz functions for semiperfect gases, variable specific heats. Gas tables, Gibbs-Dalton Law and its application to partial pressures, volumetric composition, internal energy, enthalpy, entropy of gas mixtures. Heats of reaction, heats of formation, adiabatic temperature rise with no dissociation.

Notation

A_o Avogadro number
c velocity
C_p specific heat at constant pressure
C_v specific heat at constant volume
e specific internal energy
E internal energy
f degrees of freedom
g specific Gibbs free energy function
g_o specific Gibbs free energy function at absolute zero
g° specific Gibbs free energy function at temperature T and unit pressure
G Gibbs free energy function
h specific enthalpy
h_f heat of formation
h_o specific enthalpy at absolute zero
H enthalpy
I mass momentum of inertia
k Boltzmann constant
m molecular weight
M number of mols
N number of molecules
p pressure
p_i partial pressure
p_o reference or unit pressure
P_i probability
q_x quantum number
Q heat transfer to system
Q_p heat of reaction at constant pressure
Q_v heat of reaction at constant volume
R gas constant
R_{mol} universal gas constant
s specific entropy
s_o specific entropy at absolute zero
S entropy
T temperature absolute
t degree of freedom variable
u specific internal energy in the absence of motion, gravity, etc.
u_o specific internal energy in the absence of motion, gravity, etc. at absolute zero
U internal energy in the absence of motion, gravity, etc.
V volume
v specific volume
w mass
W' shaft work
x mol-fraction
ε instantaneous energy or time average energy
Ω number of accessible quantum states
ω angular velocity
ΔE_o heat of reaction at constant volume at absolute zero
ΔH_o heat of reaction at constant pressure at absolute zero
ΔS entropy of mixing

In the previous chapter we discussed the general thermodynamic relations for systems of constant chemical composition. It was pointed out that these may be applied to a wide range of engineering applications. A volume of this size precludes examination of all but a few of these applications. In this chapter we will be mainly concerned with the thermodynamic properties of gases. We will consider at first systems containing single gases, then systems containing mixtures of gases of constant chemical composition. Finally we will examine the combustion process and calculate "flame" temperatures, assuming that all the reactants are transferred to products of combustion. Within certain limitations these problems cover a wide range of engineering applications such as the compression ignition engine, the gas turbine, steam generator plant, gas heating units, etc. In the next chapter we will examine certain limitations in the methods we have developed and more exact methods will be discussed to take into account dissociation. In the final chapter, a brief discussion on the thermodynamics of simple elastic, electrical, magnetic and thermoelectric systems will be given. As before we will begin with a revision of the important ideas upon which the analysis will be made.

In many engineering applications the operating conditions with real gases are far from the critical pressure and temperature and the state equation is the same as the ideal gas. The thermodynamic properties such as the internal energy and enthalpy are therefore functions of temperature only. In systems with reacting gases, for example in the combustion chamber of gas turbines in the regions of excess air, we can consider the initial reactants to be a mixture of constant composition and the final products to be of similar form. We can therefore examine the thermodynamics of the gas turbine by replacing the products of combustion by an equivalent gas of constant composition. There is a limitation to this procedure which we will discuss in the next chapter.

It is proposed first of all to discuss the general state equation for ideal gases, then to examine the quantitative determination of the specific heats, the internal energy, the enthalpy and entropy of ideal gases. The specific Gibbs free energy function will be evaluated in terms of the enthalpy and entropy data. The properties of ideal gas mixtures will be examined including the internal energy, enthalpy, and entropy.

Finally, the adiabatic temperature rise during combustion and the calorific values of fuel-air mixtures will be discussed.

State Equation for all Ideal Gases

Let us consider a vessel, volume V, containing a gas A at a pressure p and temperature T. The ideal gas equation is

$$pV = wR_aT \tag{3.1}$$

where p is in lbf per ft^2 or newton per square metre, N/m^2, V is in ft^3, or cubic metres, m^3, T is in ^oR (or ^oK), R_a is the gas constant ft lbf per lb per ^oR (or ^oK) or Joule per kg per ^oK, w is in lb or kilogramme, kg.

Let w equal the molecular weight of the gas and $\overline{V}$ the volume of the vessel for this special case, then

$$\overline{V} = \frac{mR_aT}{p}. \tag{3.2}$$

By Avogadro's Hypothesis equal volumes of all gases contain the same number of molecules at the same pressure and temperature, hence $\overline{V}$ in equation (3.2) is the same for all gases at pressure p and temperature T. This volume is called the molar or molecular volume.

Consider two gases A and B at the same pressure p and temperature T. If m_a, m_b are the molecular weight of the gases, R_a, R_b the gas constants, then

$$p\frac{\overline{V}}{T} = m_aR_a = m_bR_b \tag{3.3}$$

The product mR is, therefore, the same for both gases, and since we have not specified the gases A and B it is clear that mR is the same for all gases. The product mR is called the Universal Gas Constant R_{mol}.

It is convenient for our subsequent discussion to introduce the mol or pound-molecule or kilogramme-molecule (the pound molecule is strictly the pound mass molecule). This is the volume occupied by the molecular weight of a gas. Since this volume is the same for all gases at the same pressure and temperature the mol can be thought of as a "mass". For example, one mol or kilogramme-molecule of CO_2 will have a mass of 44 kg and we can replace the actual mass in kg by the "mass" in mols. A volume of gas containing 22 kg of CO_2 can be considered to have a "mass" of one-half mol (or half kilogramme-molecule). If we have a system containing M mols of gas A of molecular weight m_a then the mass of the system w is equal to Mm_a. Equation (3.1) can therefore be written in the form

$$pV = Mm_aR_aT = MR_{mol}T.$$

The expression

$$pV = MR_{mol}T \tag{3.4}$$

is called the General Gas Law, where p is the pressure, V is the volume, M is the number of mols, R_{mol} is the Universal Gas Constant, T is the temperature.

The state equation for all ideal gases is then

$$pv = R_{mol}T \tag{3.5}$$

where v is the specific volume or the volume per unit mol.

The universal gas constant R_{mol} is a fundamental property of matter. From Avogadro's hypothesis the number of molecules per unit volume at a given pressure and temperature is fixed. If A_o is the number of molecules per mol (called the Avogadro number) then the ratio

$$\frac{R_{mol}}{A_o} = k$$

is a constant for all substances; k is called Boltzmann's constant and occurs frequently in statistical thermodynamics. The values of R_{mol}, A_o, k will depend on the system of units. Typical values for R_{mol} are:

1545	ft-lbf	per lb mol per ^oR
2782	ft-lbf	per lb mol per ^oK
1.985	Btu	per lb mol per ^oR
1.985	CHU	per lb mol per ^oK
1.985	kcal	per kg mol per ^oK
8314	Joules	per kg mol per ^oK

In thermodynamic calculations it is important to use the correct system of units for R_{mol}. In the subsequent discussions we will refer to R_{mol} without the units being specified.

The Internal Energy and Enthalpy of an Ideal Gas

For an ideal gas the specific heats at constant pressure and constant volume are independent of pressure and volume; hence

$$C_v = \left(\frac{\partial u}{\partial T}\right)_v = \frac{du}{dT} \tag{3.6}$$

$$C_p = \left(\frac{\partial h}{\partial T}\right)_p = \frac{dh}{dT}. \tag{3.7}$$

For an ideal gas[†]

$$C_p - C_v = R_{mol} \tag{3.8}$$

Integration of equations (3.6), (3.7) yields

$$h = \int_0^T C_p \, dT + h_o \tag{3.9}$$

$$u = \int_0^T C_v \, dT + u_o \tag{3.10}$$

where h_o, u_o are constants of integration at $T = 0$.

Now by definition

$$h = u + pv.$$

For an ideal gas

$$pv = R_{mol}T,$$

hence

$$h = u + R_{mol}T. \tag{3.11}$$

At the absolute zero, $T = 0$, $h = h_o$, $u = u_o$ and

$$h_o = u_o. \tag{3.12}$$

Classical thermodynamics, the thermodynamics of macroscopic systems, does not supply analytical information on the functional relationship between C_p and T or C_V and T. Equations (3.9) and (3.10) cannot therefore be directly evaluated. To obtain this information, using classical thermodynamics, one must rely on experimental calorimetric measurements. Under certain conditions it is not possible to obtain accurate data

†See Chapter 2, p.45.

and alternative methods must be sought. The basis for these alternative methods is the science of statistical thermodynamics. In statistical thermodynamics a study is made of the particles (molecules, atoms or electrons) within the system, in particular the study of the motions of these particles and the associated energy. The expressions derived from the analytical methods used in statistical thermodynamics are compared with the classical expressions representing the macroscopic properties of the system and certain of the parameters are identified with one another; this correlation then leads to the development of the remaining macroscopic properties in terms of the statistical thermodynamic expressions. For a particular molecular model it is possible by this method to derive analytical expressions for the specific heats in terms of the temperature and many other properties. A number of methods may be used in statistical thermodynamics depending on the system under investigation. One generalized method however, can be used for all studies; this is called quantum mechanics. Some of the important ideas in this theory will be discussed, although no attempt will be made to develop in a rigorous or detailed manner the basic conclusions.

In classical mechanics we are concerned with the study of the motion of particles which have fixed positions in say an x-y-z coordinate system. In quantum mechanics we are concerned in rather a general way with the study of motion of particles whose positions are not fixed, but are probably in certain positions in the x-y-z coordinate system. The reason for this approach is based on a well-known principle† which states it is impossible to locate accurately an infinitesimal particle since by virtue of our observation of the particle we change its velocity and hence its position. The probability of finding a particle in a certain position or strictly speaking within a small region near that position can be expressed by a well-known equation called the Schrödinger Wave Equation.†† This will give us a mathematical expression of the probability of finding a particle at any point x, y, z. It can be shown that for each probability distribution the particle has a fixed value for its energy.††† There may, however, be a number of solutions to the equation each corresponding to a fixed or discrete, but not necessarily the identical, value for the energy of the particle. Hence the energy of a particle is not a continuous function of its position. It is a discontinuous function or, in the terminology of the theory, the energy is quantized. We can

†This is the Uncertainty Principle of Heisenberg.

††It can be shown that the Heisenberg uncertainty principle follows from Schrödinger's equation. For these small fundamental particles we say they may act sometimes as a particle and sometimes as a wave.

†††Although we refer to fixed value of energy, in fact the "uncertainty principle" asserts that we cannot strictly give a fixed value for the energy but some value within a small range of the fixed value.

obtain a clear picture of the physical significance of this if we refer once again to classical mechanics.

In dynamics the direct linear motion of a body is called the translational motion. In an x-y-z coordinate system there are three components of this motion in the directions $\vec{x}$, $\vec{y}$, $\vec{z}$. Associated with each component is one <u>degree of freedom</u>, that is, a particle is <u>free</u> to move in either the $\vec{x}$, $\vec{y}$ or $\vec{z}$ direction and the total degrees of freedom for translation is three. Similarly if a body rotates about axes parallel to the x,y,z axes the body will have <u>three degrees of rotational freedom</u>. In an x-y-z coordinate system a <u>rigid</u> body will have a <u>maximum</u> of six degrees of freedom; N bodies will have a <u>maximum of 6N</u> degrees of freedom. The kinetic energy of a body for each degree of freedom can be represented by At^2 where A is a constant and t a variable associated with the degree of freedom called the <u>degree of freedom variable</u>. For example, the kinetic energy of translation of a body, mass w moving with velocity c is $\frac{1}{2}\,wc^2$, here A equals $\frac{1}{2}$ w and t equals c. The kinetic energy of rotation is $\frac{1}{2}\,I\omega^2$ where A equals $\frac{1}{2}$ I, and t equals ω. (I is the mass moment of inertia, ω is the angular velocity.) In classical mechanics the energy (for example, the kinetic energy) is a continuous function of the position of a particle, that is, if $\varepsilon = \frac{1}{2}\,wc^2$. ε can have any value for a given mass depending only on the velocity c. In quantum mechanics a body can only have discrete values of energy ε and it can only have discrete values of velocity c (in classical terms).

Let us now examine a system of constant energy (in macroscopic terms). If we could examine the system microscopically we would find the molecules are in a constant state of movement. Assuming we can isolate any one molecule and determine its energy it will be found to consist of three parts. The kinetic energy associated with its motion, the potential energy due to the attraction forces between molecules, and the internal energies within the molecular structure. If, for the purpose of our present discussion, we consider the molecule to be a rigid structure we can neglect the intramolecular energies. This then leaves only the kinetic and potential energy. When the dimensions and the number of molecules are such that the volume occupied by the molecules is small compared with the volume of the system the forces of attraction are negligible and the energy of the molecules is associated with their motion only (i.e. kinetic energy); furthermore, the total energy is the sum of the kinetic energy of each molecule. This condition corresponds to the ideal gas. (It will be shown later that the assumption of a rigid body model conforms to a gas with constant specific heats.)

Since the laws of quantum mechanics apply to the system it is not possible to determine the total energy simply by integrating a function representing the energy distribution over the whole system. The location of the molecule is uncertain and the velocities are discontinuous. We can, however, for the special case of the perfect gas solve the Schrodinger wave equation for the boundary conditions of fixed energy and volume.† When we perform the analysis the answer is in the form of a wave function. It is found, however, that a number of numerical solutions of the wave function are possible (indeed in many cases a large number) and each of these solutions defines a probability of a given state of the system. Each of these states is called a quantum state. For convenience each solution is numbered; this number being called the quantum number. The wave function for the special case of an ideal gas can be separated into wave functions for each degree of freedom for an individual molecule. Once again a number of numerical solutions can be derived corresponding as before to the number of quantum states. Associated with each state is a discrete value of energy ε. For an ideal gas, if q_x is the quantum number for the xth degree of freedom (i.e. the degree of freedom in the x direction) and ε_x is the discrete value of energy for the corresponding quantum state, it can be shown that ε_x is proportional to q_x^2. The quantum number q_x can therefore be associated with the velocity in the classical sense.†† The total energy for the three translational degrees of freedom will be

$$\varepsilon_T = \varepsilon_x+\varepsilon_y+\varepsilon_z = a_x q_x^2+a_y q_y^2+a_z q_z^2$$

where a_x, a_y, a_z are constants. When figures are inserted into the expression it can be shown that it is possible for a molecule to be in a number of quantum states (different q's) yet have the same energy; in these cases it is usual to refer to a molecule "occupying" a given energy level or state.†††

Since there is, in general, a larger number of quantum states than molecules, it is possible for a molecule to occupy a range of quantum states or energy levels over a finite interval of time. The instantaneous total energy of a system is equal to the sum of the molecular energies at a particular instant of time. For a macroscopic system, for example, in thermal equilibrium with its surroundings, it will take a finite time to measure the equilibrium state of the system. During this time interval the system will pass through a large number of quantum states each giving about the same total energy. In macroscopic thermodynamics we measure the time average property of the molecules; the magnitude of this value will depend on the time each molecule is in a particular quantum state or energy level.

---

†This solution is obtained because the potential energy due to the attraction forces between molecules is neglected.

††In classical mechanics $\varepsilon \propto c^2$, for the perfect gas in quantum mechanics $\varepsilon \propto q^2$, hence c and q are analogous.

†††If $a_x=a_y=a_z$, $q_x=1$, $q_y=5$, $q_z=7$, $\varepsilon_T=75a_x$; if $q_x=5$, $q_y=5$, $q_z=5$, $\varepsilon_T=75a_x$. Thus we have two quantum states with the same value for ε_T.

To obtain a numerical value for the energy we must resort to statistical methods. The time a molecule is in a particular state divided by the total time we are making the observation is called the thermodynamic probability† P_i of the molecule being in state i. For example, if a molecule occupies two energy levels $\varepsilon_1 = 1$ and $\varepsilon_2 = 3$ energy units, and if over a time interval of one second it occupies state ε_1 for one-quarter of a second and ε_2 for three-quarters of a second, then the average energy is $\frac{1}{4}\varepsilon_1 + \frac{3}{4}\varepsilon_2 = \frac{1}{4} + \frac{9}{4} = 2\frac{1}{2}$ units. The probability of the molecule being in state 1 is therefore 0.25 and in state 2, 0.75, or $P_1 = 0.25$, $P_2 = 0.75$. In general, therefore, ε_T the "time average" energy is $\varepsilon_T = \Sigma P_i \varepsilon_i$ and the sum of the probabilities is unity, $\Sigma P_i = 1$.

To determine the thermodynamic probability we must consider the macroscopic system. For an ideal gas we assume that a molecule can occupy any state and its probability of being in a particular state is of equal weight, that is, it is not biased to one state or another. There are, however, many more states than there are molecules. For an isolated system (constant internal energy), since each state is equally probable, the probability is just the reciprocal of the total number of states. It is impossible, however, to determine this number. We therefore look at a more useful system. This consists of a vessel immersed in a large constant-temperature bath. Whilst the system (vessel plus bath) is of constant internal energy in the macroscopic sense, the contents of the vessel in the microscopic sense are not at constant energy, since the vessel is in thermal contact with the large bath. For this condition it can be shown that the thermodynamic probability P_i is a function of the energy ε_i (of that state). Using this relationship we now can determine the total energy of the vessel. The energy levels ε_i are governed by the probable distribution of the molecules in the vessel and this depends on the geometry only. For an ideal gas the difference between each energy level is small compared with kT, where k is the Boltzmann constant, and we can replace the discrete energy levels by a continuous function. The integration of this function enables us to obtain an analytical function for the probability for each degree of freedom. It can be shown that for these conditions the "time average" energy per molecule for each degree of freedom in which the freedom variable is squared is equal to $\frac{1}{2}$ kT.

---

†The term probability is used in two contexts. Up to this point we have been concerned with the probable distribution of the particles in the box, this is given by the wave equation. For each distribution there will be a fixed value for the energy of the particles. Since there are many energy or quantum states some may be more probable than others. The probability of a particle is in one energy state or another over a period of time is the second use of the term. This is usually called the thermodynamic probability.

For an ideal gas in which the molecule is considered as a rigid body each degree of freedom has a squared variable, each will therefore contribute the same quantity of energy, namely, $\frac{1}{2}$ kT per molecule. Since the potential energy is neglected the total internal energy or the macroscopic internal energy is therefore

$$U = \tfrac{1}{2} NfkT \tag{3.13}$$

where N = number of molecules in the system, f = degrees of freedom.

If A_o is Avogadro's number, i.e. the number of molecules per mol, then the universal gas constant (R_{mol}) is $A_o k$ and the number of molecules N equals the number of mols M times A_o, i.e. $N = MA_o$.

Then

$$\frac{U}{M} = u = \tfrac{1}{2} fR_{mol}T \tag{3.14}$$

Equations (3.13) and (3.14) are mathematical formulations of the principle of the equipartition of energy. This states that the total energy of a system is equally shared amongst the degrees of freedom. This strictly applies only to the special case of the ideal gas in which the potential energy is neglected and the molecules are considered to be rigid bodies. Differentiating equation (3.14) with respect to T we obtain an expression for the specific heat at constant volume.

$$C_v = \left(\frac{\partial u}{\partial T}\right)_v = \tfrac{1}{2} fR_{mol} \tag{3.15a}$$

and

$$C_p = C_v + R_{mol} = (\tfrac{1}{2} f+1)R_{mol}. \tag{3.15b}$$

For these particular cases, therefore, the specific heats are independent of the temperature.†

For a monatomic gas the molecule consists of a single atom with only three translational degrees of freedom. Hence f = 3 and

$$C_v = \frac{3}{2} R_{mol}, \quad C_p = \frac{5}{2} R_{mol},$$

$$\gamma = \frac{C_p}{C_v} = 1.67.$$

†As pointed out in the footnote to p.28 (Chapter 2) in some texts the "ideal" gas with constant specific heats is called the "perfect" gas, i.e. the gas is calorically perfect.

The atoms of a diatomic gas can be considered to be two rigid bodies joined by a massless rod. Assuming the molecule cannot rotate about an axis parallel to the rod, but can rotate about the other two axes, five degrees of freedom are possible and

$$C_v = \frac{5}{2} R_{mol}, \quad C_p = \frac{7}{2} R_{mol}, \quad \gamma = 1.4.$$

The atoms of a triatomic gas can be considered to be three rigid bodies joined by massless rods in triangular formation. Six degrees of freedom (three translational and three rotational) are possible and

$$C_v = 3R_{mol}, \quad C_p = 4R_{mol}, \quad \gamma = 1.33.$$

The models of rigid bodies joined by massless rods indicate that the specific heats are independent of temperature. In practice this holds only for a moderate temperature range corresponding to normal room temperatures. In fact the specific heats increase with temperature. The assumption of rigid bodies and massless rods is, therefore, not satisfactory. Within the molecule there will be intra-molecular vibrations associated with the atomic structure. Furthermore, within the atom there will be electronic vibrations associated with the electron orbits and within the nucleus there will be spins. Little is known of the effect of the latter on the macroscopic properties of systems. For an ideal gas we can assume that the intra-molecular phenomena are uncoupled. The energy level of each form of motion is independent of the other and a probability distribution can therefore be derived for each form. The "time average" molecular energy will be the sum of each energy contribution. If we let

ε_T = the total "time average" energy

ε_{tr} = "time average" translational energy

ε_r = "time average" rotational energy

ε_v = "time average" vibrational energy

ε_e = "time average" electronic energy

then

$$\varepsilon_T = \varepsilon_{tr} + \varepsilon_r + \varepsilon_v + \varepsilon_e;$$

all except the first term on the right are non-linear functions of temperature T.

If we write

$$\varepsilon_{tr} = aT$$

and the remainder as $\varepsilon = \varepsilon(T)$ with the appropriate suffix we have

$$C_v = \left(\frac{\partial \varepsilon_T}{\partial T}\right)_v = a + \varepsilon_r'(T) + \varepsilon_v'(T) + \varepsilon_e'(T) \qquad (3.17)$$

where the prime represents differentiation with respect to T.†

The second and third terms come more into prominence when the temperature T is increased; the fourth term is only of significance at very high temperatures. The functional relationships in (3.17) are complex and it is more convenient to express the whole expression in an empirical form

$$C_v = A + BT + CT + DT \qquad (3.18)$$

where A, B, C, D are derived from experiments.

The last three terms in expression (3.17) do not provide a direct theoretical calculation for the specific heats. Certain unknowns are included which can only be obtained from spectrographic measurements. These now form the basis of the latest information for specific heats.

For an <u>ideal</u> gas the specific heats are related by

$$C_p - C_v = R_{mol}.$$

In place of equation (3.18) we can obtain a similar relation for the specific heats at constant pressure.

In practice we are generally interested in the internal energy or enthalpy of ideal gases. It is therefore not convenient to present the data in the form shown in equation (3.18).

The internal energy is given by

$$du = C_v \, dT \qquad (3.19)$$

since u is a function of T only (see Chapter 2). Now C_v is a function of T and we can write (3.19) as

---

†An ideal gas whose specific heat is a function of temperature is called in certain texts an imperfect gas (see footnote to p.28, Chapter 2). It is implied that the gas is calorically imperfect. The major statistical assumptions for an ideal gas are (1) the degrees of freedom are uncoupled, (2) the energies are uncoupled.

$$du = C_v(T)\ dT \tag{3.20}$$

where $C_v(T)$ indicates that C_v is a function of T. The problem when integrating equation (3.20), in order to obtain numerical values for the internal energy at any temperature T, is the determination of the constant of integration, or, in physical terms to establish whether or not there is a state of zero internal energy. The evaluation of the specific heats based on the quantum theory refers to the change in internal energy above the ground or zero energy state. This may be used as a basis for the calculation of the macroscopic internal energy. If we let u_o be the internal energy of a substance at the absolute zero corresponding to the ground state then integration of equation (3.20) will give

$$u = \int_o^T C_v(T)\ dT + u_o \tag{3.21a}$$

or

$$u = u(T) + u_o \tag{3.21b}$$

where

$$u(T) = \int C_v(T)\ dT.$$

To calculate the change in internal energy of a gas of constant composition the constant u_o is not required, on the other hand to calculate the change in internal energy of a mixture of gases in which there is a change in composition a knowledge of u_o may be required. This arises because the value of u_o is not the same for all gases. If the composition of the mixture remains constant, the changes in internal energy do not involve the constants u_o and these will cancel out. The important term in these cases is the first term on the right-hand side of equation (3.21b). If we are concerned with gas mixtures in which the composition can vary over the temperature range then, although we cannot neglect the constant u_o, we will find that it is not necessary to know its value but only the differences in u_o for the gases. These can be determined by calorimetric or spectrographic experiments as will be described later.

By selecting the constant u_o at absolute zero $u_o = h_o$ and the enthalpy relation as a function of temperature is

$$dh = C_p\ dT = C_p(T)\ dT$$

$$h = \int C_p(T)\ dT + h_o$$

$$h = h(T) + h_o = h(T) + u_o \tag{3.22}$$

where

$$h(T) = \int C_p(T)\ dT.$$

Data to compute enthalpy or internal energy changes are generally presented in the form of tables or charts. There are a number of alternative methods. These will be discussed.

The first method is a direct presentation of the function h(T) or u(T) in polynomial form, for example,

$$\frac{h(T)}{R_{mol}T} = a_1 + a_2T + a_3T^2 + a_4 T^3 + a_5T^4$$

or

$$h(T) = R_{mol} (a_1T + a_2T^2 + a_3T^3 + a_4T^4 + a_5T^5)$$

$$h(T) = R_{mol} \sum_{j=1}^{j=5} a_j T^j$$

Typical values for the constant a_j are given in Table A.2.†

The second method, which we will discuss with reference to the enthalpy data only, although the same remarks apply to internal energy data, uses a modified approach.

For any substance the change in enthalpy between temperatures T_1 and T_2 is

$$h_2-h_1 = \int_{T_1}^{T_2} C_p(T)\, dT = h(T_2)-h(T_1)$$

since the constant of integration is eliminated. We can rewrite the above expression in the form

$$\overline{C}_p(T_2-T_1) = h(T_2)-h(T_1) \qquad (3.23)$$

where $\overline{C}_p$ is the <u>mean</u> specific heat over the temperature range T_1 to T_2. From tabulated <u>mean</u> specific heats over a range of temperatures the changes in enthalpy can be calculated directly from equation (3.23). The mean specific heats will vary with temperature. It is usual to specify the mean specific heat at a particular temperature. This is the <u>mean</u> specific heat over the range from the reference temperature (T_0) to the temperature (T) at which the value is given. Once again it is possible to write the <u>mean</u> specific heat at constant pressure in the form

---

†See Appendix, pp. 272-308.

$$\overline{C}_p = a + bT + cT + \ldots \tag{3.24}$$

The constants a, b, c, etc., may be tabulated. It should be emphasized that the constants a, b, c, etc., apply over the range specified and that if the enthalpy change between two temperatures say, T_1 to T_2, is required then

$$h_2 - h_1 = (\overline{C}_p)_{T_2}(T_2 - T_o) - (\overline{C}_p)_{T_1}(T_1 - T_o) \tag{3.25}$$

where $(\overline{C}_p)_{T_2}$ is the mean specific heat over the temperature range T_o to T_2 and so on.

In practice one need only calculate the enthalpy change from tables or charts and the internal energy can then be found by deducting $R_{mol}(T_2 - T_1)$. If only internal energy data are available then the enthalpy change can be calculated by adding $R_{mol}(T_2 - T_1)$.

We have discussed two alternative methods of presenting specific heat or internal energy data. In some texts the reference temperature may not be the absolute zero. This does not create any difficulty provided calculations are not carried out below the reference temperature. For mixtures of gases of constant composition no problem arises with this formulation. If there is a change in composition of the gases, due to say combustion, then the difference in the internal energies of the component gases is required at the reference temperature. This can be computed if calorimetric data are provided. Examples of the application of these data are given at the end of the chapter.

So far we have discussed the variation in u or h with temperature for an ideal gas. For real gases as the temperature approaches absolute zero the enthalpy is not

$$h = \int C_p(T)\, dT + h_o.$$

Due to phase changes the gas liquefies and may eventually solidify. The latent heats of vaporization, fusion and sublimation must be included in the calculations.

In some texts an equivalent h is defined to allow for the deviation from the ideal gas relationship. Space precludes a discussion of this point.† The published enthalpy data for

---

†See Tables of Properties of Gases with Dissociation Theory and its Applications, by E.W. Geyer and E.A. Bruges, published by Longmans (1948).

gases at all temperatures include corrections for the liquid and solid phases.

Entropy of an Ideal Gas - The Third Law of Thermodynamics

For nearly all cycle analysis and advanced combustion calculations we require to know the entropy of a gas.

We know from Chapter 2 that the entropy s is related to h, p and T by

$$ds = \frac{dh}{T} - R_{mol}\frac{dp}{p}. \qquad (3.26)$$

Since h is a function of temperature, we can directly integrate equation (3.26)

$$s = \int_{T_o}^{T} \frac{dh}{T} - R_{mol}\ln\frac{p}{p_o} + s_o. \qquad (3.27)$$

Two major problems arise in using (3.27) to establish numerical data for the entropy of a substance. The first is the integration of the first term and the second the evaluation of s_o. As before it is convenient to take the lower limit T_o as the absolute zero. In general we are concerned with gases. However, as all substances (except helium) approach the absolute zero they solidify and we cannot simply write the relationship $dh = C_p dT$ in equation (3.27). We must also include the latent heats of vaporization and fusion (or sublimation). This raises no difficulty until we finally reach an expression for the entropy of the solid state. The enthalpy will be in the form

$$dh = C_p dT = C_p(T)dT \text{ and } \frac{dh}{T} = \frac{C_p(T)dT}{T}.$$

It is clear that at the lower limit, $T_o = 0$, there is a possibility of the function having an infinite value if $C_p(T)>0$. Below a certain temperature it can be shown that the specific heat of a solid may be represented by the law $C_p = aT$ where a is a constant. If it is <u>assumed</u> that this law applies as we <u>approach</u> the absolute zero, then we can integrate the first term. This term is therefore a direct function of the temperature and can be replaced by the symbol s(T), where

$$s(T) = \int_{T_o}^{T} \frac{dh}{T}. \tag{3.28}$$

The entropy is therefore

$$s = s(T) - R_{mol} \ln \frac{p}{p_o} + s_o. \tag{3.29}$$

If we make p_o equal to unit pressure (e.g. one atmosphere) then

$$s = s(T) - R_{mol} \ln p + s_o \tag{3.30}$$ †

where p is the pressure in the same units (e.g. atmospheres absolute).

The constant of integration s_o raises a different problem. For changes in entropy of single substances, the constant s_o will cancel out. Similarly for mixtures of substances of constant composition calculations of entropy changes will not involve s_o. For mixtures in which there may be changes in composition a knowledge of s_o would appear to be necessary in order to calculate the entropy change. By the methods of statistical thermodynamics, in association with our ideas of entropy from the second law, it is possible to correlate the entropy of a substance with the number of accessible quantum states of a substance. This is usually expressed in the form

$$s = k \ln \Omega \tag{3.31}$$

where s is the entropy, k the Boltzmann constant and Ω the number of accessible quantum states. For determining the number of quantum states we must consider not only the molecular, intra-molecular and electronic states but also the nuclear states (nuclear spins). At the time of writing complete knowledge on this subject is not available and it is not possible using (3.31) to evaluate the minimum value of the entropy of a substance. For a number of years the problem of absolute entropy has been the subject of intense research by physicists and others. These studies have been directed to the formulation of an axiom which is called the Third Law of Thermodynamics. A number of versions of the third law have been

† The usual convention is to specify the unit pressure as one atmosphere absolute, all pressures are then in atmospheres absolute. The new international unit of pressure is the bar this is equal to 10^5 newtons/m^2 or 10^6 dynes/cm^2 or 14.696 lbf/in^2 or 1.01325 atms. In Table A.2 the values for s(T) are based on p_o = 1.01325 bars.

produced and each in turn subjected to analytical discussion and subsequent reformulation. The most satisfactory formulation at present is that due to Fowler and Guggenheim.† The following is Denbigh's†† statement of the axiom based on Fowler and Guggenheim: "For an isothermal process involving only phases in internal equilibrium the change in entropy approaches zero at the absolute zero. By internal equilibrium it is implied that the state of the phase is determined entirely by its temperature, pressure and composition." This statement enables entropy changes to be calculated for mixtures whose composition does not remain constant, provided that the components of the mixture exist at very low temperatures as pure crystals or liquids. The difference in the constant s_0 for the various components of the mixture may be considered to be zero. For a system at constant pressure the only term required in entropy change calculation is therefore s(T), and s_0 may be taken as zero.

Not all the entropy data are published from a reference temperature at 0 °R or 0 °K. In these cases s_0 will refer to the entropy at the reference temperature T_0 and pressure p_0.

The Gibbs Free Energy Function for an Ideal Gas

We will consider only one more property of state of a substance before examining mixtures. With a knowledge of the entropy and the enthalpy of a substance we can calculate any of the other thermodynamic properties through the relations given in Chapter 2. Let us examine the Gibbs free energy function g.

By definition

$$g = h - Ts. \qquad (3.32)$$

Now from (3.22)

$$h = \int C_p(T)dT + h_0 = h(T) + h_0$$

and from (3.30) at a pressure p_0 = unit pressure (usually 1 atmosphere absolute).

$$s = s(T) - R_{mol} \ln p + s_0.$$

Substituting in (3.32) we have

---

†Fowler and Guggenheim, Statistical Thermodynamics (Cambridge) 1949.

††K. Denbigh, The Principles of Chemical Equilibrium (Cambridge) 1961.

$$g = (h_o - Ts_o) + (h(T) - Ts(T)) + R_{mol}T \ln p. \quad (3.33)$$

If we let

$$g_o = h_o - Ts_o \quad (3.34)$$

$$g(T) = h(T) - Ts(T) \quad (3.35)$$

then

$$g = g(T) + R_{mol}T \ln p + g_o. \quad (3.36)$$

It is more usual to combine g_o and $g(T)$ as

$$g^o = g(T) + g_o \quad (3.37)$$

then

$$g = g^o + R_{mol}T \ln p. \quad (3.38)^\dagger$$

g^o is the specific Gibbs function at temperature T and pressure of 1 atmosphere†† and is a function of temperature only.

The Gibbs function is not an intensive property. In the next chapter it will be shown that the specific Gibbs function is equal to an intensive property called the chemical potential.††† The same difficulty arises in the computation of an absolute value of g as in the calculation of entropy, internal energy and enthalpy. The constant g_o is not required for changes in Gibbs function for pure substances or mixtures of constant composition. For mixtures of varying composition, where a knowledge of the change in Gibbs function is required, the constant g_o must be known. This can be evaluated through calorimetric or spectrographic measurements. In practice the numerical value of the specific Gibbs function is quoted as a chemical potential. It will be shown later that the difference in the Gibbs functions for a reacting mixture is associated with equilibrium data.

Mixtures of Ideal Gases

In engineering problems we are usually concerned with systems containing a mixture of gases rather than a single gas. In studying the thermodynamics of these systems we are thrown once again upon actual experimental observation to determine the

---

†The full form of equation (3.38) is $g = g^o + R_{mol}T \ln(p/p_o)$ where g^o is the Gibbs function at the reference pressure p_o.

††See footnote to p.80.

†††See footnote to p.126.

criteria upon which these studies can be made. A large number of experiments with real gas mixtures indicate that, over the range of pressures and temperatures of interest, the gas mixtures approximate to mixtures of ideal gases. We will concern ourselves with these types of mixtures. There are a number of ways of defining a mixture of ideal gases. It is proposed to use the following definitions:

(a) The gas mixture as a whole obeys the equation of state $pV = MR_{mol}T$ where M is the total number of mols of all kinds.

(b) The total pressure of the mixture is the sum of the pressures which each component would exert if it alone occupied the whole volume of the mixture at the same temperature.

(c) The internal energy, enthalpy and entropy of the mixture are respectively equal to the sums of the internal energies, enthalpies and entropies which each component of the mixture would have if each alone occupied the whole volume of the mixture at the same temperature.

The first condition implies that the gas mixture acts as if it were a single component ideal gas. The last two conditions are the well-known Gibbs-Dalton Laws.† The first two conditions both indicate that the gas molecules are considered to move independently of one another in the whole volume of the system.

Let us consider a mixture of ideal gases A, B, C at temperature T in a container of volume V.

Now the total number of mols is

$$M = M_a + M_b + M_c.$$

The ratios of the number of mols of a component gas to the total number of mols of the gas mixture is called the mol fraction x. Thus

$$x_a = \frac{M_a}{M}, \quad x_b = \frac{M_b}{M}, \quad x_c = \frac{M_c}{M} \tag{3.39}$$

and

$$x_a + x_b + x_c = 1.0. \tag{3.40}$$

---

†J.H. Keenan, Thermodynamics, published by J. Wiley & Sons, 1940, and S.R. Montgomery, The Second Law of Thermodynamics, published by Pergamon Press, 1965.

We define the partial pressure p_i of a gas i as

$$p_i = x_i p. \tag{3.41}$$

Then

$$p_a = x_a p, \quad p_b = x_b p, \quad p_c = x_c p. \tag{3.42}$$

Now from (a) the total pressure p is given by

$$p = M \frac{R_{mol} T}{V}. \tag{3.43}$$

From (3.42) and (3.43)

$$p_a = x_a p = x_a M \frac{R_{mol} T}{V}.$$

Therefore

$$p_a = M_a \frac{R_{mol} T}{V}. \tag{3.44a}$$

This is the original statement of Dalton, namely, each gas acts as if it alone occupied the whole vessel.

In general for an <u>ideal</u> gas the partial pressure p_i is given by

$$p_i = M_i \frac{R_{mol} T}{V}. \tag{3.44b}$$

The total pressure p is given by

$$p = M \frac{R_{mol} T}{V} = (M_a + M_b + M_c) \frac{R_{mol} T}{V}$$

$$= (x_a + x_b + x_c) M \frac{R_{mol} T}{V}$$

$$= x_a p + x_b p + x_c p$$

$$p = p_a + p_b + p_c. \tag{3.45}$$

Thus the total pressure is the sum of the partial pressures.

Both the mol fraction x_i of gas i in a mixture of gases and the partial pressure p_i can be obtained directly from the volumetric analysis of the mixture. A volumetric analysis is usually carried out at a constant pressure p and temperature T. The volume v_i of gas i will be

$$v_i = \frac{M_i R_{mol} T}{p}$$

and the total volume of the mixture will be

$$V = \frac{M R_{mol} T}{p}.$$

The volumetric analysis at p and T will then be

$$\%v_i = 100 \frac{V_i}{V} = 100 \frac{M_i}{M} = 100 x_i$$

or

$$x_i = \frac{\%v_i}{100} \tag{3.46}$$

And from (3.41) the partial pressure p_i is

$$p_i = x_i p$$

then

$$p_i = \frac{\%v_i}{100} p. \tag{3.47}$$

The condition (c) in the Gibbs-Dalton Law implies no heat of mixing. If one uses the notation e, h, for the specific internal energy, enthalpy per unit mol, then we have for the gas mixture

$$E = Me = \sum M_i e_i \tag{3.48}$$

$$H = Mh = \sum M_i h_i \tag{3.49}$$

where e, h are the specific internal energy and the enthalpy for the mixture and e_i, h_i are the specific internal energy and the enthalpy of the component gas i. These may be directly evaluated in terms of the mol fractions.

From (3.48)

$$e = \frac{1}{M} \sum M_i e_i = \sum \frac{M_i}{M} e_i.$$

Hence

$$e = \sum x_i e_i \tag{3.50}$$

and similarly

$$h = \sum x_i h_i. \tag{3.51}$$

If we consider a system in the absence of motion, gravity, electricity, magnetism and capillarity,

then
$$e = u$$

and
$$u = \sum x_i u_i . \tag{3.52}$$

For an ideal gas
$$h = u + R_{mol}T.$$

For a mixture of ideal gases

$$h = \sum x_i h_i$$

$$h = \sum x_i (u + R_{mol}T)_i$$

$$h = \sum (x_i u_i) + R_{mol}T \sum x_i$$

$$= \sum (x_i u_i) + R_{mol}T \tag{3.53}$$

$$h = u + R_{mol}T \tag{3.54}$$

Heats of Reaction or Calorific Values and Adiabatic Combustion

The methods discussed above can be used to determine the heats of reaction for combustion processes and the calculation of gas temperatures.

In Fig. 3.1 a calorimeter is diagrammatically represented by a control volume V. The reactants (air plus fuel) enter the calorimeter at temperature T and pressure p. The fuel is ignited and the products of combustion leave the calorimeter at a temperature slightly above temperature T and pressure p. The heat transferred between the products and the water-cooled jacket is equal to the heat of reaction at constant pressure.

Let suffix R refer to reactants and P refer to products.

For the control volume the heat exchanged equals the change in enthalpy of the gas mixture. Thus

$$Q = H_P - H_R . \tag{3.55}^\dagger$$

†For processes with fluid motion H_P, H_R are the stagnation enthalpies.

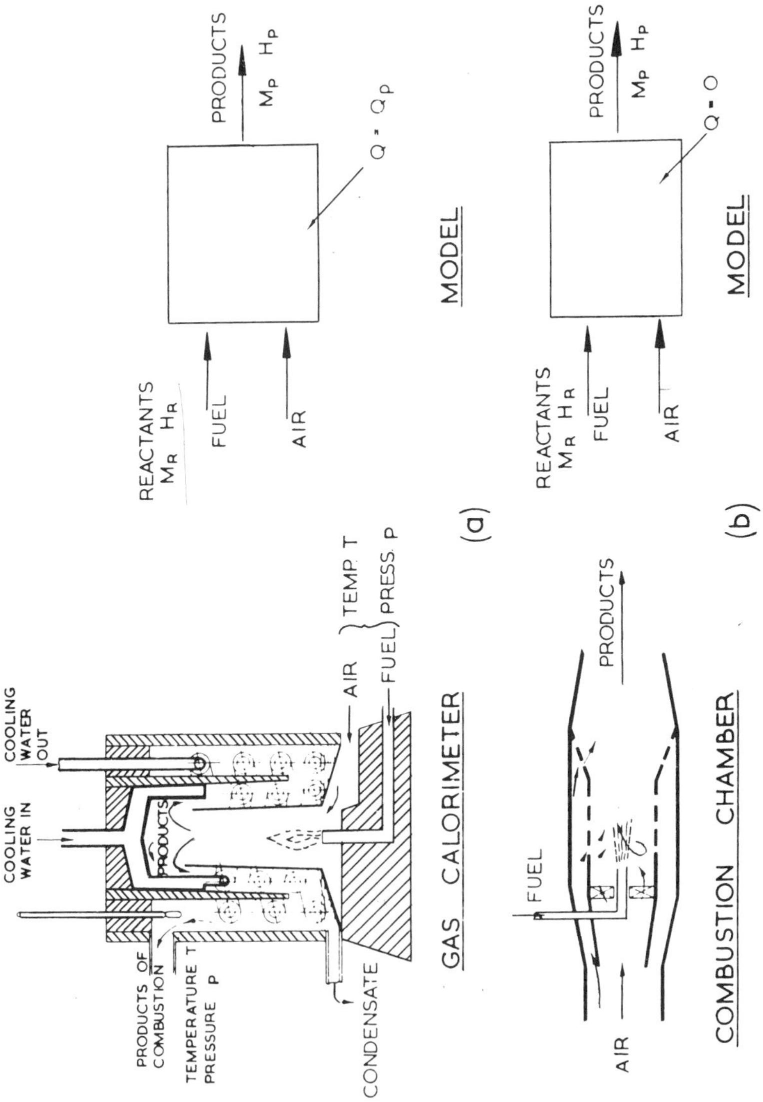
COOLING WATER IN
COOLING WATER OUT
PRODUCTS
PRODUCTS OF COMBUSTION
TEMPERATURE T
PRESSURE p
CONDENSATE
AIR
FUEL
TEMP. T
PRESS. p
GAS CALORIMETER
(a)
REACTANTS
M_R H_R
FUEL
AIR
Q = Q_p
PRODUCTS
M_P H_P
MODEL
FUEL
AIR
PRODUCTS
COMBUSTION CHAMBER
(b)
REACTANTS
M_R H_R
FUEL
AIR
Q = 0
PRODUCTS
M_P H_P
MODEL

Fig. 3.1.

Now

$$H_P = M_P h_P \tag{3.56}$$

where

$$h_P = \sum (x_i h_i)_P \tag{3.57}$$

and

$$h_i = h_i(T) + h_{io}. \tag{3.58}$$

Also

$$H_R = M_R h_R \tag{3.59}$$

$$h_R = \sum (x_i h_i)_R. \tag{3.60}$$

We can write H_P as

$$H_P = M_P(h_P(T) + h_{oP}) \tag{3.61}$$

where

$$h_P(T) = \sum (x_i h_i(T))_P \tag{3.62}$$

$$h_{oP} = \sum (x_i h_o)_P. \tag{3.63}$$

Similarly

$$H_R = M_R(h_R(T) + h_{oR}) \tag{3.64}$$

where

$$h_R(T) = \sum (x_i h_i(T))_R \tag{3.65}$$

$$h_{oR} = \sum (x_i h_o)_R \tag{3.66}$$

The heat exchanged Q is equal to the heat of reaction at constant pressure. This is equal to Q_p. (Since heat is lost from the system Q_p will be negative for an exothermic reaction.)

For the constant pressure calorimeter we have

$$Q_P = M_P h_P(T) + M_P h_{oP} - M_R h_R(T) - M_R h_{oR}$$

$$Q_p = M_P h_P(T) - M_R h_R(T) + \Delta H_o \tag{3.67}$$

where

$$\Delta H_o = M_P h_{oP} - M_R h_{oR} \tag{3.68}$$

If we plot the enthalpies of the reactants and products as a function of temperature T on a single diagram we would obtain graphs of the form shown in Fig. 3.2.

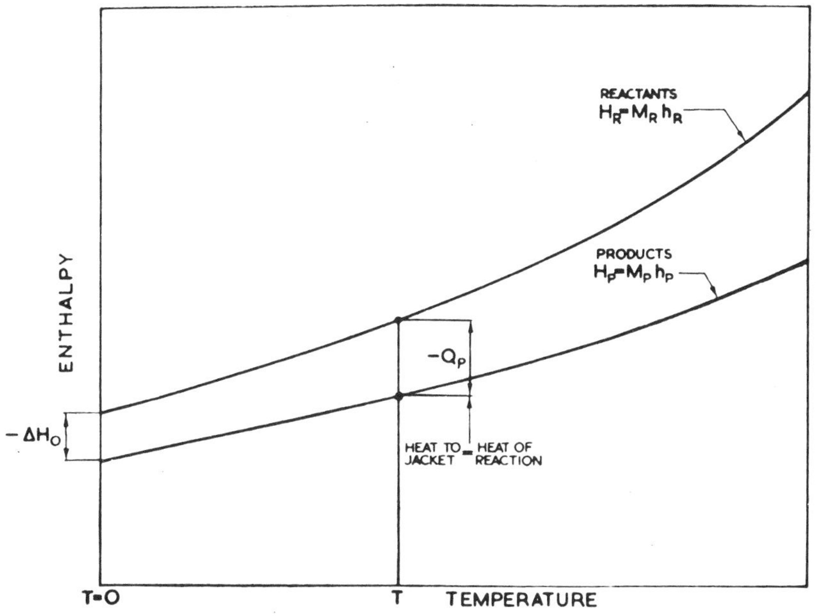

FIG. 3.2. Enthalpy–temperature diagram for constant pressure calorimeter.

It will be observed that since heat is lost in the calorimeter, H_R is greater than H_P at constant T and the enthalpy curve for the reactants is above the corresponding curve for the products. The value ΔH_0 is the difference in enthalpy at absolute zero for the combustion process. The heat of reaction is equal to the vertical displacement between the curves at constant temperature.

The heats of reaction are negative for exothermic reactions. For endothermic reactions, in which heat is absorbed, the heats of reaction are positive. It is clear from expression (3.67) that the heats of reaction are dependent on temperature. It is usual to quote the standard values at a temperature of 25°C and pressure of 1 atmosphere. The heats of reaction at constant pressure are sometimes called the calorific value at constant pressure. From the heats of reaction and a knowledge of the enthalpy data, in the form h(T), calculations of temperature changes for processes involving combustion can be made without a knowledge of the enthalpy constant h_0.

Consider the system shown in Fig. 3.1(b) in which Q = 0. There is no heat lost and the combustion process is adiabatic. The temperature of the products T_P is now greater than the temperature of the reactants T_R. The steady flow energy equation for the control volume is

$$0 = H_P - H_R.$$

Substitution of equations (3.61) to (3.68) gives

$$0 = M_P h_P(T)_P - M_R h_R(T)_R + H_0. \qquad (3.69)$$

In this case $h_P(T)_P$ and $h_R(T)_R$ refer to the numerical value of these functions at temperature T_P and T_R respectively. Let the calorific value or heat of reaction $(Q_p)_s$ data be available at some standard temperature T_s. Then from equation (3.67) we have

$$-\Delta H_0 = M_P h_P(T)_s - M_R h_R(T)_s - (Q_p)_s. \qquad (3.70)$$

Substituting into equation (3.69) we obtain

$$0 = M_P(h_P(T)_P - h_P(T)_s) - M_R(h_R(T)_R - h_R(T)_s) + (Q_p)_s. \qquad (3.71)$$

For a reaction with no dissociation the following values are known:

$$M_P,\ M_R,\ h_P(T)_s,\ h_R(T)_s,\ h_R(T)_R,\ (Q_p)_s.$$

The only unknown is $h_P(T_P)$ and from this we can calculate T_P. We can represent this process on an enthalpy-temperature diagram as shown in Fig. 3.3.

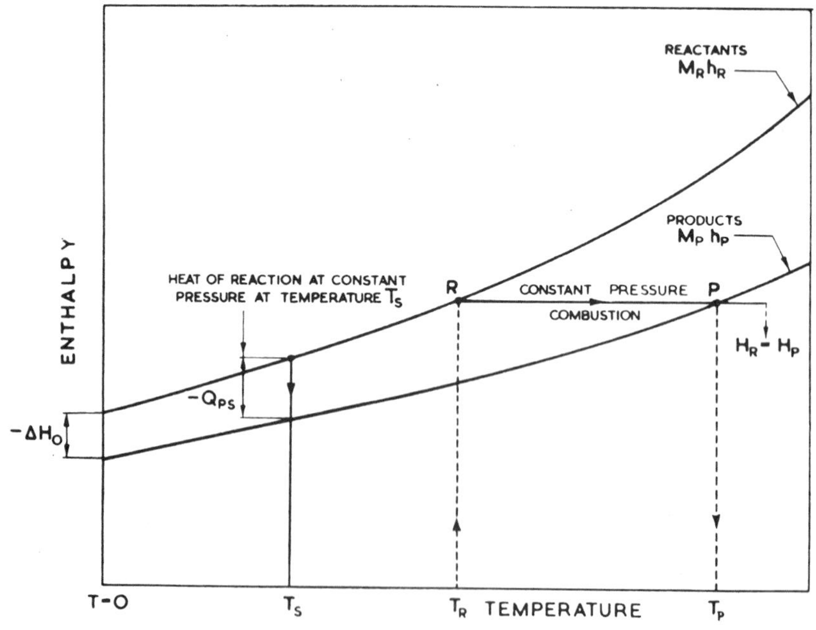

FIG. 3.3. Enthalpy-temperature diagram for adiabatic combustion at constant pressure.

Many combustion processes are at constant volume. Calorimetric experiments at constant volume include the well-known bomb calorimeter. Let us consider a closed system containing reactants (fuel plus air). After combustion the temperatures of products are cooled to the same temperature as the reactants. The heat lost by the gases in the system equals the heat of reaction of the fuel plus air at constant volume. For this case from the first law

$$Q = E_P - E_R \tag{3.72}$$

Q is equal to Q_v the heat of reaction at constant volume (where Q_v is negative for an exothermic reaction).

$$Q_v = E_P - E_R. \tag{3.73}$$

As before we can write

$$E_P = M_P(e_P(T) + e_{oP}) \tag{3.74}$$

$$E_R = M_R(e_R(T) + e_{oR}) \tag{3.75}$$

$$e_P(T) = \sum(x_i e(T))_P \tag{3.76}$$

$$e_{oP} = \sum(x_i e_o)_P \tag{3.77}$$

$$e_R(T) = \sum(x_i e(T))_R \tag{3.78}$$

$$e_{oR} = \sum(x_i e_o)_R \tag{3.79}$$

and the heat of reaction, Q_v, is

$$Q_v = M_P e_P(T) - M_R e_R(T) + \Delta E_o \tag{3.80}$$

$$\Delta E_o = M_P e_{oP} \cdot M_R e_{oR}. \tag{3.81}$$

The combustion process is represented on the internal energy-temperature diagram shown in Fig. 3.4.

The same remarks about ΔE_o apply as ΔH_o. Since h_o equals e_o for all ideal gases (see p.68), ΔE_o, equals ΔH_o.

For an adiabatic combustion process at constant volume, $E_P = E_R$ and the calculation procedure is the same as for a constant enthalpy. This is represented in diagrammatic form in Fig. 3.5.

The heat of reaction at constant volume is dependent on the temperature. The two heats of reaction Q_p and Q_v are related through the following expressions.

From equations (3.67) and (3.80) we have

$$Q_p - Q_v = M_P(h_P(T) - e_P(T)) - M_R(h_R(T) - e_R(T)). \tag{3.82}$$

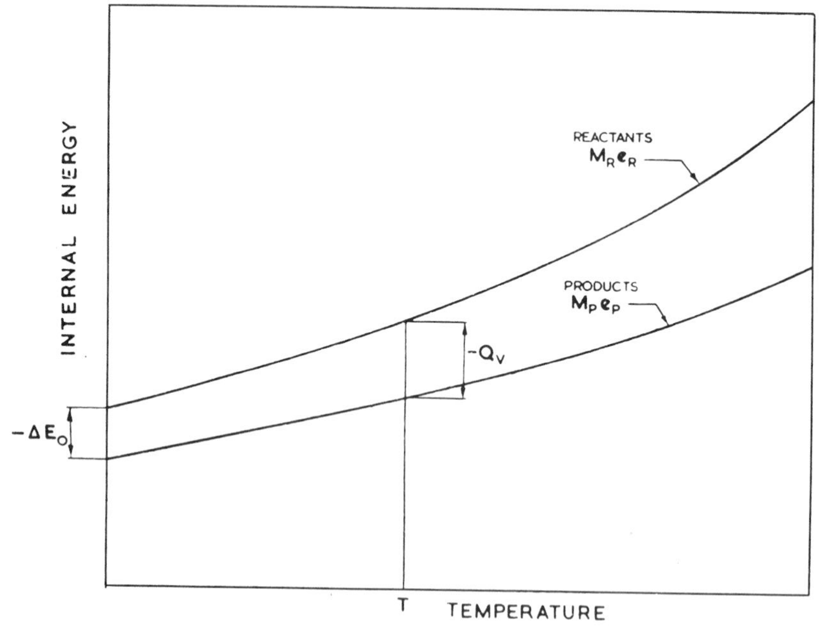

FIG. 3.4. Internal energy-temperature diagram for constant volume calorimeter.

Since $\Delta E_o = \Delta H_o$ and $h_P(T)-e_P(T) = R_{mol}T$ for an ideal gas equation (3.82) reduces to

$$Q_p-Q_v = M_pR_{mol}T-M_RR_{mol}T = (M_P-M_R)R_{mol}T. \quad (3.83)$$

If the number of mols of the reactants equal the number of mols of the products then the two heats of reaction are the same. From (3.83) it will be seen that, if we know the change in the number of mols during combustion, we require to know only one heat of reaction to calculate the other.

In our discussions we have not indicated whether the heat of reaction is independent of the fuel-air ratio (or strictly the fuel-oxidant ratio). If we neglect dissociation, then for oxidation combustion processes, the excess oxygen above that required for complete combustion will pass into the products of combustion. The reaction will therefore be

$$F+A_c+A_x = P+A_x \quad (3.84)$$

where F is the fuel, A_c the correct air (or oxidant), P the products of complete combustion and A_x the excess air (or oxidant).

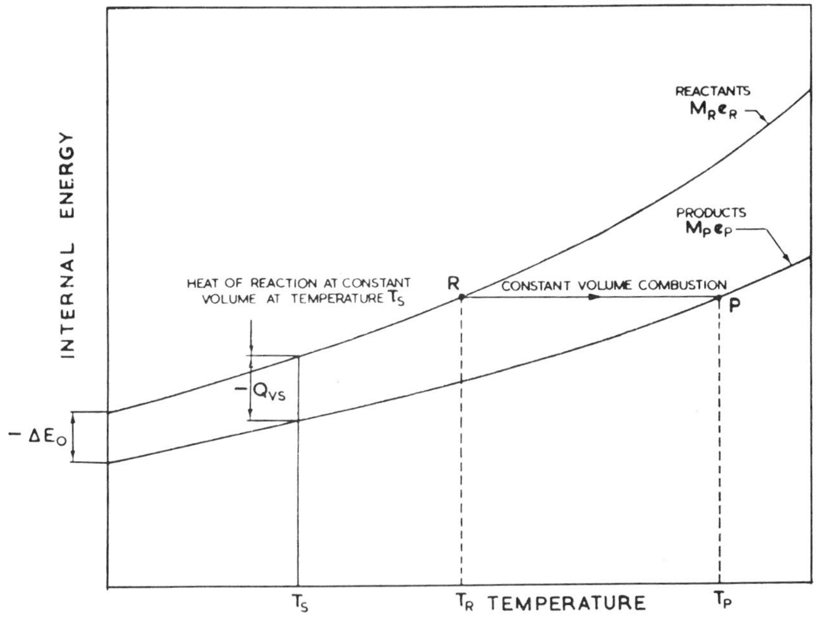

FIG. 3.5. Internal energy–temperature diagram for adiabatic combustion at constant volume.

For a calorimetric experiment at constant pressure

$$Q_p = H_{product} - H_{reactants} \tag{3.85}$$

$$H_{products} = M_P h_P + M_{Ax} h_{Ax} \tag{3.86}$$

$$H_{reactants} = M_F h_F + M_{Ac} h_{Ac} + M_{Ax} h_{Ax}. \tag{3.87}$$

Hence

$$Q_p = M_P h_P - M_F h_F - M_{Ac} h_{Ac} \tag{3.88}$$

and the heat of reaction is <u>independent</u> of the excess <u>air</u>.

For excess fuel we must take dissociation into account in <u>all</u> cases. We may, therefore, summarize for combustion reactions neglecting dissociation.

(1) The heats of reaction at constant pressure and constant volume are dependent on the temperature of the reactants.

(2) The heats of reaction at constant pressure and constant volume are equal if there is no change in the number of molecules before and after combustion.

(3) The difference between the heat of reaction at constant pressure and the heat of reaction at constant volume is equal to $(M_P - M_R)R_{mol}T$.

(4) The heats of reaction are independent of the excess air (or oxidants) in the fuel-air (or oxidant) mixture.

(5) The adiabatic temperature rise during combustion corresponds to either a constant enthalpy or a constant internal energy processes.

In the case of processes in which both work and heat transfer take place we can apply the general steady flow energy equation

$$\dot{Q} - \dot{W}' = \dot{M}_e h_{eo} - \dot{M}_i h_{io} \tag{3.89}$$

where $\dot{Q}$ = heat transfer rate, $\dot{W}'$ = rate of shaft work, $\dot{M}_i$ = inlet molar flow rate, $\dot{M}_e$ = exit molar flow rate, h_{io} = stagnation enthalpy of inlet gas, h_{eo} = stagnation enthalpy of outlet gas.

The stagnation enthalpy is given by

$$h + \frac{c^2}{2}. \tag{3.90}$$

Care should be taken in correcting the units for the velocity term $c^2/2$. The static enthalpy h is equal to the expressions given in equation (3.54), if the gas is a mixture of ideal gases.

Heats of Formation and Hess's Law

From the engineering viewpoint data on the heats of reaction are most important. To reduce the quantity of data necessary and to provide information for the range of possible reactions, we may use a well-known law of chemistry called Hess's Law. Two formulations will be given. The first formulation† states: "If a reaction at constant pressure or constant volume is carried out in stages, the algebraic sum of the amounts of heat evolved in the separate stages is equal to the total evolution of heat when the reaction occurs directly." The second formulation is "The heat liberated by a reaction is independent of the path of the reaction between the initial and final states". Both statements are identical and are a consequence of the law of conservation of energy. Using Hess's Law we can calculate the heat of reaction either from reaction

†Page 394 - College Course of Inorganic Chemistry, J.R.Partington, Macmillan, 1945.

data or from heats of formation. The heat of formation of a compound is the quantity of heat absorbed during the formation of a compound from its elements†. This may be obtained from calorimetric data either directly or indirectly. Let us consider a reaction

$$A + B = C + D$$

where A, B, C, D are compounds.

If the compounds A and B are reduced to their original elements then heat will be evolved equal to the algebraic sum of the heats of formation. If we now combine these elements to form C and D heat is absorbed equal to the algebraic sum of the heats of formation. The heat of reaction is then by Hess's Law the sum of the heats of formation of the final compounds minus the sum of the heats of formation of initial reactants. Or if we let

$$(h_f)_P = \sum Mh_f \text{ products}$$

$$(h_f)_R = \sum Mh_f \text{ reactants}$$

then

$$Q_p = (h_f)_P - (h_f)_R \qquad (3.91)$$

for the heat of reaction Q_p. If the heats of formation h_f are known, we can calculate the heat of reaction.

Let us consider a simple example. Consider the reaction

$$CO + \tfrac{1}{2}O_2 = CO_2$$

It is required to calculate the heat of reaction at constant pressure at 25°C. The heat of formation (h_f) for carbon monoxide is -112.2 kJ/g mol obtained from the reaction $C + \frac{1}{2}O_2 = CO$. The heat of formation for carbon dioxide is -393.6 kJ/g mol obtained from the reaction $C + O_2 = CO_2$. Using Hess's Law

$$(Q_p) = (h_f)_P - (h_f)_R.$$

The only reaction compound is CO and $(h_f)_R = -112.2$; the only product compound is CO and $(h_f)_P = -393.6$. Substituting into equation (3.91) we have

†We use the same sign convention for heats of formation as for heats of reaction, exothermic processes having negative h_f, endothermic processes having positive h_f. Hence for the general case (positive h_f) heat is absorbed during the formation of a compound from its elements. (In some texts the heat of formation is defined as the quantity of heat evolved during the formation of its elements.)

$$Q_p = -393.6 - (-112.8) = -280.8 \text{ kJ/g mol.}$$

Notice the sign convention. Heats of formation have a negative sign if heat is released and a positive sign if heat is absorbed.

Let us consider another reaction:

$$CH_4 + 2O_2 = 2\ H_2O + CO_2.$$

This reaction is the correct combustion of methane CH_4. The heats of formation (h_f) are

Methane	CH_4	=	-74.9	kJ/g mol.
Water	H_2O	=	-286	kJ/g mol.
Steam	H_2O	=	-242	kJ/g mol.
Carbon dioxide	CO_2	=	-393.6	kJ/g mol.

All the above data refer to a pressure of 1.01325 bar and 25°C.

To determine the heat of reaction at constant pressure we must stipulate whether the final state of the H_2O is water or steam. Let us consider the former case. The heats of formation are:

$$(h_f)_{product} = -2 \times 286 - 393.6 = -965.6 \text{ kJ/g mol.}$$

$$(h_f)_{reactant} = -74.9 \text{ kJ/g mol.}$$

Notice again the only reactant we consider is the compound CH_4, and we multiply the heat of formation for water by the number of mols.

$$Q_p = (h_f)_P - (h_f)_R$$

$$= -965.6 - (-74.9) = -890.7 \text{ kJ/g mol.}$$

If we consider the final state of the H_2O to be steam then

$$(h_f)_P = -2 \times 242 - 393.6 = -877.6 \text{ kJ/g mol.}$$

and

$$Q_p = -877.6 - (-74.9) = -802.7 \text{ kJ/g mol.}$$

The first value is called the higher calorific value or heat of reaction. The second value is the lower calorific value or heat of reaction. Whenever hydrogen is present in a fuel we will have two calorific values. In calculating temperature changes the lower value is used in the expression for Q_p, because steam rather than water is usually one of the products.

Heats of formation can be calculated from Hess's equation if the heats of reaction are known. In practice this is one method of establishing these data.

Entropy of Ideal Gas Mixtures

The entropy of a mixture of ideal gases is equal to the sum of the entropies each component of the mixture would have if each alone occupied the whole volume of the mixture at the same temperature.

Let us consider two gases A and B. The total number of mols of the mixture (M_T) is given by

$$M_T = \sum M = M_a + M_b$$

and the entropy balance is

$$M_T s_T = \sum Ms = M_a s_a + M_b s_b \tag{3.92}$$

From (3.29), if the reference pressure p_o is equal to unit pressure,†

$$s = s(T) - R_{mol} \ln p + s_o$$

and

$$M_T s_T = (M_a s_a(T) + M_b s_b(T)) - R_{mol}(M_a \ln p_a + M_b \ln p_b) + (M_a s_{oa} + M_b s_{ob}) \tag{3.93}$$

Dividing by M_T

$$s_T = \left(\frac{M_a}{M_T} s_a(T) + \frac{M_b}{M_T} s_b(T)\right) - R_{mol}\left(\frac{M_a}{M_T} \ln p_a + \frac{M_b}{M_T} \ln p_b\right) + \left(\frac{M_a}{M_T} s_{oa} + \frac{M_b}{M_T} s_{ob}\right) \tag{3.94}$$

†All pressure in the subsequent analysis will be referred to the unit pressure p_o.

where p_a, p_b are the partial pressures of A and B in the gas mixture. The ratio M/M_T is the mol fraction x. The terms in the first bracket can therefore be written as

$$x_a s_a(T) + x_b s_b(T) = \Sigma_n x_i s_i(T). \qquad (3.95)$$

The terms in the last bracket can be expressed as

$$x_a s_{oa} + x_b s_{ob} = \Sigma_n x_i s_{oi}. \qquad (3.96)$$

The pressure terms can be rewritten in the following manner:

$$x_a \ln p_a + x_b \ln p_b = x_a \ln \frac{p_a}{p_T} p_T + x_b \ln \frac{p_b}{p_T} p_T.$$

Now from (3.41)

$$\frac{p_a}{p_T} = x_a, \quad \frac{p_b}{p_T} = x_b$$

and

$$\begin{aligned} x_a \ln p_a + x_b \ln p_b &= x_a \ln x_a p_T + x_b \ln x_b p_T \\ &= x_a \ln x_a + x_b \ln x_b + x_a \ln p_T + x_b \ln p_T \\ &= x_a \ln x_a + x_b \ln x_b + (x_a + x_b) \ln p_T \\ &= x_a \ln x_a + x_b \ln x_b + \ln p_T \\ &= \Sigma_n x_i \ln x_i + \ln p_T. \end{aligned} \qquad (3.97)$$

Substituting (3.95), (3.96), (3.97) into (3.94) we have for the specific entropy of the gas mixture

$$s_T = \Sigma_n x_i s_i(T) + \Sigma_n x_i s_{oi} - R_{mol} \Sigma_n x_i \ln x_i - R_{mol} \ln p_T. \qquad (3.98)$$

The notation Σ_n indicates the summation over all the components in the gas mixture, p_T is the total pressure of the mixture, $s(T)$, s_o are the specific entropies, at unit pressure p_o (e.g. one atmosphere), at temperature T and at the reference temperature T_o (or the absolute zero) respectively.

If we consider gases which exist as pure crystalline substances near the absolute zero, the difference in the entropies of these gases at the absolute zero is zero. We can therefore assume s_{oi} to be zero. Hence the term $\Sigma_n x_i s_{oi}$ will be zero. Equation (3.98) then becomes

$$s_T = \Sigma_n x_i s_i(T) + R_{mol}\Sigma_n x_i \ln \frac{1}{x_i} - R_{mol} \ln p_T. \qquad (3.99)\dagger$$

The first term is the usual summation expression for gas mixtures. The second term is the <u>entropy of mixing</u>. This is due to the gases occupying the whole volume at partial pressures different to the reference pressure of one atmosphere. This term is associated with the irreversibility of a gas mixing process.

Let us consider two gases A and B, each initially at a pressure of one atmosphere and temperature T, occupying a vessel volume V, but separated by a partition. Let us open the partition and let the gases diffuse at constant temperature T. After a period of time we have a homogeneous mixture at temperature T. Let the final pressure be p_T. The initial conditions are

$$V_a = \frac{M_a R_{mol} T}{p_a} = M_a R_{mol} T$$

$$V_b = \frac{M_b R_{mol} T}{p_b} = M_b R_{mol} T$$

since $p_a = p_b = 1.0$ atm.

The initial entropies of A and B are

$$M_a s_a + M_b s_b = M_a s_a(T) + M_a s_{oa} + M_b s_b(T) + M_b s_{ob}. \qquad (3.100)$$

Notice the pressure term $R_{mol} \ln p$ is zero, since $p_a = p_b =$ 1.0 atm. After mixing

$$M_T = M_a + M_b$$

$$p_T = \frac{M_T R_{mol} T}{V}.$$

the initial volume is

$$V = V_a + V_b = (M_a + M_b)\, R_{mol} T$$

---

†If the reference temperature is <u>not</u> at absolute zero then the $\Sigma_n x_i s_{oi}$ term must be included.

and

$$p_T = \frac{(M_a+M_b)R_{mol}T}{(M_a+M_b)R_{mol}T} = 1.0 \text{ atm.}$$

The mol fractions of A and B in the gas mixtures are

$$x_a = \frac{M_a}{M_T}, \quad x_b = \frac{M_b}{M_T}.$$

The entropy of the mixture from (3.98) is, therefore,

$$M_T s_T = M_T(x_a s_a(T)+x_b s_b(T)+x_a s_{oa}+x_b s_{ob})$$
$$-R_{mol}M_T(x_a \ln x_a+x_b \ln x_b)-M_T R_{mol}\ln 1.0$$
$$= (M_a s_a(T)+M_b s_b(T)+M_a s_{oa}+M_b s_{ob})$$
$$-(R_{mol}M_a \ln x_a + R_{mol}M_b \ln x_b). \qquad (3.101)$$

The entropy change due to mixing is the expression (3.101) minus expression (3.100).

Hence $$\Delta s = -(R_{mol}M_a \ln x_a + R_{mol}M_b \ln x_b) \qquad (3.102)$$

The mol fractions x_a and x_b are less than unity and Δs is greater than zero. There is, therefore, an increase in entropy on mixing. This is called the entropy of mixing.

In this example we did not distinguish between the gases A and B. If we were to consider both gases to be identical it would appear from (3.102) that for mixing of two identical gases there would be an increase in entropy. This is not correct. We must therefore re-examine this special case. In this case, since we are dealing with a single gas, the initial pressure before mixing must be equal to the partial pressure after "mixing", because the "mixture" is a single component gas. Reverting to the basic expression (3.93). For an initial pressure equal to the partial pressure of the component gas the entropy summation term $\Sigma M_i s_i$ is the same before and after mixing, but the pressure terms will cancel out.† In applying the

---

†Let 'A' and 'B' gases be identical,

Initial entropy

$$M_a s_a+M_b s_b = M_a s_a(T)+M_b s_b(T)+M_a s_{oa}+M_b s_{ob}-R_{mol}M_a \ln p_a+R_{mol}M_b \ln p_b$$

Final entropy

$$M_a s_a+M_b s_b = M_a s_a(T)+M_b s_b(T)+M_a s_{oa}+M_b s_{ob}$$
$$-R_{mol}M_a \ln \frac{p_a}{p_o} -R_{mol}M_b \ln \frac{p_b}{p_o} -(M_a+M_b)R_{mol} \ln p_o.$$

But $p_a=p_b=p_o$ and the pressure terms cancel out. The entropy of mixing $\Delta S = 0$.

the entropy expressions care must be taken to distinguish between the component gases. This example of the "mixing" of two identical component gases and the apparent "entropy of mixing" is called Gibbs' Paradox.

It is possible for the components of a gas mixture to have the same partial pressures after mixing as the initial pressure of the gas. If all the component gases have the same partial pressures in the mixture as their initial pressures, then there will be a reduction in the system volume after mixing. There will be no entropy change, but the increase in entropy due to mixing will exactly equal the decrease in entropy due to the change in volume.

The Gibbs function for a mixture of gases can be examined in the same manner as the entropy, internal energy, etc. We will discuss this in the next chapter when examining more complex combustion phenomena.

Exercises

1. Gas Air Mixture

A closed vessel of 5 ft^3 capacity contains a mixture of methane (CH_4) and air, the air being 20 per cent in excess of that required for chemically correct combustion. The pressure and temperature in the vessel before combustion are respectively 50 p.s.i. and 200°F. Determine:

(a) the individual partial pressures and the weights of methane, nitrogen and oxygen present before combustion,

(b) the individual partial pressures of the burnt products, on the assumption that these are cooled to 200°F without change of volume and that all the vapour of combustion is condensed. (Air contains 79 per cent N_2 by volume.)

2. An engine runs on a rich mixture of methyl and ethyl alcohol and air. At a pressure of 14.5 p.s.i.a. and 50°F the fuel is completely vaporized. Calculate the air-fuel ratio by volume under these conditions, and the percentage of ethyl alcohol in the fuel by weight. If the total pressure of the exhaust gas is 14.5 p.s.i.a., calculate the dew point of the water vapour in the exhaust and the percentage by volume of carbon monoxide in the dry exhaust gas assuming all the hydrogen in the fuel forms water vapour.

Vapour pressure at 50°F, methyl alcohol (CH_4O) 1.08 p.s.i.a. and ethyl alcohol (C_2H_6O) 0.45 p.s.i.a. (Air contains 21 per cent oxygen by volume.)

(Univ. Liv.)

3. An engine working on the constant volume cycle has a compression ratio of 6.5 to 1, and the compression follows the law $PV^{1.3}$ = constant, the initial pressure and temperature being 14.0 p.s.i.a. and 110°F. The specific heats per lb throughout compression and combustion are 0.23 + 0.00001T and 0.16 + 0.00001T respectively, where T is the absolute temperature in degrees Fahrenheit.

Find (a) the change in entropy per lb during compression; (b) the heat rejected per lb during compression; (c) the heat rejected per lb during combustion if the maximum pressure is 625 p.s.i. and the energy liberated by the combustion is 920 Btu per lb.

4. A compression-ignition engine runs on a fuel of the following analysis by weight: carbon 84 per cent, hydrogen 16 per cent. If the pressure at the end of combustion is 800 p.s.i.a., the volume ratio of expansion 15 to 1, the pressure and temperature at the end of expansion 25 p.s.i.a. and 600°C respectively, calculate

(a) the variable specific heat at constant volume for the products of combustion; and

(b) the change in entropy during the expansion stroke per pound molecule.

The expansion may be assumed to follow the law $pv^n = C$ and there is 60 per cent excess air. The specific heats at constant volume in C.H.U./lb mol/°C between 600°C and 2400°C are

O_2	7.66 + 0.0006T;	N_2	7.22 + 0.0006T;
H_2O	8.12 + 0.02T;	CO_2	12.1 + 0.0016T,

where T is in °C absolute. The water vapour (H_2O) may be considered to act as a perfect gas. Air contains 79 per cent nitrogen by volume.

(Univ. Liv.)

5. The exhaust gases of a compression-ignition engine are to be used to drive an exhaust gas turbo-supercharger. Estimate the mean pressure ratio of expansion and the isentropic heat drop per pound molecule of gas in the turbine if the mean exhaust temperature is 600°C and the isentropic temperature drop is 100°C. The composition of the exhaust gas by volume is CO_2, 8 per cent; H_2O, 9.1 per cent; O_2, 7.5 per cent; N_2, 75.4 per cent.

The specific heats at constant volume in C.H.U. per pound-molecule per degree centigrade are

O_2	7.66 + 0.0006T;	N_2	7.22 + 0.0006T;
H_2O	8.12 + 0.002T;	CO_2	12.1 + 0.0016T,

where T is in degrees Centigrade absolute. The water vapour (H_2O) may be considered to act as a perfect gas.

(Univ. Liv.)

6. The exhaust gas from a two-stroke cycle compression-ignition engine is exhausted at an elevated pressure into a large chamber. The gas from the chamber is subsequently expanded in a turbine. If the mean temperature in the chamber is 1460°R and the pressure ratio of expansion in the turbine is 4 to 1, calculate the isentropic heat drop in the turbine per pound of gas.

Specific heat of gas at constant volume is 0.16 + 0.00001T Btu/lb°R where T is in °R; mean molecular weight of gas 30 lb.

(Univ. Liv.)

7. The following data refer to an analysis of a dual combustion cycle with a gas having specific heats varying linearly with temperature:

Pressure and temperature at end of compression 450 p.s.i.a. and 440°F respectively; maximum pressure 900 p.s.i.a.; maximum temperature 3140°F, temperature at the end of expansion 2270°F. The entropy increase during constant volume and constant pressure combustion is 0.211 and 0.346 Btu/lb°F respectively. Molecular weight of gas 30.5.

Calculate the equations of the specific heats and the volume ratio for isentropic expansion.

(Univ. Liv.)

8. A correct mixture of carbon monoxide and oxygen is ignited in a closed vessel and cooled to the original temperature. If the initial pressure is 5 atms absolute and temperature 140°F calculate the change in entropy per pound-molecule of original mixture.

The absolute entropy at 140°F and 1 atm. abs. for CO_2 = 52.1; O_2 = 49.8; CO = 48.1; Btu/lb mol°F.

(Univ. Liv.)

9.(1) Two moles of an ideal gas at temperature T and pressure p are contained in a compartment. In an adjacent compartment is one mole of an ideal gas at temperature 2T and pressure p. The gases mix adiabatically but do not react chemically when a partition separating the compartments is withdrawn. Show that, as a result of the mixing process, the entropy increases by

$$R\left(\ln\frac{27}{4} + \frac{\gamma}{\gamma-1}\ln\frac{32}{27}\right)$$

provided that the gases are different and that γ, the ratio of specific heats, is the same for both gases and remains constant in the temperature range T to 2T.

(2) What would be the entropy change if the mixing gases were of the same species?

(Univ. Manch.)

10. In an experiment to determine the calorific value of octane (C_8H_{18}) with a bomb calorimeter the weight of octane was 0.001195 lb, the water equivalent of the calorimeter including water 5.9 lb, and the corrected temperature rise in the water jacket 4.2 °F. Calculate the lower heat of reaction of octane, in Btu/lb mol at 60°F. The latent heat at constant volume of water vapour at this temperature is 18,000 Btu/lb mol.

If the initial pressure and temperature were 25 atm absolute and 60°F respectively and there was 400 per cent excess oxygen, estimate the maximum pressure and temperature reached immediately after ignition assuming no heat losses to water jacket during this time. No air was present in the bomb.

Internal energies of octane, oxygen, carbon dioxide and water vapour in Btu/lb mol/°F above 0°R are:

Temperature (°R)	520	4,000	4,500	5,000
C_8H_{18}	13,280			
O_2	2,480	24,890	28,500	32,190
CO_2		41,370	47,750	54,200
H_2O		32,730	38,110	43,660

Neglect dissociation and latent heat of fuel.

(Univ. Liv.)

11. A vessel contains a mixture of ethylene (C_2H_4) and air, the air being 100 per cent in excess of that required for complete combustion. If the initial pressure and temperature are 5 atm abs and 800°R, calculate the adiabatic temperature rise and maximum pressure when the mixture is ignited.

If the products of combustion are cooled until the water vapour is just about to condense, calculate the final temperature, pressure and heat loss per pound molecule of original mixture.

The enthalpy of combustion at the absolute zero is -569,936 Btu/lb mol ethylene.

ABSTRACT FROM GAS TABLES

Temperature °R	Internal energy above 0°R Btu/lb mol			Specific volume of water vapour ft^3/lb
	Air	Ethylene	Production of combustion	
600			3 035	123
610			3 087	97
620			3 139	77.3
630			3 191	62.1
800	4490	7106		
3700			23,058	
3750			23,423	
3800			23,788	
3850			24,155	

(Univ. Liv.)

12. Calculate the lower heat of reaction at constant volume for benzene C_6H_6 at 25°C. The heat of formation at 25°C for benzene, C_6H_6, 19.2 kcal/g mol; water vapour, H_2O, -57.8 kcal/g mol; carbon dioxide, CO_2, -94.0 kcal/g mol.

A mixture of one part by volume of vaporized benzene to fifty parts by volume of air is ignited in a cylinder and adiabatic combustion ensues at constant volume. If the initial pressure and temperature of the mixture is 10 atm abs, and 500°K calculate the maximum pressure and temperature neglecting dissociation.

INTERNAL ENERGIES ABOVE 298°K kcal/g mol

T °K	C_6H_6 (Vap)	O_2	N_2	CO_2	H_2O
500	4.343	0.467	0.421	0.991	0.666
2 500		13.772	12.804	24.292	18.496
2 600		14.503	13.480	25,580	19.574

$$R_{mol} = 1.985 \times 10^{-3} \text{ kcal/g mol°K}$$

(Air contains 21 per cent oxygen by volume)

(Univ. Manch.)

13. The heat of reaction of methane (CH_4) is determined in a constant pressure calorimeter. The temperature rise of the cooling water is 3.96°F, the water flow rate is 10 lb/min, the mean gas temperature (inlet to outlet) is 77°F and the gas flow rate 2.4 ft^3/hr. Calculate the higher and lower heats of

reaction at constant volume and constant pressure in Btu/lb mol if the gas pressure is 14.7 p.s.i.a.

A spark ignition engine runs on a mixture of one part by volume of methane to twelve parts by volume of air. If the temperature at the end of the compression stroke is 1040°R and the combustion is at constant volume, calculate the adiabatic temperature rise during combustion neglecting dissociation.

Air contains oxygen 21 per cent by volume.
Latent heat of H_2O at 77°F is 1050 Btu/lb.
Universal gas constant is 1.985 Btu/lb mol/°R.
Specific heats at constant pressure Btu/lb mol/°R.
(T in degrees Rankine)

Methane	CH_4	8.485
Carbon dioxide	CO_2	$5.32 + 14.29 \times 10^{-3}T$
Water vapour	H_2O	$6.97 + 3.46 \times 10^{-3}T$
Nitrogen	N_2	$6.53 + 1.49 \times 10^{-3}T$
Oxygen	O_2	$6.73 + 1.51 \times 10^{-3}T$

(Univ. Manch.)

14. Derive an expression connecting the lower heat of reaction at constant pressure with the lower heat of reaction at constant volume. Show how the lower heat of reaction at any temperature T_2 may be estimated from the lower heat of reaction at another temperature T_1.

A gas of volumetric composition 50 per cent methane (CH_4) and 50 per cent carbon monoxide is mixed with the correct quantity of oxygen for complete combustion at 240°F and 14.7 p.s.i.a. in a vessel 0.1 ft³ capacity. The mixture is ignited and cooled until the temperature is again 240°F. The heat released is 20.25 Btu.

Calculate: (a) the lower heat of reaction per pound molecule of gas at constant volume and constant pressure respectively at 240°F; (b) the lower heat of reaction per pound molecule of gas at constant volume and constant pressure respectively at 1040°F; (c) the lower heats of reaction of 1 ft³ of gas at 240°F and 1040°F respectively at constant volume and 14.7 p.s.i.a.

The mean specific heats at constant volume in Btu/lb mol/°F between 240°F and 1040°F are: CH_4, 10.4; CO, 5.3; O_2, 5.7; CO_2, 9.3; H_2O, 6.75.

(Univ. Liv.)

15. A gas engine is to run on a correct mixture of methane and air. If the volume ratio of compression is 10 to 1, the initial temperature and pressure 140°F and 14.0 p.s.i. absolute respectively, calculate the maximum temperature and pressure

reached during combustion. Assume (a) that 10 per cent of the heat released is lost during combustion and (b) that compression is isentropic.

Combustion takes place at constant volume; γ = 1.4 for unburnt methane (CH_4)-air mixture; heat of reaction at constant volume for methane - 344,000 Btu/lb mol.

INTERNAL ENERGY ABOVE 0°R IN BTU/LB MOL

Temperature °R	600	4000	4500	5000	5500
CO_2	3,434	41,367	47,746	54,197	60,703
H_2O	3,615	32,727	38,113	43,656	49,314
N_2	2,979	23,423	26,793	30,193	33,618

Air contains 21 per cent oxygen by volume.

(Univ. Liv)

16. A mixture of octane vapour (C_8H_{18}) and air is contained in an engine cylinder at a pressure and temperature of 100p.s.i. and 700°F the amount of air being 10 per cent in excess of that required for theoretically correct combustion. Assuming that combustion takes place at constant volume without heat-loss, and neglecting dissociation estimate the final temperature and pressure of the products of combustion, making use of the data given below:

Mean C_v per mol. between 32°F and 700°F: octane vapour 53.8 air 5.09.

Lower heat of reaction of octane vapour at constant volume = -2,182,000 Btu per mol C_8H_{18}

(assumed to be independent of temperature).

Internal energy of 1 mol of products of combustion at temperature T°F above 32°F in Btu.

T°F	4,200	4,400	4,600	4,800	5,000
E	29,400	31,000	32,400	33,900	35,800

17. Distinguish between an "ideal" and a "perfect" gas and show that in both cases the entropy s is given by

$$s = s_o + \int_{T_o}^{T} \frac{dh}{T} - R_{mol} \ln \left(\frac{p}{p_o}\right)$$

Two streams A and B of perfect gas mix adiabatically at constant pressure and without chemical change to form a third stream. The molal specific heat at constant pressure C_p of the gas in stream A is equal to that in stream B. A flows at M mols per second and is at a temperature T_1. B flows at one mol per second and is at a temperature nT_1. Assuming that the gases in streams A and B are different, show that the rate of entropy increase is

$$C_p \ln\left\{\frac{1}{n}\left(\frac{M+n}{M+1}\right)^{M+1}\right\} - R_{mol} \ln\left\{\frac{1}{M}\left(\frac{M}{M+1}\right)^{M+1}\right\}$$

How is the above expression modified if the gases in streams A and B are the same?

For the case n = 1, evaluate the rate of entropy increase
(a) when different gases mix, and
(b) when the gas in each stream is the same.

(Univ. Manch.)

18. An ideal gas has a specific heat at constant volume C_v given by

$$C_v = a + bT + cT^2$$

where a, b and c are constants and T is the absolute temperature.

The gas expands in a thermally insulated cylinder over a volume ratio $v_2/v_1 = 16$. If the actual expansion follows the law

$$pv^{1.25} = \text{constant}$$

prove that the entropy increase, $s_2 - s_1$ is

$$R_{mol} \ln 16 - a \ln 2 - \frac{b}{2} T_1 - \frac{3}{8} cT_1^2$$

where T_1 is the initial temperature.

For the case of isentropic expansion over the same volume ratio from an initial temperature of 1800°K, show that the final temperature would be approximately 850°K for a gas in which

$$a = 20{,}000 \text{ J/(kg mol deg.K)}$$
$$b = 6 \text{ J/(kg mol deg.K}^2\text{)}$$
$$c = 1.85 \times 10^3 \text{ J/(kg mol deg.K}^3\text{)}$$

Calculate the ratio of the work done by this gas in the actual expansion, to that done in the isentropic expansion, if the initial temperature is 1800°K in both cases.

(Univ. Manch.)

19. The molal specific heat at constant volume, C_v of an ideal gas is given by

$$C_v = 2.27\, R_{mol} - 6.3T + 1.2 \times 10^{-2} T^2$$

where T is the absolute temperature and R_{mol} is the Universal gas constant, 8,314 J/(kg mol deg.K). A sample of the gas at a temperature of 750°K is contained in a cylinder closed by a frictionless piston. The gas first expands adiabatically to eight times the original volume. The expansion stroke is followed by a compression stroke back to the initial volume but with heat exchange to the surroundings, so that the compression follows the law

$$pv^{4/3} = \text{constant.}$$

Show that the temperature after the expansion is approximately 300°K and calculate

(a) the temperature at the end of the compression stroke
(b) the heat exchange per mol of gas, and
(c) the entropy change during the compression stroke.

(Univ. Manch.)

20. A piston engine runs on a mixture of gaseous fuel and air, the fuel:air ratio being 1:24 by volume. The specific heats at constant volume, C_v, are:

for the air $C_v = 20.0 \times 10^3 + 2.5T$ J/(kg mol °K)

for the fuel $C_v = 12.5 \times 10^3 + 40T$ J/(kg mol °K)

where T is the temperature in degrees Kelvin.

Give an expression for C_v of mixture of fuel and air.

At the start of the compression stroke, the mixture is at a temperature of 15°C. The mixture is then compressed to one eighth of its initial volume. Show that the gases would reach a temperature of approximately 645°K if the compression were isentropic.

In the actual compression, the final temperature is 720°K. Part of this temperature increase is accounted for by heat gains of 600×10^3 J per kg mol of gas mixture, flowing to the gases from the piston and cylinder walls.

Calculate (a) the work done on the gas during the actual compression, expressed as a ratio of the isentropic compression work,

and (b) the entropy rise per kg mol of gas mixture during the actual compression.

Treat the gases as ideal gases. (Univ. Manch.)

21. Gases in a diesel engine exhaust manifold have a specific heat C_p which varies with the absolute temperature T according to the relationship

$$C_p = 500 + 0.625T \text{ J/(kg K)}$$

The molecular weight of the gas mixture is 29.

The gases are at an average temperature of 800K when they enter the turbine of a turbocharger, which expands to a pressure of 1 bar absolute. If the isentropic work done by the turbine is to be 80 kJ/kg, show that the isentropic turbine outlet temperature must be approximately 718°K and calculate the pressure ratio across the turbine. Assume that the mixture behaves as an Ideal Gas and that the temperatures and pressures are stagnation values.

(Univ. Manch.)

22. A jet engine burns a weak mixture of octane (C_8H_{18}) and air, the quantity of air being double the stoichiometric value. The products of combustion, in which dissociation may be neglected, enter the nozzle with negligible velocity at a temperature of 1,000°K. The gases leave the nozzle at atmospheric pressure with an exit velocity of 500 metres per second. The nozzle may be considered to be adiabatic and frictionless. The gases behave as ideal gases.

Determine:

(a) The specific heat at constant pressure C_p, of the products as a function of temperature.

(b) The molecular weight of the products.

(c) The temperature of the products at the nozzle exit.

(d) The pressure of the products at the nozzle inlet.

C_p data:

CO_2	$21 \times 10^3 + 34.0\ T$	
H_2O	$33 \times 10^3 + 8.3\ T$	
O_2	$28 \times 10^3 + 6.4\ T$	
N_2	$29 \times 10^3 + 3.4\ T$	J/(kg.mol.deg.K),

where T is the temperature in degrees Kelvin.

(Univ. Manch.)

23. The products of combustion of a jet engine have a molecular weight of 30 and a molal specific heat at constant pressure given by $C_p = 3.3 \times 10^4 + 15T$ J/(kg mol°K) where T°K is the gas temperature. When the jet pipe stagnation temperature is 1200°K the gases leave the nozzle at a relative speed of 600 m/s. Show that the static temperature of the gas at the nozzle exit will be approximately 1,092°K and estimate the total to static pressure ratio across the nozzle. Assume that the products of combustion behave as an ideal gas and that the flow is isentropic.

In a frictional nozzle producing the same mean outlet speed from the same inlet gas temperature, how would (a) the mean outlet static temperature and (b) the total to static pressure ratio be affected?

(Univ. Manch.)

CHAPTER 4

THERMODYNAMIC PROPERTIES OF GAS MIXTURES WITH VARIABLE COMPOSITION

More advanced thermodynamics of gas mixtures: chemical potential, chemical equilibrium, equilibrium constant, calculation of dissociation products, adiabatic temperature rise, discussion of processes involving dissociation - Lighthill ideal dissociating gas, extension to ionization and real gas effects, "frozen" flow and equilibrium flow.

Notation

- c velocity
- C_p specific heat at constant pressure
- C_v specific heat at constant volume
- D heat of dissociation
- e specific internal energy or exponential
- E internal energy
- E_P internal energy of products
- E_R internal energy of reactants
- f specific Helmholtz free energy function
- F Helmholtz free energy function
- g specific Gibbs free energy function
- g_o specific Gibbs free energy function at absolute zero
- g specific Gibbs free energy function at temperature T and unit pressure
- G Gibbs free energy function
- h specific enthalpy
- h_o specific enthalpy at absolute zero
- H enthalpy
- k_b backward reaction rate velocity constant
- k_d velocity constant for dissociation
- k_f forward reaction rate velocity constant
- k_D velocity constant for dissociation
- k_r velocity constant for recombination
- k_R velocity constant for recombination
- K_C equilibrium constant
- K_P equilibrium constant
- m molecular weight
- M number of mols
- n number of excess mols
- p pressure
- p_i partial pressure
- Q_p heat of reaction at constant pressure
- Q_v heat of reaction at constant volume
- R gas constant
- R_f one-way equilibrium rate
- R_{mol} universal gas constant
- s specific entropy
- S entropy
- T temperature absolute
- u specific internal energy in the absence of motion, gravity, etc.
- u_o specific internal energy in the absence of motion, gravity, etc.at absolute zero
- U internal energy in the absence of motion, gravity, etc.
- V volume
- v specific volume
- x_i mol fraction
- x length, degrees of richness or weakness (mixture strength)
- α degree of dissociation
- ε degree of reaction
- ρ density
- ρ_d characteristic density
- ΔE_o heat of reaction at constant volume at absolute zero
- ΔH_o heat of reaction at constant pressure at absolute zero
- ΔG^o_T the Gibbs free energy of formation at temperature T and unit pressure
- θ_D characteristic temperature for dissociation
- θ_I characteristic temperature for ionization
- μ chemical potential
- μ^o chemical potential at temperature T and unit pressure
- ν stoichiometric coefficient

In certain engineering applications it is found that the temperature changes are somewhat different to those predicted by the methods outlined in the previous chapter. For example, in a spark ignition engine, operating over the normal range of fuel-air ratio, it is observed that the maximum temperature is less than that calculated assuming the combustion reaction proceeds directly from the initial reactants to the final products of combustion. In rocket nozzles it is found that the measured thrust is different from the calculated thrust, assuming a gas of constant composition, when the nozzle gases operate at elevated temperatures. In both these cases dissociation is present. Experiments show that in a reaction between two or more elements or compounds the rate of conversion of the initial reactants to the final products is retarded by the dissociation of some of the final products to the initial reactants. Equilibrium is established when for the dissociating elements or compounds the speed of the forward reaction equals the speed of the backward reaction (this is called the Law of Mass Action). If the initial forward reaction is exothermic then the dissociation is endothermic. In practical terms this implies for an adiabatic combustion process the temperature rise will be less with dissociation than with no dissociation. For given initial reactants the degree of dissociation will increase with increase in temperature. For a given temperature the degree of dissociation will depend on the composition of the initial reactants.

Dissociation is not only present during a combustion process. Many pure substances, in the form of elements or compounds, may only exist at certain elevated temperatures in dissociated form, for example, free oxygen exists in atomic form (O) in equilibrium with molecular oxygen (O_2). Above certain temperatures gases may also ionize, that is, the atoms become dissociated shedding free electrons.

A finite time is required for equilibrium to be established in all dissociation processes. This time is called the relaxation time. If a system passes through a rapid change in pressure or temperature then it is possible that there will be insufficient time for equilibrium to be established. Calculations of the composition of the gases in these circumstances require a knowledge of reaction kinetics. In certain cases, for example, hypersonic wind tunnels and hypersonic ram jets, the "time" for a change in state of the system can be so short compared with the relaxation time that the gas is "frozen" at a fixed composition depending on the stagnation pressure, stagnation temperature and dimensions of the nozzle.

Calculations of flows between the limits of equilibrium flow are complex and can only reasonably be carried out with high-speed computers. Furthermore, there is, at the present time, a lack of important experimental data on reaction rate kinetics for the temperature range of the gases of interest to engineers. For these reasons preliminary design studies are carried out using the two extreme cases of "equilibrium" flow and "frozen" flow. For more detailed qualitative studies an ideal dissociating gas is used. In addition to the relaxation phenomenon for dissociating molecules, a similar phenomenon can be observed in ionized gas flow (that is in dissociating atoms) and it is possible in some problems to have "frozen" ionized flow.

In all except "frozen" flows the dissociated products recombine when the temperature of a gas mixture drops. Some of the available "chemical" energy is recovered in the recombination process. In "frozen" flow, as the name implies, there is no recombination.

In order to establish the equilibrium conditions in the systems involved in the processes described it is necessary to examine both the chemistry of the process (i.e. the stoichiometry) as well as the thermodynamics. We will first discuss the thermodynamics of systems of variable composition and then proceed with some elementary stoichiometry to the establishment of the equilibrium criteria. Examples will be given of typical reactions met in the mechanical engineering field. Methods will be described for calculating temperature and pressure changes of gas mixtures with variable composition. A brief treatment of the ideal dissociating gas will follow and some elementary expressions will be developed for the ionized gas. Finally some of the basic ideas of "non-equilibrium" flow will be discussed. Since the literature field in the latter topics is in a state of flux only one or two fairly elementary ideas will be discussed.

Thermodynamic Relations for Mixtures of Variable Composition

In Chapter 2, p.35 we defined the specific internal energy, enthalpy, Helmholtz function and Gibbs function in the form

$$du = T\,ds - p\,dv \quad \text{or} \quad u = u(s,v) \tag{4.1}$$

$$dh = T\,ds + v\,dp \quad \text{or} \quad h = h(s,p) \tag{4.2}$$

$$df = -p\,dv - s\,dT \quad \text{or} \quad f = f(v,T) \tag{4.3}$$

$$dg = v\,dp - s\,dT \quad \text{or} \quad g = g(p,T). \tag{4.4}$$

For a system containing M mols we obtain the following extensive equations corresponding to (4.1) to (4.4):

$$dU = T\,dS - p\,dV \quad \text{or} \quad U = U(S,V) \tag{4.5}$$

$$dH = T\,dS + V\,dp \quad \text{or} \quad H = H(S,p) \tag{4.6}$$

$$dF = -p\,dV - S\,dT \quad \text{or} \quad F = F(V,T) \tag{4.7}$$

$$dG = V\,dp - S\,dT \quad \text{or} \quad G = G(p,T). \tag{4.8}$$

All the above relations describe the extensive properties of a system as a function of two other properties (either extensive or intensive). Relations (4.5) to (4.8) hold only for systems of constant composition.

Let us consider a system containing a mixture of substances 1, 2, 3,...,k. If we add some further quantities of the same substance to the system then the energy of the system will increase. At any time the composition (by mass or other unit of quantity) of the system will depend on the proportions of the substance added. Let us consider again the expression (4.5); this shows that for a system of constant composition the change in internal energy of the system depends only on changes in the extensive properties of entropy and volume. If we add a substance A to the system† then there will be a change in the internal energy of the system. The magnitude of the change will depend on the quantity of substance added. If other substances are added there will be further changes in the internal energy. Thus for a system of variable composition the internal energy depends not only on the entropy and volume, but also on the number of mols of the various constituents of the system (the quantity of a substance being represented by the number of mols). Mathematically we can express this fact by the functional relationship

$$U = U(S,V,M_1,M_2,M_3,\ldots,M_k) \tag{4.9}$$

where $M_1,M_2,M_3,\ldots,M_k$ are the number of mols of substances $1,2,3,\ldots,k$.

We are primarily interested in systems in which the composition is changing either due to addition or to chemical reactions taking place within the system. In both cases there will be a change in the number of mols of substances $1,2,3,\ldots,k$. To obtain the change in internal energy of the system we require to know the change in composition and the change in volume and entropy. If we consider the function (4.9) to be continuous we can write, for the change in internal energy,

---

†If the substance is added to the system then we have an "open" system. It will be shown shortly that the relations developed equally apply to "closed" systems.

$$dU = \left(\frac{\partial U}{\partial S}\right)_{V,M_1,M_2,\ldots,M_k} dS + \left(\frac{\partial U}{\partial V}\right)_{S,M_1,M_2,\ldots,M_k} dV$$
$$+ \left(\frac{\partial U}{\partial M_1}\right)_{S,V,M_2,M_3,\ldots,M_k} dM_1$$
$$+ \left(\frac{\partial U}{\partial M_2}\right)_{S,V,M_1,M_3,\ldots,M_k} dM_2$$
$$+\ldots+ \left(\frac{\partial U}{\partial M_k}\right)_{S,V,M_1,M_2,\ldots,M_{k-1}} dM_k. \qquad (4.10)$$

Expression (4.10) gives the change in internal energy due to the change in entropy, volume and concentrations of substances 1,2,3,...,k. Let us examine the separate terms on the right-hand side of (4.10).

(1) The first term is the change in internal energy of the system due to change in entropy, when the volume is constant and the number of mols of all the constituents are constant, or the change in internal energy due to the change in entropy of a system of <u>constant composition</u> with the volume held constant.

From equation (4.5) for this case we have

$$dU = T\,dS \qquad (4.11)$$

Hence

$$T = \left(\frac{\partial U}{\partial S}\right)_{V,M_1,M_2,\ldots,M_k} \qquad (4.12)$$

(2) The second term is the change in internal energy of the system due to the change in volume, when the entropy and the number of mols of all the constituents are constant, or the change in internal energy due to the change in volume of a system of <u>constant</u> composition with the entropy held constant. From equation (4.5) for this case

$$dU = -pdV. \qquad (4.13)$$

Hence

$$-p = \left(\frac{\partial U}{\partial V}\right)_{S,M_1,M_2,\ldots,M_k} \qquad (4.14)$$

(3) The third term is the change in internal energy of the system due to the change in number of mols of constituent 1, when the volume and entropy of the system are constant and the number of mols of the remaining constituents remain constant.

(4) The fourth and subsequent terms have the same physical interpretation as (3) replacing 1 by 2 and so on.

Equation (4.10) is quite general and applies to closed and open systems. Let us consider an example. In a closed isolated system a chemical reaction takes place in which there is a change in the number of mols of the individual constituents. Equation (4.10) shows that, provided the partial derivatives $(\partial U/\partial M_i)$ have finite values, the chemical changes may produce changes in the pressure, temperature or entropy of the system since the net change in internal energy must be zero ($dU = 0$). Whether or not all three properties p, T and S change or only one or two will depend on the constraints on the system. We will show, shortly, that $\partial U/\partial M_i$ is a property of the constituent i and has a finite value. In general, therefore, changes in the concentrations of the constituents in a closed isolated system will produce changes in the usual thermodynamic properties (pressure, temperature, entropy and volume in this case).

If there are changes in the molar concentrations of the constituents of the system, even though there is no change in the total number of mols of the system, there may still be changes in pressure, temperature, entropy and volume, the magnitude of which will depend, amongst other things, on the $(\partial U/\partial M_i)$ terms.

It is convenient to introduce a special notation to simplify the analysis. We let M_i be the number of mols of the constituent i, M_j the number of mols of all the constituents except i, and k the number of the constituents.

With this notation the third and remaining terms in (4.10) can be represented by

$$\sum_{i=1}^{i=k} \left(\frac{\partial U}{\partial M_i}\right)_{S,V,M_j} dM_i$$

and (4.10) becomes

$$dU = T\,dS - p\,dV + \sum_{i=1}^{i=k} \left(\frac{\partial U}{\partial M_i}\right)_{S,V,M_j} dM_i. \qquad (4.15)$$

If the number of mols of <u>each</u> constituent remains constant then we have a mixture of constant composition and equation (4.15) reduces to the well-known relation

$$dU = T\,dS - p\,dV.$$

Let us consider a system containing a mixture of substances which can react with each other. Initially let there be k_1 substances. A reaction takes place in which it is observed that k_2 substances are present. The additional substances are equal to $k_2 - k_1$. If we require to evaluate the properties of the system it is necessary to include in equation (4.15) the partial derivatives of all the possible constituents (i.e. k_2 in number), for if we leave some of the partial derivatives out of this expression it will not be possible to ascertain the contribution of those constituents to the change in internal energy. Of course until these additional constituents, not present in the original system, are present their contribution to the change in internal energy will be zero. Hence we must consider, in any analysis of systems of variable composition, both the actual and possible constituents.

To illustrate this point let us consider a mixture of two substances A and B. Let the two substances react to form a third substance C. Initially we have M_a mols of A, M_b mols of B. During the reaction we have a change in the number of mols of A and B, given by dM_a, dM_b and some mols of C are produced, given by dM_c. Then, in order to calculate the change in internal energy of the system we have

$$dU = T\,dS - p\,dV + \left(\frac{\partial U}{\partial M_a}\right)_{S,V,M_b,M_c} dM_a$$

$$+ \left(\frac{\partial U}{\partial M_b}\right)_{S,V,M_c,M_a} dM_b + \left(\frac{\partial U}{\partial M_c}\right)_{S,V,M_a,M_b} dM_c .$$

Until some of the mols of C appear the last term is zero, but once they do appear the last term will have a finite value. The partial derivatives $(\partial U/\partial M_a)$, $(\partial U/\partial M_b)$, $(\partial U/\partial M_c)$, are important variables associated with the energy changes in a chemical reaction. They also have significance in other processes in which there are some forms of mass transfer (for example, diffusion processes).

The partial derivative $(\partial U/\partial M_i)_{S,V,M_j}$ is called the chemical potential μ_i

$$\mu_i = \left(\frac{\partial U}{\partial M_i}\right)_{S,V,M_j} \tag{4.16}$$

The chemical potential is the energy added to a system per unit mass (or mol) transferred into the system, when the volume and entropy of the whole system are constant and the mass (or mols) of all the other constituents is held constant.

Equation (4.15) in terms of the chemical potential becomes

$$dU = T\,dS - p\,dV + \sum_{i=1}^{i=k} \mu_i\,dM_i. \qquad (4.17)$$

By the same reasoning as above the enthalpy H, the Helmholtz function F and the Gibbs function G for open systems may be expressed in the form

$$H = H(S,p,M_1,M_2,M_3,\ldots,M_k) \qquad (4.18)$$

$$F = F(V,T,M_1,M_2,M_3,\ldots,M_k) \qquad (4.19)$$

$$G = G(p,T,M_1,M_2,M_3,\ldots,M_k). \qquad (4.20)$$

We will examine (4.18) in the same manner as (4.9) and leave the reader to derive the differential relations for F and G. Equation (4.18) in differential form is

$$dH = \left(\frac{\partial H}{\partial S}\right)_{p,M_1,\ldots,M_k} dS + \left(\frac{\partial H}{\partial p}\right)_{S,M_1,\ldots,M_k} dp + \sum_{i=1}^{i=k} \left(\frac{\partial H}{\partial M_i}\right)_{S,p,M_j} dM_i. \qquad (4.21)$$

The first two terms correspond to systems of <u>constant composition</u>. Hence from (4.6) we have

$$dH = T\,dS + V\,dp + \sum_{i=1}^{i=k} \left(\frac{\partial H}{\partial M_i}\right)_{S,p,M_j} dM_i. \qquad (4.22)$$

Deducting d(pV) from both sides of (4.22)

$$d(H-pV) = T\,dS + V\,dp - p\,dV - V\,dp + \sum_{i=1}^{i=k} \left(\frac{\partial H}{\partial M_i}\right)_{S,p,M_j} dM_i. \qquad (4.23)$$

Now

$$d(H-pV) = dU$$

and

$$dU = T\,dS - p\,dV + \sum_{i=1}^{i=k} \left(\frac{\partial H}{\partial M_i}\right)_{S,p,M_j} dM_i. \qquad (4.24)$$

Comparing (4.24) and (4.17), it will be seen that

$$\mu_i = \left(\frac{\partial H}{\partial M_i}\right)_{S,p,M_j} \tag{4.25}$$

By similar reasoning it can be shown that

$$\mu_i = \left(\frac{\partial U}{\partial M_i}\right)_{S,V,M_j} = \left(\frac{\partial H}{\partial M_i}\right)_{S,p,M_j}$$

$$= \left(\frac{\partial F}{\partial M_i}\right)_{T,V,M_j} = \left(\frac{\partial G}{\partial M_i}\right)_{p,T,M_j} \tag{4.26}$$

and the group of differential equations are:

$$dU = T\ dS - p\ dV + \sum_{i=1}^{i=k} \mu_i\ dM_i \tag{4.17}$$

$$dH = T\ dS + V\ dp + \sum_{i=1}^{i=k} \mu_i\ dM_i \tag{4.27}$$

$$dF = -p\ dV - S\ dT + \sum_{i=1}^{i=k} \mu_i\ dM_i \tag{4.28}$$

$$dG = V\ dp - S\ dT + \sum_{i=1}^{i=k} \mu_i\ dM_i. \tag{4.29}$$

It will be seen that equations (4.5) to (4.8) are special cases of equations (4.17), (4.27), (4.28), (4.29) when the dM_i's are zero. Equations (4.17), (4.27), to (4.29) are the <u>state</u> equations for systems of variable composition.

<u>The Chemical Potential</u>

Re-examining equation (4.26) we note three important facts:

1. The chemical potential μ is a function of properties, hence it is itself a <u>property</u>.
2. The numerical value of μ is <u>independent</u> of the <u>type of property</u>.

3. The numerical value of μ is independent of the size of the system.

For example,

$$\mu = \left(\frac{\partial U}{\partial M_i}\right)_{S,V,M_j} = \frac{\partial(M_T u)}{\partial(M_T x_i)} = \left(\frac{\partial u}{\partial x_i}\right)_{S,V,M_j}$$

$$\mu = \left(\frac{\partial G}{\partial M_i}\right)_{p,T,M_j} = \frac{\partial(M_T g)}{\partial(M_T x_i)} = \left(\frac{\partial g}{\partial x_i}\right)_{p,T,M_j}$$

where M_T is the total number of mols, x_i is the mole-fraction of the substance i, and u and g are the specific internal energies and Gibbs free energy functions for the mixture.†

From (2) and (3) it will be seen that the chemical potential is an intensive property. That is, for a particular substance the chemical potential is the same throughout the system and independent of the quantity of the substance in the same way as temperature is independent of the quantity of a substance.

Since the chemical potential is a thermodynamic property of a system it is possible to define any other property of a system in terms of the chemical potential and a second property. For example, for a system of constant composition we can write

$$p = p(\mu, T)$$

in the same way as one may write

$$p = p(V, T)$$

although the functional relationship will be different for the two expressions.

For mixtures of substances any functional relationship containing the chemical potential must include all the chemical potentials of all the possible substances as well as the actual substances. This follows from the previous discussion on the

---

†The derivatives $(\partial u/\partial x_i)_{S,V,M_j}$ and $(\partial g/\partial x_i)_{p,T,M_j}$ refer to the changes in the specific internal and free energies of the mixture due to the change in the concentration of the substance i when all the other variables are constant. Hence μ_i is associated with the substance i and is a property of i alone.

development of equation (4.17).

For a single substance the chemical potential is a function of any two properties. Thus, for example,

$$\mu = \mu(p,T).$$

If the pressure and temperature of a system containing the single substance is not changed the chemical potential is constant. If we have a mixture of substances, and if the pressure, temperature and composition of the mixture remains constant, then the chemical potentials of all the substances remain constant. Now the pressure, the temperature and the chemical potential are all intensive properties. Hence if <u>two intensive</u> properties of a system are constant, the <u>third</u> will be constant.

Consider a system with 1,2,...,k substances. Let us represent the thermodynamic state equation by the expression

$$dG = V\,dp - S\,dT + \sum_{i=1}^{i=k} \mu_i\,dM_i. \tag{4.29}$$

Initially the Gibbs function for the system is equal to G_I. The system is enlarged by adding substances 1,2,...,k in the same proportions as present initially in the system. The process is to be carried out without any change in the intensive properties (p,T,μ_i) of the system. The final Gibbs function for the system can be obtained from integration of (4.29).

Since $dp = dT = 0$, μ, p and T are constant and

$$dG = \sum_{i=1}^{i=k} \mu_i\,dM_i$$

which on integration becomes

$$G_{II} - G_I = \sum_{i=1}^{i=k} (M_{II} - M_I)_i\,\mu_i. \tag{4.30}$$

The suffixes I and II refer to the initial and final conditions of the system.

Consider the <u>specific</u> Gibbs function g in terms of v, p, s and T. We have from (4.4)

$$dg = v\,dp - s\,dT.$$

In a system subjected to a change in which the pressure and temperature are held constant, dg is equal to zero, and the specific Gibbs function is constant.

For the system under discussion

$$G = M_T g$$

$$G_I = (M_I)_T g_I$$

$$G_{II} = (M_{II})_T g_{II}$$

$$g_I = g_{II} \quad \text{(since } dp = dT = 0\text{)}$$

and

$$\frac{G_{II}}{G_I} = \frac{(M_{II})_T}{(M_I)_T}. \tag{4.31}$$

We have specified for each substance that the quantity added was such that the proportion of each substance in the system was the same as initially in the system. Thus the ratio of the number of mols of substance 1 before and after the addition is the same as the ratio of the number of mols of substance 2 before and after the addition and so on, or

$$\frac{(M_{II})_1}{(M_I)_1} = \frac{(M_{II})_2}{(M_I)_2} = \cdots \frac{(M_{II})_i}{(M_I)_i} = \frac{(M_{II})_T}{(M_I)_T}$$

Substituting the above into (4.31) we have:

$$\frac{G_{II}}{G_I} = \frac{(M_{II})_T}{(M_I)_T} = \frac{(M_{II})_i}{(M_I)_i}. \tag{4.32}$$

If we let

$$\beta = \frac{G_{II}}{G_I}$$

then from (4.32)

$$\beta = \frac{(M_{II})_i}{(M_I)_i}$$

and

$$G_{II} - G_I = (\beta-1)\, G_I$$

$$(M_{II})_i - (M_I)_i = (\beta-1)\, (M_I)_i.$$

Substitution into (4.30) gives

$$(\beta-1)\, G_I = \sum_{i=1}^{i=k} (\beta-1)\, (M_I)_i \mu_i$$

or

$$G_I = \sum_{i=1}^{i=k} (M_i \mu_i)_I. \qquad (4.33)$$

Since state I was not specified we can write

$$G = \sum_{i=1}^{i=k} (M_i \mu_i) \qquad (4.34)$$

For a single component system i = k = 1 hence

$$G = M\mu$$

or

$$\mu = \frac{G}{M} = g. \qquad (4.35)$$

Hence the chemical potential is equal in magnitude to the specific Gibbs function at a given temperature and pressure.†

From (4.33) the Gibbs function for a mixture of gases is equal to the sum of the products of the number of mols and the chemical potential of each constituent substance.

The specific Gibbs function g was defined in equation (3.38) as

$$g = g^o + R_{mol} T \ln p \qquad (3.38)$$

where g^o is a function of temperature at reference pressure of one atmosphere and

$$g^o = g(T) + g_o. \qquad (3.37)$$

---

†There is a clear thermodynamic distinction between the chemical potential, μ, an intensive property and the specific Gibbs function, g, a specific quantity. If we consider a system containing M_i mols of a single substance i, the Gibbs free energy, G, is equal to $M_i g_i = M_i \mu_i$, but the chemical potential of the system is μ_i, not $M_i \mu_i$, since the chemical potential is an intensive property and is uniform throughout the system.

To determine the chemical potential, g is replaced by μ in (3.38) and (3.37) and the following important relations are obtained:

$$\mu = \mu^o + R_{mol} T \ln p \tag{4.36}$$

$$\mu^o = \mu(T) + \mu_o. \tag{4.37}$$

μ_o is a constant and depends on the reference temperature T_o; this might be the absolute zero or some standard temperature. μ^o is a function of temperature and is the value of μ when the pressure p is equal to one atmosphere.

In Chapter 2 (p.35) we showed that

$$g = h - Ts.$$

We can replace g by μ and obtain the following relationship for μ:

$$\mu = h - Ts. \tag{4.38}$$

The Maxwell relations IV (b) gave

$$s = -\left(\frac{\partial g}{\partial T}\right)_p. \tag{4.39}$$

Substituting for s in (4.38) and g in (4.39) we have

$$h = \mu - T\left(\frac{\partial \mu}{\partial T}\right)_p. \tag{4.40}$$

It is more convenient to express the relationship between h, and T in another form.

Rearranging equation (4.40) we obtain

$$\frac{T\left(\frac{\partial \mu}{\partial T}\right)_p - \mu}{T^2} = -\frac{h}{T^2}. \tag{4.41}$$

Consider the expression μ/T. If we differentiate this expression with respect to T, keeping p constant, we obtain

$$\left(\frac{\partial\left(\frac{\mu}{T}\right)}{\partial T}\right)_p = \frac{T\left(\frac{\partial \mu}{\partial T}\right)_p - \mu}{T^2};$$

this is the left-hand side of (4.41). Hence equation (4.41) can be written as

$$h = -T^2 \left(\frac{\partial \left(\frac{\mu}{T} \right)}{\partial T} \right)_p . \tag{4.42}$$

This expression will be used later in developing equations for the computation of equilibrium data.

Chemical Stoichiometry and Dissociation

Before proceeding to consider the application of the relations developed above to the thermodynamics of chemical reactions a brief revision of the stoichiometry of some elementary chemical reactions of interest will be given.

Let us consider the chemical reaction

$$1 \text{ mol } CO + \tfrac{1}{2} \text{ mol } O_2 = 1 \text{ mol } CO_2 . \tag{4.43}$$

This equation corresponds to the complete combustion of carbon monoxide in oxygen. The equation implies that one mol of carbon monoxide combines with half a mol of oxygen to form one mol of carbon dioxide. If this reaction proceeded to completion then no carbon monoxide or oxygen would be present after combustion. The number of mols of the carbon monoxide, oxygen and carbon dioxide are such to give exact chemical balance, i.e., the mass of carbon originally in the carbon monoxide is equal to the mass of carbon in the carbon dioxide and the mass of oxygen in the carbon monoxide plus the free molecular oxygen is equal to the mass of oxygen in the carbon dioxide. Equation (4.43) is the stoichiometric equation for the reaction. The numbers, corresponding to the number of mols of each constituent in the stoichiometric equation, are called the stoichiometric coefficients. The stoichiometric coefficients for CO, O_2, and CO_2 in equation (4.43) are 1, ½ and 1 respectively.

In any general reaction between say elements or compounds A and B in which elements or compounds C and D may be formed the stoichiometric equation will be

$$\nu_a A + \nu_b B = \nu_c C + \nu_d D \tag{4.44}$$

where ν_a, ν_b, ν_c, ν_d are the stoichiometric coefficients of A, B, C and D.

The stoichiometric equation will be represented in the text in the form shown in (4.44) with the equality sign between the initial reactants and the final products and except for unity (1) the stoichiometric coefficients will be placed directly before the chemical symbol. Thus (4.43) would be represented by

$$CO + 0.5\ O_2 \quad = \quad CO_2$$

and

$$\nu_{CO} = 1.0, \quad \nu_{O_2} = 0.5, \quad \nu_{CO_2} = 1.0.$$

In order to distinguish between the stoichiometric equation and the actual direction of a chemical reaction the equality sign will be replaced by an arrow indicating the direction of the reaction. The stoichiometric coefficients will be replaced by the actual number of mols taking part in the reaction.

In combustion problems the special case of a chemical reaction in which the actual number of mols for each reactant is in the same ratio as the stoichiometric coefficients is called the correct mixture for combustion. If the number of mols of oxygen in the mixture is greater than the correct number the mixture is considered to be weak. If the number of mols of oxygen is less than the correct number the mixture is considered to be rich.

Let us consider some examples of this procedure. In all calculations one starts with the stoichiometric equation. We will first consider a correct mixture of carbon monoxide and oxygen. The stoichiometric equation is

$$CO + \tfrac{1}{2}\ O_2 \quad = \quad CO_2.$$

The chemical reaction with no dissociation is written in the form

$$CO + \tfrac{1}{2}\ O_2 \quad \longrightarrow \quad CO_2.$$

Let us consider a weak mixture of carbon monoxide and oxygen. For this purpose let there be n mols of excess oxygen. We then proceed as follows:

The stoichiometric equation is:

$$CO + \tfrac{1}{2}\ O_2 \quad = \quad CO_2.$$

The chemical reaction with no dissociation is:

$$CO + \left(\tfrac{1}{2} + n\right)O_2 \longrightarrow CO_2 + nO_2.$$

This expression assumes that all the carbon monoxide burns to form carbon dioxide, CO_2. Finally, let us consider a rich mixture with n mols of carbon monoxide in excess of the correct value.

The stoichiometric equation is:

$$CO + \tfrac{1}{2}\,O_2 \;=\; CO_2.$$

The chemical reaction with no dissociation is:

$$(1+n)CO + \tfrac{1}{2}\,O_2 \rightarrow CO_2 + nCO.$$

Here we have assumed that all the oxygen combines with CO to form CO_2.

Let us re-examine the chemical reaction for a correct mixture of carbon monoxide and oxygen.

The stoichiometric equation is:

$$CO + \tfrac{1}{2}\,O_2 \;=\; CO_2. \tag{4.45}$$

The chemical reaction with no dissociation is:

$$CO + \tfrac{1}{2}\,O_2 \rightarrow CO_2.$$

If in practice we were to examine the products of this reaction, we would find that there was always some carbon monoxide and oxygen present in addition to the carbon dioxide. The true equation for the chemical reaction is therefore

$$CO + \tfrac{1}{2}\,O_2 \rightarrow aCO_2 + bCO + dO_2 \tag{4.46}$$

where a, b and d are the numbers of mols of CO_2, CO and O_2 when the reaction is "complete". Closer examination of the reaction would show that for each mol of CO_2 present in a reaction without dissociation some α mols of CO_2 would decompose to form the mols b and d of CO and O_2. This decomposition or dissociation process is governed by an equilibrium reaction

$$\alpha CO_2 \rightleftharpoons \alpha(CO + \tfrac{1}{2}\,O_2). \tag{4.47}$$

This implies that as fast as the CO_2 is dissociated into CO and O_2 some of the constituents recombine to form CO_2, the system being in equilibrium. (This is called the Law of Mass Action.) For the equilibrium condition we are in fact saying that there is a continuous interchange between some of the mols of CO, O_2 and CO_2 represented by

$$\alpha(CO + \tfrac{1}{2}\,O_2) \rightarrow \alpha CO_2 \rightarrow \alpha(CO + \tfrac{1}{2}\,O_2).$$

Notice these reactions take place in the ratios of the stoichiometric coefficients, that is, the number of mols dissociating and recombining satisfy the stoichiometric equation (4.45).

Reverting now to our original reaction. We observe that at any temperature we have present in the "products" the gases CO_2, CO and O_2. Now for the mass balance of (4.46) the number of mols of carbon is given by

$$1 = a + b,$$

the number of mols of oxygen is given by

$$1 = a + \frac{b}{2} + d.$$

Therefore,

$$a = (1-b)$$

$$d = \frac{b}{2}.$$

Equation (4.46) can be rewritten as:

$$CO + \tfrac{1}{2}\,O_2 \longrightarrow (1-b)CO_2 + bCO + \frac{b}{2}\,O_2. \qquad (4.48)$$

It will be seen from this expression that only b mols of CO_2 have been dissociated into b mols of CO and b/2 mols of O_2. The number of mols "b" therefore correspond to the number of mols "α" of CO_2 in (4.47). This leads to an alternative method of obtaining (4.48). For the dissociation process, when equilibrium is reached, the law of mass action gives:

$$CO_2 \rightleftharpoons CO + \tfrac{1}{2}\,O_2.$$

For the case of α mols of CO_2 we have

$$\alpha CO_2 \rightleftharpoons \alpha CO + \frac{\alpha}{2}\,O_2.$$

We can represent the process in a number of steps as follows:

1. Forward reactions

$$CO + \tfrac{1}{2}\,O_2 \longrightarrow CO_2.$$

2. Backward reactions

$$\alpha CO_2 \longrightarrow \alpha CO + \frac{\alpha}{2}\,O_2.$$

3. Dissociation of CO

$$CO_2 \longrightarrow (1-\alpha)CO_2 + \alpha CO_2 \longrightarrow (1-\alpha)CO_2 + \alpha CO + \frac{\alpha}{2}\,O_2.$$

4. The final reaction

$$CO + \tfrac{1}{2}\,O_2 \longrightarrow (1-\alpha)CO_2 + \alpha CO + \frac{\alpha}{2}\,O_2. \qquad (4.49)$$

When we compare (4.49) with (4.48) it will be seen that α and b are identical.

For this reaction we have α mols of carbon dioxide dissociating into α mols of CO and $\alpha/2$ mols of O_2. This dissociation reaction follows the expression (4.47). It will be observed that the molar quantity actually involved in the dissociation process, corresponding to the stoichiometric ratios, appears in the final products. That is the ratio of the number of mols of CO_2 dissociated (α) to the number of mols of CO (α) and O_2 ($\alpha/2$) present in the "products" is the same as the ratio of the stoichiometric coefficients for CO_2, CO and O_2. This only applies to a correct mixture. For other mixtures the molar quantities can be obtained if the correct mixture equation is set up "within" the main reaction equations and the procedure outlined above carried out. In some cases this may not be possible and the method outlined to develop (4.48) should be used. The symbol α is called the degree of dissociation.

Before proceeding further let us examine a few reactions.

Example 1

Consider a weak mixture of CO and O_2. Determine the final composition of the mixture at equilibrium.

Original mixture:

$$CO + (\tfrac{1}{2} + n)O_2.$$

In this case n is the number of excess mols of O_2 in the mixture. Let α be the degree of dissociation.

(a) Reaction without dissociation:

$$CO + (\tfrac{1}{2} + n)O_2 \rightarrow CO_2 + nO_2.$$

(b) Dissociation of CO_2:

$$\alpha CO_2 \rightarrow \alpha CO + \frac{\alpha}{2}\, O_2.$$

$$CO_2 \rightarrow (1-\alpha)CO_2 + \alpha CO_2 \rightarrow (1-\alpha)CO_2 + \alpha CO + \frac{\alpha}{2}\, O_2.$$

(c) Overall reaction:

$$CO + (\tfrac{1}{2} + n)\, O_2 \rightarrow (1-\alpha)CO_2 + \alpha CO + \left(n + \frac{\alpha}{2}\right) O_2.$$

The final composition is, therefore,

$$(1-\alpha)CO_2 + \alpha CO + \left(n + \frac{\alpha}{2}\right) O_2.$$

Notice the increase in the number of mols of O_2 due to the dissociation of CO_2.

Example 2

Consider a rich mixture of CO and O_2. Determine the final composition of the mixture at equilibrium.

Original mixture:

$$(1+n)CO + O_2.$$

In this case n is the number of mols of excess CO in the mixture. Let α be the degree of dissociation.

(a) Reaction without dissociation:

$$(1+n)CO + \tfrac{1}{2}\, O_2 \rightarrow nCO + CO_2.$$

(b) Dissociation of CO_2:

$$\alpha CO_2 \rightarrow CO + \frac{\alpha}{2}\, O_2$$

$$CO_2 \rightarrow (1-\alpha)CO_2 + \alpha CO_2 \rightarrow (1-\alpha)CO_2 + \alpha CO + \frac{\alpha}{2}\, O_2.$$

(c) Overall reaction:

$$(1+n)CO + \tfrac{1}{2}\, O_2 \rightarrow (1-\alpha)CO_2 + (n+\alpha)CO + \frac{\alpha}{2}\, O_2.$$

The final composition is, therefore,

$$(1-\alpha)CO_2 + (n+\alpha)CO + \frac{\alpha}{2}\, O_2.$$

Example 3

Consider a mixture of 1 mol of CO_2, 1 mol of O_2 and 1 mol of CO. Determine the final composition of the mixture.

Original mixture:

$$CO_2 + O_2 + CO.$$

(a) Reaction without dissociation:

$$CO_2 + O_2 + CO \rightarrow CO + \tfrac{1}{2}\, O_2 + \tfrac{1}{2}\, O_2 + CO_2 \rightarrow 2CO_2 + \tfrac{1}{2}\, O_2.$$

(b) Dissociation of CO_2:

$$\alpha CO_2 \rightarrow \alpha CO + \frac{\alpha}{2}\, O_2$$

$$2CO_2 \rightarrow 2(1-\alpha)CO_2 + 2\alpha CO_2 \rightarrow 2(1-\alpha)CO_2 + 2\alpha CO + \alpha O_2.$$

(c) Overall reaction:

$$CO_2 + O_2 + CO \rightarrow 2(1-\alpha)CO_2 + 2\alpha CO + (\alpha + \tfrac{1}{2})\, O_2.$$

Example 4

Consider a correct mixture of benzene C_6H_6 and oxygen. Determine the final composition of the mixture after combustion.

(a) Reaction without dissociation:

$$C_6H_6 + 7.5\ O_2 \rightarrow 6\ CO_2 + 3H_2O.$$

Now the dissociation process involves both CO_2 and H_2O. We will use α_1 and α_2 for the degrees of dissociation of CO_2 and H_2O respectively.

(b) Dissociation of CO_2

$$\alpha_1 CO_2 \rightarrow \alpha_1 CO + \frac{\alpha_1}{2} O_2$$

$$6CO_2 \rightarrow 6(1-\alpha_1)CO_2 + 6\alpha_1 CO_2 \rightarrow 6(1-\alpha_1)CO_2 + 6\alpha_1 CO + 3\alpha_1 O_2.$$

(c) Dissociation of H_2O:

$$\alpha_2 H_2O \rightarrow \alpha_2 H_2 + \frac{\alpha_2}{2} O_2.$$

(In this case the equilibrium reaction is $H_2O \rightleftharpoons H_2 + \tfrac{1}{2}O_2$.)

$$3H_2O \rightarrow 3(1-\alpha_2)H_2O + 3\alpha_2 H_2O \rightarrow 3(1-\alpha_2)H_2 + 3\alpha_2 H_2 + \frac{3}{2}\alpha_2 O_2.$$

(d) Overall reaction

$$C_6H_6 + 7.5O_2 \rightarrow 6(1-\alpha_1)CO_2 + 6\alpha_1 CO + 3(1-\alpha_2)H_2O + 3\alpha_2 H_2 + \frac{3}{2}(2\alpha_1 + \alpha_2)O_2.$$

The final composition after combustion is given by the terms on the right-hand side of the equation.

In all the above examples we have specified the dissociation process. This is necessary in all problems involving dissociation. If there are other products of dissociation these should be included if accurate results are required in calculations. In the above examples it is possible to include the additional products of dissociation O, H, OH.

Chemical Equilibrium

To evaluate the degrees of dissociation (α_1, α_2, etc.) it is necessary to establish the criterion for chemical equilibrium in the chemical reactions given above. It is usual to consider the dissociation process to take place at constant temperature and pressure. For these conditions the system will be in stable equilibrium when the Gibbs function for the reaction reaches its minimum value. It is convenient for the development of the criterion for equilibrium of a chemical reaction to refer to the "nominal" reactants and the "nominal" products of a reaction. The "nominal" products are the products of a reaction without dissociation. For example, in the CO-O_2 reaction the "nominal" products are CO_2, the "nominal" reactants are CO and O_2. During the course of the reaction there will be present both the "nominal" products and the "nominal" reactants. For example, if the reaction at any stage is given by

$$CO + \tfrac{1}{2}\,O_2 \rightarrow aCO + bCO_2 + dO_2$$

we have present at any stage a mols of CO, b mols of CO_2, d mols of O_2. The Gibbs function for the mixture, G, at any stage will be

$$a\mu_{CO} + b\mu_{CO_2} + d\mu_{O_2} = G$$

since the specific Gibbs function g equals the chemical potential μ.

At equilibrium the value of G will be a minimum. In this simple reaction, a, b, d will be continually varying until G has reached a minimum. We have already shown that a, b, d are related at equilibrium by the degree of dissociation α. We would therefore expect that a, b and d are also connected during the reaction by some variable; this variable is called the degree of reaction ε. At equilibrium ε will equal $(1-\alpha)$ but will vary throughout the reaction. At any stage of the reaction $CO + \frac{1}{2}O_2$ we can replace α by $(1-\varepsilon)$ and we have

$$CO + \tfrac{1}{2}\,O_2 \rightarrow \varepsilon CO_2 + (1-\varepsilon)CO + \tfrac{1}{2}(1-\varepsilon)O_2.$$

(The reader will observe that ε can have any numerical value between 0 and 1 in the above equation.) The Gibbs function for the mixture is then

$$G = \varepsilon\mu_{CO_2} + (1-\varepsilon)\mu_{CO} + \tfrac{1}{2}(1-\varepsilon)\mu_{O_2}.$$

We have simplified the expression for G which is now a function of ε only. At equilibrium $(\partial G/\partial\varepsilon)_{p,T}$ will equal zero, since at this the point the Gibbs function will be a minimum. The value of $\alpha = (1-\varepsilon)$ can then be determined.

In the above example we have taken a special case (i.e. a correct mixture) and briefly indicated the method of approach. We will now consider the general case of CO-O_2-CO_2 reaction.

Let us consider a mixture of

$$x\left\{\nu_{CO}CO + \nu_{O_2}O_2\right\} + yO_2 + zCO_2$$

where $x\nu_{CO}$, $\left\{x\nu_{O_2}+y\right\}$, z are the number of mols of CO, O_2 and CO_2 present. The stoichiometric coefficients are ν_{CO}, ν_{O_2}. At any stage of the reaction we will have

$$x(1-\varepsilon)\nu_{CO}CO + x(1-\varepsilon)\nu_{O_2}O_2 + x\nu_{CO_2}\varepsilon CO_2 + yO_2 + zCO_2,$$

where ε is the degree of reaction. When $\varepsilon = 0$ the reaction has not started. If there were no dissociation ε would go to unity. Notice we associate the degree of reaction with the stoichiometric proportions of the initial reactants, excess O_2 (y mols) and CO_2 (z mols) taking no direct part in the chemical reaction. (This remark refers to the detailed balancing of the mols taking part in the reaction. The excess mols of oxygen and carbon monoxide do influence the reaction and limit the degree of dissociation, but as far as the detailed balancing of the various constituents they are considered to be "inert".)

The total number of mols at any stage will be

$$\left.\begin{aligned} M_{CO} &= x(1-\varepsilon)\nu_{CO} \\ M_{CO_2} &= \varepsilon x\nu_{CO_2} + z \\ M_{O_2} &= x(1-\varepsilon)\nu_{O_2} + y. \end{aligned}\right\} \qquad (4.52)$$

When $\varepsilon = 0$ we have the "nominal" reactants and when $\varepsilon = 1$ we have the "nominal products. The rate of change of the individual concentrations with the degree of reaction is obtained by direct differentiation. Thus

$$
\begin{aligned}
dM_{CO} &= -x\nu_{CO}\, d\varepsilon \\
dM_{CO_2} &= x\nu_{CO_2}\, d\varepsilon \\
dM_{O_2} &= -x\nu_{O_2}\, d\varepsilon
\end{aligned}
\tag{4.53}
$$

or

$$
-\frac{dM_{CO}}{\nu_{CO}} = -\frac{dM_{O_2}}{\nu_{O_2}} = \frac{dM_{CO_2}}{\nu_{CO_2}} = x\, d\varepsilon. \tag{4.54}
$$

This equation is called the <u>equation of constraint</u> of the chemical system. Notice that the change in the number of mols of any substance during the reaction is proportional to the stoichiometric coefficients.

With the aid of equation (4.54) we can obtain the equilibrium conditions for the reaction.

From equation (4.29)

$$
dG = V\, dp - S\, dT + \sum_{i=1}^{i=k} \mu_i\, dM_i.
$$

For a constant pressure and temperature reaction $dp = dT = 0$ and

$$
dG = \sum_{i=1}^{i=k} \mu_i\, dM_i.
$$

Substituting for dM_i from (4.54) we have

$$
dG)_{p,T} = (-\nu_{CO}\,\mu_{CO} - \nu_{O_2}\mu_{O_2} + \nu_{CO_2}\,\mu_{CO_2})x\, d\varepsilon.
$$

For equilibrium at constant pressure and temperature

$$
\left(\frac{dG}{\partial\varepsilon}\right)_{p,T} = 0.
$$

Hence

$$
\nu_{CO_2}\mu_{CO_2} = \nu_{CO}\mu_{CO} + \nu_{O_2}\mu_{O_2} \tag{4.55}
$$

since x is not equal to zero.

For an ideal gas

$$\mu = \mu^{o} + R_{mol}T \ln p^{\dagger}.$$

Therefore at equilibrium

$$\nu_{CO_2} \ln p_{CO_2} - \nu_{CO} \ln p_{CO} - \nu_{O_2} \ln p_{O_2}$$

$$= \frac{1}{R_{mol}T} \left[\nu_{CO}\mu^{o}_{CO} + \nu_{O_2}\mu^{o}_{O_2} - \nu_{CO_2} \mu^{o}_{CO_2}\right].$$

Substituting the stoichiometric coefficients for the CO-O_2-CO_2 reaction

$$\nu_{CO_2} = 1, \quad \nu_{CO} = 1, \quad \nu_{O_2} = \tfrac{1}{2}$$

we obtain on rearrangement

$$\ln K_p = \ln\left[\frac{p_{CO_2}}{p_{CO}\sqrt{p_{O_2}}}\right]$$

where

$$\ln K_p = \frac{1}{R_{mol}T} \left[\mu^{o}_{CO} + \tfrac{1}{2} \mu^{o}_{O_2} - \mu^{o}_{CO_2}\right].$$

Finally the relationship between the partial pressures of CO_2, CO and O_2 at equilibrium is given by

$$K_p = \frac{p_{CO_2}}{p_{CO}\sqrt{p_{O_2}}} . \qquad (4.56)^{\dagger\dagger}$$

Since μ^{o} is a function of temperature only, K_p will also be a function of temperature and the relationship between the partial pressures of CO_2, CO and O_2 in the mixture will depend only on the temperature. Furthermore, as the partial pressure is proportional to the mole fraction it is possible from (4.56) to evaluate the proportion of CO_2, CO, and O_2 in the final products at equilibrium.

---

† See footnote to p.82 (Chapter 3).

†† It is important to remember that since we have defined μ^{o} in terms of unit pressure (atm abs) all partial pressures must be in atm abs.

We have evaluated at some length the conditions for equilibrium for the special case of the CO-O_2-CO_2 reaction. It is proposed now to consider the general case of a reaction between the substances A, B, C and D.

It will be recalled in the special case discussed above that the main reaction proceeds between the stoichiometric proportions of the nominal reactants and the nominal products. For the general case we define the stoichiometric equation as

$$\nu_a A + \nu_b B = \nu_c C + \nu_d D,$$

the equilibrium equation as

$$\nu_a A + \nu_b B \rightleftharpoons \nu_c C + \nu_d D,$$

and the reaction at any stage is represented by

$$\nu_a A + \nu_b B \rightarrow (1-\varepsilon)\nu_a A + (1-\varepsilon)\nu_b B + \varepsilon\nu_c C + \varepsilon\nu_d D \qquad (4.57)$$

where as before ε is the degree of reaction. At any stage of the reaction the number of mols will be

$$\left.\begin{aligned} M_a &= (1-\varepsilon)\,\nu_a + C_1 \\ M_b &= (1-\varepsilon)\,\nu_b + C_2 \\ M_c &= \varepsilon\nu_c + C_3 \\ M_d &= \varepsilon\nu_d + C_4. \end{aligned}\right\} \qquad (4.58)$$

The constants C_1, C_2, C_3, C_4 are any number greater than or equal to zero necessary for chemical balance. These are included in (4.58), since up to this stage we have only been concerned with the stoichiometric proportion of the "nominal" reactants and "nominal" products. The reader should compare (4.58) with (4.52) to evaluate the constants for the CO-O_2-CO_2 reaction.

The <u>equation of constraint</u> is the differential form of (4.58)

$$-\frac{dM_a}{\nu_a} = -\frac{dM_b}{\nu_b} = \frac{dM_c}{\nu_c} = \frac{dM_d}{\nu_d} = d\,\varepsilon. \qquad (4.59)$$

As before the Gibbs function in differential form is given by (4.29)

$$dG = V\,dp - S\,dT + \sum_{i=1}^{i=k} \mu_i\, dM_i \tag{4.29}$$

or

$$dG = V\,dp - S\,dT + \mu_a dM_a + \mu_b dM_b + \mu_c dM_c + \mu_d dM_d.$$

Substituting for dM_a, dM_b, dM_c, dM_d from the equation of constraint we obtain

$$dG = V\,dp - S\,dT + (-\nu_a\mu_a - \nu_b\mu_b + \nu_c\mu_c + \nu_d\mu_d)d\varepsilon. \tag{4.60}$$

At equilibrium, at constant pressure and temperature,

$$\left(\frac{\partial G}{\partial \varepsilon}\right)_{p,T} = 0. \tag{4.61}$$

Differentiating equation (4.60) with respect to ε at constant temperature and pressure and equating to zero

$$\nu_a\mu_a + \nu_b\mu_b = \nu_c\mu_c + \nu_d\mu_d. \tag{4.62}$$

Expression (4.62) gives the conditions for chemical equilibrium in terms of the stoichiometric coefficients and the chemical potentials.

For an ideal gas the chemical potential is

$$\mu = \mu^o + R_{mol}T \ln p. \qquad (4.36)^\dagger$$

Substituting for μ in (4.62) we obtain after rearrangement,

$$\nu_a\mu_a^o + \nu_b\mu_b^o - \nu_c\mu_c^o - \nu_d\mu_d^o = R_{mol}T\,(\nu_c \ln p_c + \nu_d \ln p_d - \nu_a \ln p_a - \nu_b \ln p_b). \tag{4.63}$$

The left-hand side of (4.63) is the difference in the standard

† See footnote to p.82 (Chapter 3).

chemical potentials (referred to one atmosphere) and is usually expressed as:

$$-\Delta G_T = \nu_a\mu_a^o + \nu_b\mu_b^o - \nu_c\mu_c^o - \nu_d\mu_d^o. \tag{4.64}$$

The right-hand side of (4.63) is in terms of the partial pressures and is simplified to

$$R_{mol}T \ln \frac{p_c^{\nu_c}\, p_d^{\nu_d}}{p_a^{\nu_a}\, p_b^{\nu_b}}$$

Equation (4.63) can be expressed in the form

$$\ln \frac{p_c^{\nu_c}\, p_d^{\nu_d}}{p_a^{\nu_a}\, p_b^{\nu_b}} = -\frac{\Delta G_T^o}{R_{mol}T} = \ln K_p \tag{4.65}$$

or

$$K_p = \frac{p_c^{\nu_c}\, p_d^{\nu_d}}{p_a^{\nu_a}\, p_b^{\nu_b}} \tag{4.66}$$

K_p is called the <u>equilibrium constant</u>.

From (4.65)

$$\ln K_p = -\frac{\Delta G_T^o}{R_{mol}T}. \tag{4.67}$$

Since ΔG_T^o is a function of <u>temperature only the equilibrium constant</u> is <u>a function of temperature only</u>.

The reader should compare equations (4.66) and (4.56). The partial pressures p_a, p_b, p_c, p_d can be expressed in terms of the mol-fractions x_a, x_b, x_c, x_d and the total pressure p.

Thus

$$p_a = x_a p, \quad p_b = x_b p, \quad p_c = x_c p, \quad p_d = x_d p.$$

Equation (4.66) can then be written in the form

$$K_p = \left(\frac{x_c^{\nu_c}\, x_d^{\nu_d}}{x_a^{\nu_a}\, x_b^{\nu_b}}\right) p^{\nu_c+\nu_d-\nu_a-\nu_b}.$$

The molar concentration is defined as the number of mols per unit volume (M/V) and is represented by square brackets [ ] thus:

$$[A] = \frac{M_a}{V}, \quad [B] = \frac{M_b}{V}, \quad \text{etc.}$$

Now from the general gas law

$$pV = MR_{mol}T$$

and the molar concentration is

$$[\;] = \frac{M}{V} = \frac{p}{R_{mol}T}.$$

For example,

$$[A] = \frac{M_a}{V} = \frac{p_a}{R_{mol}T}$$

and

$$[B] = \frac{p_b}{R_{mol}T}, \quad [C] = \frac{p_c}{R_{mol}T}, \quad [D] = \frac{p_d}{R_{mol}T}.$$

Substituting for p_a, p_b, p_c, p_d in (4.66) we obtain

$$K_p = \frac{[C]^{\nu_c}[D]^{\nu_d}}{[A]^{\nu_a}[B]^{\nu_b}}(R_{mol}T)^{\nu_c+\nu_d-\nu_a-\nu_b}.$$

The equilibrium constant, K_c, in terms of the concentration is

$$K_c = \frac{[C]^{\nu_c}[D]^{\nu_d}}{[A]^{\nu_a}[B]^{\nu_b}}. \tag{4.68a}$$

K_c and K_p are related by

$$K_c = K_p(R_{mol}T)^{\nu_a+\nu_b-\nu_c-\nu_d}. \qquad (4.68b)^\dagger$$

† The units of R_{mol} are most important in this expression. These will depend on the units of K_p and K_c.

Let us re-examine the simple case of the correct mixture of CO and O_2:

$$CO + \tfrac{1}{2} O_2 \rightarrow (1-\alpha) CO_2 + \alpha CO + \frac{\alpha}{2} O_2.$$

The partial pressures at equilibrium are

$$p_{CO_2} = \frac{1-\alpha}{1+\alpha/2} p$$

$$p_{CO} = \frac{\alpha}{1+\alpha/2} p$$

$$p_{O_2} = \frac{\alpha}{2(1+\alpha/2)} p.$$

The $(1+\alpha/2)$ term in the denominator is equal to the total number of mols in the final products.

For this reaction

$$K_p = \frac{p_{CO_2}}{p_{CO}\sqrt{p_{O_2}}}$$

$$K_p = \frac{(1-\alpha)}{\alpha} \sqrt{\frac{2+\alpha}{\alpha}} \sqrt{\frac{1}{p}}.$$

In the final pressure p is known, or can be determined, and the temperature is known, the degree of dissociation, α, can be calculated.

Example 5

A correct mixture of CO and O_2 is exploded in a vessel in which the initial pressure and temperature of the reactants are 1 atm. absolute and 600°R respectively. Calculate the composition of the products of combustion and the pressure at 4500°R. The equilibrium constant at 4500°R is given by

$$\frac{p_{CO_2}}{p_{CO}\sqrt{p_{O_2}}} = 26.3$$

The combustion equation is

$$CO + \tfrac{1}{2}\, O_2 \rightarrow (1-\alpha)\, CO_2 + \alpha CO + \frac{\alpha}{2}\, O_2$$

where α is the degree of dissociation.

We will define the two states by suffixes,

1 initial reactants,
2 final products.

<u>Partial Pressures at State 2</u>

$$p_{CO} = \frac{\alpha}{M_2}\, p_2; \quad p_{CO_2} = \frac{1-\alpha}{M_2}\, p_2; \quad p_{O_2} = \frac{\alpha}{2M_2}\, p_2.$$

<u>State Equations</u>

$$p_1V_1 = M_1R_{mol}T_1$$

$$p_2V_2 = M_2R_{mol}T_2.$$

Now

$$V_1 = V_2.$$

Therefore

$$\frac{p_2}{M_2} = \frac{p_1T_2}{M_1T_1}.$$

$$p_1 = 1 \text{ atm abs}, \quad M_1 = \frac{3}{2}, \quad T_2 = 4500^{\circ}R, \quad T_1 = 600^{\circ}R$$

and

$$\frac{p_2}{M_2} = \frac{2}{3} \times \frac{4500}{600} = 5.$$

<u>Equilibrium Equation</u>

$$\frac{p_{CO_2}}{p_{CO}\sqrt{p_{O_2}}} = 26.3.$$

Substituting for the partial pressures and squaring

$$\left(\frac{1-\alpha}{\alpha}\right)^2 \cdot \frac{2}{\alpha} = (26.3)^2 \times 5$$

or

$$\left(\frac{1-\alpha}{\alpha}\right)^2 \cdot \frac{1}{\alpha} = 1729$$

Solving for α we obtain

$$\alpha = 0.0789 \simeq 0.079$$

and

$$M_2 = 1 + \frac{\alpha}{2} = 1.0395.$$

The gas analysis is given by

$$\text{per cent by volume} = \frac{M}{M_2} \times 100$$

$$CO = \frac{0.079}{1.0395} \times 100 = 7.6 \text{ per cent}$$

$$CO_2 = \frac{0.921}{1.0395} \times 100 = 88.6 \text{ per cent}$$

$$O_2 = \frac{0.0395}{1.0395} \times 100 = 3.8 \text{ per cent.}$$

Now

$$\frac{p_2}{M_2} = 5,$$

therefore,

$$p_2 = 5 \times 1.0395 \simeq 5.2 \text{ atm abs.}$$

A word about sign convention; if we consider the "nominal" stoichiometric reaction

$$\nu_a A + \nu_b B = \nu_c C + \nu_d D. \qquad (4.69)$$

Then

$$K_p = \frac{\Pi(p^{\nu})_{\text{"products"}}}{\Pi(p^{\nu})_{\text{"reactants"}}} \qquad (4.70)$$

where the "products" are the nominal products and the "reactants" the nominal reactants.

In some texts the equilibrium constant is defined as the reciprocal of (4.70), i.e.

$$K_p = \frac{\Pi(p^{\nu})_{\text{"reactants"}}}{\Pi(p^{\nu})_{\text{"products"}}}. \qquad (4.71)$$

The magnitude of K_p is dependent on the sign convention and units used in the free energy change in equation (4.64). The reader should exercise care in problems to make sure that the correct expression is used for the equilibrium constant.†

The Relationship Between the Equilibrium Constant and the Heat of Reaction - Van't Hoff's Equation

The equilibrium constant K_p is given by

$$\ln K_p = \frac{-\Delta G_T}{R_{mol} T} \tag{4.67}$$

where

$$-\Delta G_T^o = \sum(\nu\mu^o)_{reactants} - \sum(\nu\mu^o)_{products} \tag{4.64}$$

The suffixes "reactants" and "products" refer to the "nominal" reactants and products in the stoichiometric equation. We will represent these suffixes by the short abbreviations R and P respectively. Thus

$$-\Delta G_T^o = \sum(\nu\mu^o)_R - \sum(\nu\mu^o)_P. \tag{4.72}$$

Slight rearrangement of (4.67) gives

$$\ln K_p = \frac{1}{R_{mol}} \left(\sum\left(\nu \frac{\mu^o}{T}\right)_R - \sum\left(\nu \frac{\mu^o}{T}\right)_P \right). \tag{4.73}$$

The standard chemical potential μ^o is a function of T only and the stoichiometric coefficient ν is independent of T. Differentiating (4.73) with respect to T,

$$\frac{d(\ln K_p)}{dT} = \frac{1}{R_{mol}} \left(\sum\left(\nu \frac{d(\mu^o/T)}{dT}\right)_R - \sum\left(\nu \frac{d(\mu^o/T)}{dT}\right)_P \right). \tag{4.74}$$

---

†In Table A.4 the equilibrium constant K_p is expressed in the form

$$K_p = \frac{\Pi((p/p_o)^\nu)_{products}}{\Pi((p/p_o)^\nu)_{reactants}}$$

for

$$p_o = 1.01325 \text{ bars}$$

The ratio μ^o/T can be eliminated by using (4.42)

$$h = -T^2\left(\frac{\partial(\mu/T)}{\partial T}\right)_p. \tag{4.42}$$

Now

$$\frac{\mu}{T} = \frac{\mu^o}{T} + R_{mol} \ln p \quad \text{from (4.36).}$$

Hence

$$\left(\frac{\partial(\mu/T)}{\partial T}\right)_p = \left(\frac{\partial(\mu^o/T)}{\partial T}\right)_p = \frac{d(\mu^o/T)}{dT}$$

since μ^o is a function of T only.

Equation (4.42) in terms of the standard chemical potential is, therefore,

$$h = -T^2\left(\frac{d(\mu^o/T)}{dT}\right). \tag{4.75}$$

Substituting (4.75) into (4.74)

$$\frac{d \ln K_p}{dT} = \frac{1}{R_{mol}T^2}\left[\sum(\nu h)_P - \sum(\nu h)_R\right]. \tag{4.80}$$

To interpret the term in the bracket let us consider the "nominal" reaction

$$\nu_a A + \nu_b B = \nu_c C + \nu_d D.$$

At temperature T the heat of reaction Q_p is given by

$$Q_p = (\nu_c h_c + \nu_d h_d) - (\nu_a h_a + \nu_b h_b)$$

or

$$Q_p = \sum(\nu h)_P - \sum(\nu h)_R.$$

The term in the bracket in (4.80) is the heat of reaction Q_p at temperature T for the "nominal" reaction. Equation (4.80) becomes on substitution of the heat of reaction:

$$\frac{d(\ln K_p)}{dT} = \frac{Q_p}{R_{mol}T^2}. \tag{4.81a}$$

This expression is called Van't Hoff's equation. The heat of reaction Q_p is for the "nominal" reactants and the "nominal" products for the "nominal" reaction.

The dissociation process is the reaction

$$\nu_c C + \nu_d D \rightarrow \nu_a A + \nu_b B.$$

The heat of reaction for this process is the negative of the heat of reaction for the "nominal" reaction.

In an alternative form of Van't Hoff's equation the equilibrium constant is expressed in terms of the heat of reaction of the dissociation process. This is usually called the heat of dissociation D. Van't Hoff's equation is then given as

$$\frac{d(\ln K_p)}{dT} = \frac{D}{R_{mol}T^2} \tag{4.81b}$$

In this case K_p will be the reciprocal of the value given by (4.81a). It is usual to use this form when considering a gaseous system which apparently is non-reacting. This will be illustrated by an example. Consider a system containing oxygen in molecular form, O_2. On heating the oxygen it is found that there is some atomic form of oxygen, O, present. The reaction is represented by

$$O_2 \rightarrow (1-\alpha)\ O_2 + 2\alpha O.$$

The equilibrium process in the system is

$$O_2 \rightleftharpoons 2O.$$

Thus for this system the equilibrium constant is given by

$$K_p = \frac{p^2{}_O}{p_{O_2}}$$

The heat of dissociation D is given by

$$D = 2h_O - h_{O_2}$$

and Van't Hoff's equation is

$$\frac{d(\ln K_p)}{dT} = \frac{D}{R_{mol}T^2} = \frac{2h_O - h_{O_2}}{R_{mol}T^2}.$$

Notice in this method we have considered the "products" to be the dissociated atoms O, and the "reactants" the undissociated molecule O_2; K_p is obtained as before from (4.70). If we had considered this problem in the "nominal" form discussed earlier we would have for the <u>equilibrium</u> process

$$2O \rightleftharpoons O_2.$$

The "reactants" would now be the atomic form of oxygen O, and the "products" the molecular form O_2.

The equilibrium constant would then be

$$K_p = \frac{p_{O_2}}{p^2_O}$$

and Van't Hoff's equation

$$\frac{d(\ln K_p)}{dT} = \frac{Q_p}{R_{mol}T^2} = \frac{h_{O_2}-2h_O}{R_{mol}T^2}.$$

The general principles and calculation procedures are the same for both cases. These may be briefly summarized:

(1) Write down the equilibrium process and specify the "reactants" and "products".

(2) Write down the equilibrium constant in terms of the partial pressures for the "reactants" and "products" as in (4.70).

(3) Calculate K_p from Van't Hoff's equation noting that $Q_p = {}^P H_P - H_R$ for the equilibrium equation of the "nominal" reaction or $D = H_P - H_R$ for the equilibrium equation of the "dissociation" reaction.

The ratio D/R_{mol} is called the characteristic temperature for dissociation (θ_D). Van't Hoff's equation in terms of the characteristic temperature is

$$\frac{d(\ln K_p)}{dT} = \frac{\theta_D}{T^2} \tag{4.82}$$

We will use Van't Hoff's equation to ascertain the effect of temperature on the degree of dissociation. We will first examine the effect of temperature on the equilibrium constant and then, in a qualitative manner, examine the effect of the equilibrium constant on the degree of dissociation.

Over a small temperature range Q_p may be considered to be constant. Integrating (4.81a) we obtain

$$\ln(K_p)_{T_2} - \ln(K_p)_{T_1} = \frac{Q_p}{R_{mol}}\left(\frac{1}{T_1} - \frac{1}{T_2}\right) = \frac{Q_p}{R_{mol}}\left(\frac{T_2-T_1}{T_1T_2}\right).$$

For an exothermic reaction Q_p is negative. If T_2 is greater than T_1 $(K_p)_{T_1}$ is greater than $(K_p)_{T_2}$, and for exothermic reactions the equilibrium constant decreases with increase in temperature.

To examine the effect of K_p on the degree of dissociation let us consider the reaction

$$CO + \tfrac{1}{2} O_2 \rightarrow (1-\alpha)CO_2 + \alpha CO + \frac{\alpha}{2} O_2$$

for this reaction

$$\frac{p_{CO_2}}{p_{CO}\sqrt{p_{O_2}}} = K_p.$$

Substituting for p_{CO_2}, p_{CO}, p_{O_2} in terms of α and p

$$(K_p)^2 = \frac{(1-\alpha)^2}{\alpha^2}\left(\frac{2+\alpha}{\alpha}\right)\frac{1}{p}.$$

In general $\alpha \ll 1$ hence

$$(K_p)^2 \propto \frac{1}{\alpha^3} \quad \text{at constant p}$$

or

$$\alpha^3 \propto \frac{1}{K_p^2}.$$

As the equilibrium constant K_p decreases the degree of dissociation increases. Now the carbon monoxide-oxygen reaction is exothermic and the equilibrium constant decreases with increase in temperature. Hence the degree of dissociation increases with increase in temperature.

Another important case of dissociation, to be discussed in detail later, is the dissociation of the diatomic molecule A_2 according to the equation

$$A_2 \rightleftharpoons 2A.$$

This type of reaction was discussed earlier in connection with the dissociation of oxygen. The form of Van't Hoff's equation (4.81b) will be used. The equilibrium constant in terms of the partial pressure is

$$K_p = \frac{p^2{}_A}{p_{A_2}} = \frac{4\alpha^2}{1-\alpha^2}\,p.$$

The reaction is endothermic and D is positive in Van't Hoff's equation. If D is assumed to be constant over a small temperature range T_1 to T_2, then integration of (4.81b) gives

$$\ln(K_p)_{T_2} - \ln(K_p)_{T_1} = \frac{D}{R_{mol}}\left(\frac{T_2 - T_1}{T_1 T_2}\right)$$

and it will be seen that K_p increases with increase in temperature. Now for $\alpha \ll 1$

$$\alpha^2 \propto K_p \quad \text{at constant } p$$

The degree of dissociation, therefore, increases with increase in K_p. Finally, the degree of dissociation increases with increase in temperature. In general for gases of interest to engineers the degree of dissociation increases with increase in temperature.

Methods for Calculating Equilibrium Constants

In the same manner as standard tables are provided for the normal thermodynamic properties of substances (h, u, s, etc.), so tables are available for equilibrium constants for different reactions over a range of temperatures. Current practice is to use the methods of statistical thermodynamics in conjunction with spectrographic data to provide the empirical coefficients to evaluate the constants used in computing these tables. Two such methods will now be discussed.

The standard chemical potential is given by

$$\mu^o = g^o = g(T) + g_o.$$

If the reference temperature T_o is equal to the absolute zero and the entropy s_o will be zero for a pure substance then

$$g_o = h_o - Ts_o = h_o$$

and

$$g(T) = g^o - h_o.$$

Now from (4.73)

$$\ln K_p = \frac{1}{R_{mol}}\left(\sum\left(\frac{\nu\mu^o}{T}\right)_R - \sum\left(\frac{\nu\mu^o}{T}\right)_P\right). \qquad (4.73)$$

Substituting for $\mu^o = g(T) + h_o$ we obtain

$$\ln K_p = \sum\left(\frac{\nu(g(T) + h_o)}{R_{mol}T}\right)_R - \sum\left(\frac{\nu(g(T) + h_o)}{R_{mol}T}\right)_P$$

$$\ln K_p = \left(\sum\left(\frac{\nu g(T)}{R_{mol}T}\right)_R - \sum\left(\frac{\nu g(T)}{R_{mol}T}\right)_P\right) + \left(\sum\left(\frac{\nu h_o}{R_{mol}T}\right)_R - \sum\left(\frac{\nu h_o}{R_{mol}T}\right)_P\right). \tag{4.82}$$

The expression in the second bracket on the right-hand side of (4.82) is equal to $-\Delta H_o/R_{mol}T$; we thus obtain

$$\ln K_p = \left(\sum\left(\frac{\nu g(T)}{R_{mol}T}\right)_R - \sum\left(\frac{g(T)}{R_{mol}T}\right)_P\right) - \frac{\Delta H_o}{R_{mol}T}. \tag{4.83}$$

The function g(T) can be evaluated using the data given in the Appendix in the following manner:

The enthalpy h, the standard chemical potential μ^o and the temperature T are related by (4.75):

$$-\frac{h}{T^2} = \frac{d(\mu^o/T)}{dT} \tag{4.75}$$

where $\mu^o = g(T) + h_o$

$h = h(T) + h_o.$

The specific enthalpy h for a substance is given in the form

$$\frac{h(T)}{R_{mol}T} = \frac{h-h_o}{R_{mol}T} = a_1+a_2T+a_3T^2+a_4T^3+a_5T^4 \tag{4.84}$$

Dividing by T we obtain

$$\frac{h(T)}{R_{mol}T^2} = \frac{a_1}{T} + a_2+a_3T+a_4T^2+a_5T^3$$

Substituting for $h(T)/R_{mol}T^2$ and μ^o/T in (4.75) the following simple differential equation is formed:

$$-\left(\frac{a_1}{T} + a_2 + a_3T + a_4T^2 + a_5T^3 + \frac{h_o}{R_{mol}T^2}\right) dT = \frac{1}{R_{mol}} d\left(\frac{g(T) + h_o}{T}\right) \tag{4.85}$$

Integration of (4.85) gives

$$-\left(a_1 \ln T + a_2T + \frac{a_3T^2}{2} + \frac{a_4T^3}{3} + \frac{a_5T^4}{4} - \frac{h_o}{R_{mol}T}\right) + A = \frac{g(T)+h_o}{R_{mol}T} \tag{4.86}$$

let $A = a_1 - a_6$

Then equation (4.86) becomes after rearrangement

$$\frac{g(T)}{R_{mol}T} = a_1(1-\ln T) - \left(a_2T + \frac{a_3T^2}{2} + \frac{a_4T^3}{3} + \frac{a_5T^4}{4}\right) - a_6$$

This can be simplified to give

$$\frac{g(T)}{R_{mol}T} = a_1(1-\ln T) - \sum_{j=2}^{j=5} \frac{a_j}{j-1} T^{j-1} - a_6 \tag{4.87}$$

The constants a_1, a_2, a_3, a_4, a_5 and a_6 are given in Table A.1 (Appendix) for a number of gases.

For any temperature T the equilibrium constant K_p can be calculated from (4.83) and (4.87) with the aid of Table A.1. The tables given at the end of the book have been calculated from these data with p_o = 1.01325 bars.

An alternative method for calculating the equilibrium constant is to use the Gibbs free energy of formation of a chemical reaction. The <u>free energy of formation</u> (ΔG_T^o) at a temperature T is

$$\Delta G_T = \sum(\nu g^o)_P - \sum(\nu g^o)_R. \tag{4.88}$$

The standard reference temperature is usually 25°C (298°K). The conventional representation of the standard free energy of formation is

$$\Delta G_{298}^o = \Delta G_T^o \text{ at } T = 298^oK.$$

Now

$$g^o = g(T) + g_o$$

and

$$\Delta G_T^o = \Delta G(T) + \Delta G_{298}^o$$

where

$$\Delta G(T) = \sum(\nu g(T))_P - \sum(\nu g(T))_R .$$

The equilibrium constant K_p is given by (4.67):

$$\ln K_p = -\frac{\Delta G_T^o}{R_{mol}T}$$

$$= -\left(\frac{\Delta G(T)}{R_{mol}T} + \frac{\Delta G_{298}^o}{R_{mol}T}\right) \qquad (4.67)$$

$$\ln K_p = -\frac{1}{R_{mol}T}\left[\sum(\nu g(T))_P - \sum(\nu g(T))_R\right] - \frac{\Delta G_{298}^o}{R_{mol}T} \qquad (4.89)$$

where $g(T)$ is measured <u>above</u> 25°C (298°K). It will be observed that $-(\Delta G_{298}^o/R_{mol}T)$ is equal to the equilibrium constant at 25°C. Tabulated data for ΔG_{298}^o are available in standard reference books.

At first sight it might appear that in order to carry out calculations involving equilibrium data a large number of tables would be required to cover a range of reactions. This is not so. For many problems involving hydrocarbon-air mixtures only the following reactions are necessary:

$$CO + \tfrac{1}{2} O_2 \rightleftharpoons CO_2$$

$$H_2 + \tfrac{1}{2} O_2 \rightleftharpoons H_2O.$$

For more detailed studies the following reactions

$$H_2O \rightleftharpoons \tfrac{1}{2} H_2 + OH$$

$$O_2 \rightleftharpoons 2O$$

$$H_2 \rightleftharpoons 2H$$

$$N_2 \rightleftharpoons 2N$$

$$N + O \rightleftharpoons NO$$

may be used.

Many complex reactions can be reduced to simple reactions

of the type shown above. This can be illustrated by a simple example. Consider the equilibrium reaction

$$CO_2 + H_2 \rightleftharpoons CO + H_2O.$$

The stoichiometric equation is

$$CO_2 + H_2 = CO + H_2O \tag{4.90a}$$

and this can be represented by

$$CO_2 = CO + \tfrac{1}{2}\,O_2 \tag{4.90b}$$

$$H_2 + \tfrac{1}{2}\,O_2 = H_2O. \tag{4.90c}$$

The two reactions, added together, give the stoichiometric equation. Thus

$$\begin{array}{rcl} CO_2 & = & CO + \tfrac{1}{2}\,O_2 \\ H_2 + \tfrac{1}{2}\,O_2 & = & H_2O \\ \hline H_2 + CO_2 & = & CO + H_2O. \end{array} \tag{4.91}$$

The mols of O_2 are cancelled out from each side of the above equations.

If we replace the equality signs in (4.90a), (4.90b) and (4.90c) by the usual reversible symbols we can write, for the three reactions,

$$K_{p_x} = \frac{p_{H_2O}\, p_{CO}}{p_{CO_2}\, p_{H_2}} \tag{4.92a}$$

$$K_{p_1} = \frac{p_{CO}\sqrt{p_{O_2}}}{p_{CO_2}} \tag{4.92b}$$

$$K_{p_2} = \frac{p_{H_2O}}{p_{H_2}\sqrt{p_{O_2}}} \tag{4.92c}$$

We observe that for the reaction in (4.90b) the equilibrium constant K_{p_1} is the inverse of the usual equilibrium constant K_p, since the "nominal" products are CO and O_2 and the "nominal" reactant is CO_2.

Eliminating the partial pressures we obtain

$$K_{p_x} = K_{p_1} \times K_{p_2}. \tag{4.93}$$

Thus if K_{p_1} and K_{p_2} are known K_{p_x} can be determined from (4.93).

We can generalize (4.93). If a reaction x is the sum of n reactions then

$$K_{p_x} = K_{p_1} K_{p_2} K_{p_3}, \ldots, K_{p_n}$$

or

$$K_{p_x} = \prod_{i=1}^{i=n} K_{p_i}. \tag{4.94}^\dagger$$

If we take the logarithm of (4.94) we obtain

$$\ln K_{p_x} = \sum_{i=1}^{i=n} \ln K_{p_i}.$$

Now

$$\ln K_p = -\frac{\Delta G_T^o}{R_{mol} T} \tag{4.67}$$

and

$$\ln K_{p_x} = -\left(\frac{\Delta G_T^o}{R_{mol} T}\right)_x = \frac{1}{R_{mol} T}\left[\sum_{i=1}^{i=n} (-\Delta G_T^o)_i\right] \tag{4.95}$$

or

$$\Delta G_T^o = \sum_{i=1}^{i=n} (\Delta G_T^o)_i. \tag{4.96}$$

<u>This is the equivalent form of Hess's Law for the free energy of formation.</u>

The sign of the free energy of formation must be noted. For exothermic reactions ΔG_T^o is negative.
The reaction

$H_2 + \frac{1}{2} O_2 \rightarrow H_2O$ is exothermic

The reaction

$CO_2 \rightarrow CO + \frac{1}{2} O_2$ is endothermic.

---

† The symbol $\prod$ is the continuous product, thus $\prod_{i=1}^{i=4} x_i = x_1 x_2 x_3 x_4$.

Hence $\Delta G^o_{H_2O}$ is negative, and $\Delta G^o_{CO_2}$ is positive.

To evaluate the equilibrium constant from (4.95) the sign of ΔG^o_T must be taken into account.

One final word of warning. It is important to ascertain the number of mols involved in a reaction when using the tabulated data for the free energy of formation. It is usual to quote the free energy of formation ΔG^o_T in terms of heat units/lb mol or heat units/kg mol. Thus ΔG^o_T for H_2O will be for the reaction

$$H_2 + \tfrac{1}{2}\, O_2 = H_2O,$$

since only 1 mol of H_2O is formed.

The Calculation of Temperature Rise in a Combustion Reaction with Dissociation

Dissociation limits the rise in temperature in a combustion reaction. Dissociation processes are endothermic and we say "heat" is absorbed (this is proportional to the heat of dissociation D). For an equilibrium process the degree of dissociation increases with increase in temperature and the heat "absorbed" increases. When the temperature of the products falls the degree of dissociation decreases and we have a recombination process. "Heat" is released in a recombination process which is exothermic. To determine the temperature rise for a process in which dissociation or recombination is taking place we must include the energy equation, the equations of state and the equilibrium equations. This is best illustrated by an example.

Consider the adiabatic combustion of carbon monoxide in air under constant volume conditions. The stoichiometric equation is

$$CO + 0.5\, O_2 + 1.88\, N_2 = CO_2 + 1.88\, N_2.$$

(Air contains 1 mol of O_2 to 3.76 mols of N_2.)

The general formula for a mixture of carbon monoxide and air can be expressed as

$$(1+x)CO + 0.5\, O_2 + 1.88\, N_2.$$

If x is positive the mixture is rich, if x is negative the mixture is weak, and if x is zero we have a correct mixture. The reaction with dissociation is given by

$$(1+x)CO + 0.5\ O_2 + 1.88\ N_2$$
$$\rightarrow (1+x-a)CO_2 + aCO + 0.5(a-x)O_2 + 1.88N_2. \qquad (4.97)^{\dagger}$$

For adiabatic combustion at constant volume the internal energy E is constant. Therefore:

$$E_1 = E_2. \qquad (4.98)$$

Suffix 1 refers to the conditions immediately <u>before</u> combustion and suffix 2 to the equilibrium conditions <u>after</u> combustion. In the absence of motion, gravity, electricity, magnetism and capillarity the specific internal energy e = u.

Hence

$$E_1 = (1+x)(u_{CO})_1 + 0.5(u_{O_2})_1 + 1.88(u_{N_2})_1 \qquad (4.99)$$

$$E_2 = (1+x-a)(u_{CO})_2 + a(u_{CO})_2 + 0.5(a-x)(u_{O_2})_2 + 1.88(u_{N_2})_2. \qquad (4.100)$$

For convenience we define the internal energy of the "nominal" reactants as E_R, thus

$$E_R = (1+x)u_{CO} + 0.5u_{O_2} + 1.88u_{N_2} \qquad (4.101)$$

$$u = u(T) + u_o$$

and

$$E_R = \mathbf{E}_R + E_{RO} \qquad (4.102)$$

where

$$\mathbf{E}_R = \sum (Mu(T))_{\text{nominal reactants}} \qquad (4.103)$$

$$E_{RO} = \sum (Mu_o)_{\text{nominal reactants}}. \qquad (4.104)$$

For the reaction $CO + \frac{1}{2}\ O_2 = CO_2$ the heat of reaction at constant volume is

$$Q_v = u_{CO_2} - u_{CO} - 0.5u_{O_2}. \qquad (4.105)$$

---

† The reader will observe that the degree of dissociation α is equal to (a-x) in this expression (see examples on p.132).

At temperature T_2 the internal energy E_2 from (4.100) is therefore

$$E_2 = (1+x-a)\,(u_{CO_2} - u_{CO} - 0.5u_{O_2})_2$$

$$+(1+x)(u_{CO})_2 + 0.5(u_{O_2})_2 + 1.88(N_2)_2. \qquad (4.106)$$

Substituting (4.101) and (4.105) in (4.106) we obtain

$$E_2 = (1+x-a)\,Q_{V2} + E_{R2} \qquad (4.107)$$

where Q_{V2} is the heat of reaction at temperature T_2 and E_{R2} the internal energy of the "nominal" reactants at T_2. Substituting for E_1 and E_2 in (4.98) we obtain

$$E_{R1} = E_{R2} + (1+x-a)\,Q_{V2} \qquad (4.108)$$

and

$$(1+x-a)\,Q_{V2} = E_{R1} - E_{R2} = E_{R1} - E_{R2}. \qquad (4.109)$$

Hence

$$a = (1+x) - \frac{E_{R1} - E_{R2}}{Q_{V2}}. \qquad (4.110)$$

An alternative form of (4.110) is to replace Q_{V2}, the heat of reaction of constant volume at T_2, by $\Delta E_o + \Delta E(T)_2$ where ΔE_o is the heat of reaction at absolute zero and $\Delta E(T)_2$ is given by

$$\Delta E(T)_2 = (u(T)_{CO_2} - u(T)_{CO} - 0.5u(T)_{O_2})_{T_2}. \qquad (4.111)$$

Then

$$a = (1+x) - \frac{E_{R1} - E_{R2}}{\Delta E_o + \Delta E(T)_2}. \qquad (4.112)$$

The numerical values of E_{R1}, E_{R2}, ΔE_o can be obtained in the manner outlined in Chapter 3 or in Table A.2 in the appendices.

In expressions (4.110) and (4.112) a is given in terms of the mixture strength x, the thermodynamic properties of the "nominal" reactants at T_1 and T_2 and the heat of reaction at constant volume at temperature T_2. For each temperature T_2, mixture strength x and initial temperature T_1 there will be only one value for a.

A physical interpretation of (4.110) may be obtained by examining the internal energy-temperature diagram Fig. 4.1. The diagram is drawn for a correct mixture, that is x = 0. For this case

$$(1-a)Q_{V2} = E_{R_1} - E_{R2}.$$

For no dissociation the state 2 is at 2' and the final temperature would be $T_{2'}$. With dissociation the final temperature T_2 will be less than $T_{2'}$. The "heat absorbed" by dissociation is $-aQ_{V2}$, the "heat released" is $-(1-a)Q_{V2}$. The maximum "heat release" occurs when a = 0.

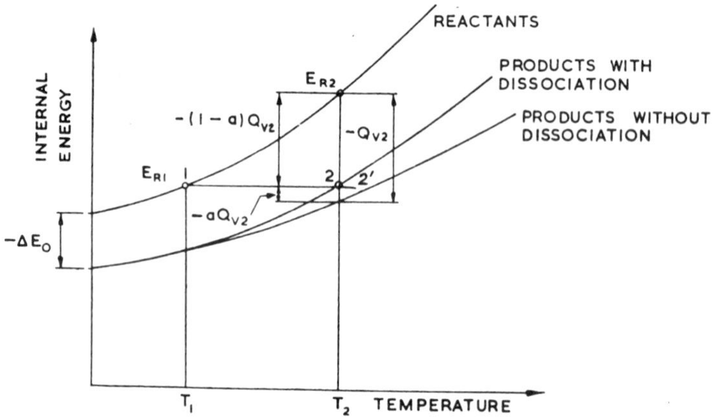

FIG. 4.1 (a). Internal energy diagram for correct mixture.

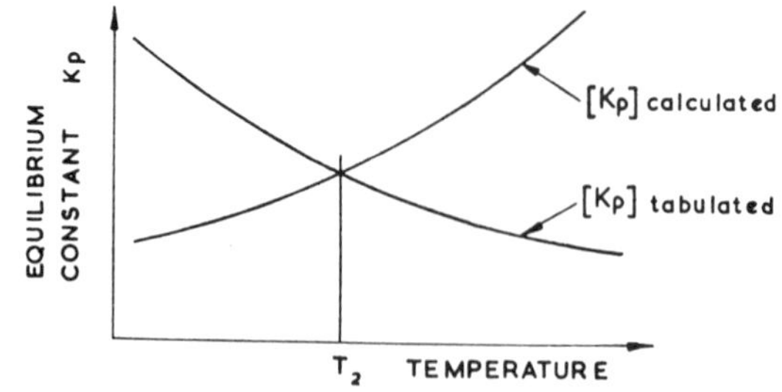

FIG. 4.1 (b).

In order to calculate the final temperature T_2 the equilibrium data are also required. These are:

(a) The equilibrium equation:

$$K_p = \frac{p_{CO_2}}{p_{CO}\sqrt{p_{O_2}}}.$$

(b) The partial pressures of carbon dioxide, carbon monoxide and oxygen:

$$p_{CO_2} = \frac{(1+x-a)}{M_2} p_2$$

$$p_{CO} = \frac{a}{M_2} p_2$$

$$p_{O_2} = \frac{0.5(a-x)}{M_2} p_2.$$

(c) The state equation:

$$\frac{p_2}{M_2} = \frac{p_1}{M_1} \times \frac{T_2}{T_1}, \quad \text{since } V_1 = V_2.$$

Combining (a), (b) and (c)

$$K_p = \left(\frac{(1+x-a)}{a}\right)\left(\frac{2}{a-x}\right)^{\frac{1}{2}} \left(\frac{M_1}{p_1} \times \frac{T_1}{T_2}\right)^{\frac{1}{2}}. \qquad (4.113)$$

The calculation procedure is to select a number of values of T_2 and calculate the value of a at each temperature from either (4.110) or (4.112). With the appropriate values of a and T_2 the values of $K_p = (K_p)_{calculated}$ are determined from (4.113) and plotted in a graph against the temperature T_2. On the same figure the true values of K_p obtained from tables or equation (4.83) are plotted against ^pT. The intersection of the two curves gives T_2. The method is illustrated in the following numerical example.

Example 6

A spark ignition engine operates with a 10 per cent rich mixture of carbon monoxide and air. At the end of the compression stroke the cylinder pressure and temperature are 8.5 bar and 600°K respectively. If combustion takes place at constant volume with no heat loss, calculate the maximum temperature and pressure. The enthalpy of reaction at absolute zero ΔH_o = -276,960 kJ/kg mol CO. The reference pressure p_o = 1.01317 bars. The internal energy u(T) above 0°K in kJ/kg mol:

Gas	Temperature °K			
	600	2800	2900	3000
N_2	12571.7	70733.4	73597.5	76468.6
O_2	12939.6	75568.6	78706.6	81863.6
CO	12626.2	71506.9	74388.9	77277.2
CO_2	17306.8	126555.6	131933.3	137320.9

The equilibrium constant for the reaction

$$CO + \tfrac{1}{2}O_2 \rightleftharpoons CO_2$$

is given by

$$K_p = \frac{\left(\frac{p}{p_o}\right)_{CO_2}}{\left(\frac{p}{p_o}\right)_{CO}\left(\frac{p}{p_o}\right)_{O_2}^{\frac{1}{2}}}$$

Temperature °K	2800	2900	3000
K_p	6.583	4.393	3.014

Solution

Correct mixture

$$CO + 0.5O_2 + 1.88N_2 \rightarrow CO_2 + 1.88N_2.$$

Dissociation

$$aCO_2 \rightarrow aCO + 0.5aO_2.$$

Rich mixture and dissociation (x = 0.1)

$$1.1CO + 0.5O_2 + 1.88N_2 \rightarrow (1.1-a)CO_2 + 0.5(a-0.1)O_2 + aCO + 1.88N_2.$$

Products of combustion

$$(1.1-a)CO_2 + 0.5(a-0.1)O_2 + aCO + 1.88N_2.$$

Number of mols

State 1 before combustion.

State 2 after combustion.

$$M_1 = 1.1 + 0.5 + 1.88 = 3.48.$$

$$M_2 = 1.1 - a + 0.5a + a + 1.88 - 0.05 = 2.93 + 0.5a.$$

Partial pressures

$$p_{CO_2} = \frac{1.1-a}{M_2} p_2$$

$$p_{CO} = \frac{a}{M_2} p_2$$

$$p_{O_2} = \frac{(a-0.1)}{2M_2} p_2.$$

Equilibrium equation

$$K_p = \frac{\left(\frac{p}{p_o}\right)_{CO_2}}{\left(\frac{p}{p_o}\right)_{CO}\left(\frac{p}{p_o}\right)_{O_2}^{\frac{1}{2}}}$$

$$= \frac{(1.1-a)}{a}\sqrt{\frac{2M_2}{(a-0.1)p_2 p_o}}$$

State equation for constant volume combustion

$$\frac{p_1}{p_2} = \frac{M_1}{M_2}\frac{T_1}{T_2}$$

$$\frac{M_2}{p_2} = \frac{600 \times 3.48}{T_2 \times 8.5} = \frac{245.65}{T_2}$$

State equation and equilibrium equation

$$K_p = \frac{1.1-a}{a}\sqrt{\frac{2 \times 245.65}{(a-0.1) \times 1.01314\ T_2}}$$

$$K_p = \frac{22.02(1.1-a)}{a}\sqrt{\frac{1}{(a-0.1)\ T_2}} \tag{i}$$

Energy balance

$$E_1 = E_2.$$

Hence from equation (4.112) a can be evaluated

$$a = (1+x) - \frac{E_{R1}-E_{R2}}{\Delta E_o+\Delta E(T)_2} \tag{4.112}$$

From (4.103)

$$E_{R2} = 1.1\big(u(T_1)\big)_{CO} + 0.5\big(u(T_1)\big)_{O_2}+1.88\big(u(T_1)\big)_{N_2}.$$

At temperature $T_1 = 600^{\circ}K$

$$E_{R1} = 1.1 \times 12626.2 + 0.5 \times 12939.6 + 1.88 \times 12571.7$$

$$= 43993.4 \text{ kJ}$$

Now

$$\Delta E_o = \Delta H_o \quad \text{(Chapter 3, p. 91)}$$

and

$$\Delta E_o = -276{,}960 \text{ kJ/kg mol CO}$$

At temperature T_2, E_{R2} and $\Delta E(T)_2$ can be evaluated from the expressions:

$$E_{R2} = 1.1\big(u(T_2)\big)_{CO} + 0.5\big(u(T_2)\big)_{O_2} +1.88\big(u(T_2)\big)_{N_2}$$

$$\Delta E(T)_2 = \big(u(T_2)\big)_{CO_2}-\big(u(T_2)\big)_{CO} - 0.5\big(u(T_2)\big)_{O_2}.$$

To calculate a from (4.112) we select a number of temperatures as shown in the table given below:

Temperature T_2 °K	2800	2900	3000
E_{R2}	249,420.7	259,544.4	269,697.7
$E_{R1}-E_{R2}$	-205,427.3	-215,551.0	-225,704.3
$\Delta E(T)_2$	17,264.4	18,191.1	19,111.9
$\Delta E(T) + \Delta E_o$	-259,694.4	-258,768.9	-257,848.1
$a = 1.1 - \dfrac{E_{R1}-E_{R2}}{\Delta E(T)_2+\Delta E_o}$	0.309	0.267	0.225

Substituting for a in (i) for each temperature, we obtain the calculated value of K_p.

For

$$T_2 = 2800^\circ K$$

$$K_p = \frac{22.02\ (1.1 - 0.309)}{0.309} \sqrt{\frac{1}{(0.309 - 0.1) \times 2800}}$$

$$= 2.33$$

and

$$T_2 = 2900^\circ K \qquad K_p = 3.12$$

$$T_2 = 3000^\circ K \qquad K_p = 4.44$$

Graphs of the calculated values of K_p and the tabulated values of K_p are plotted against T_2 in Fig. 4.2. On the same graph the calculated values of a are also drawn. The intersection of the two K_p curves gives the true temperature T_2. The true temperature T_2 is 2946°K and a is 0.246. The pressure is

$$p_2 = \frac{M_2 T_2}{245.65} = \frac{3.053 \times 2946}{245.65} = 36.61 \text{ bars}$$

For more complex combustion processes it is usual to use either charts† or a computer program. One example of the

† H.C. Hottel, G.C. Williams and C.N. Satterfield, Thermodynamic Charts for Combustion Processes, John Wiley, New York, 1956.

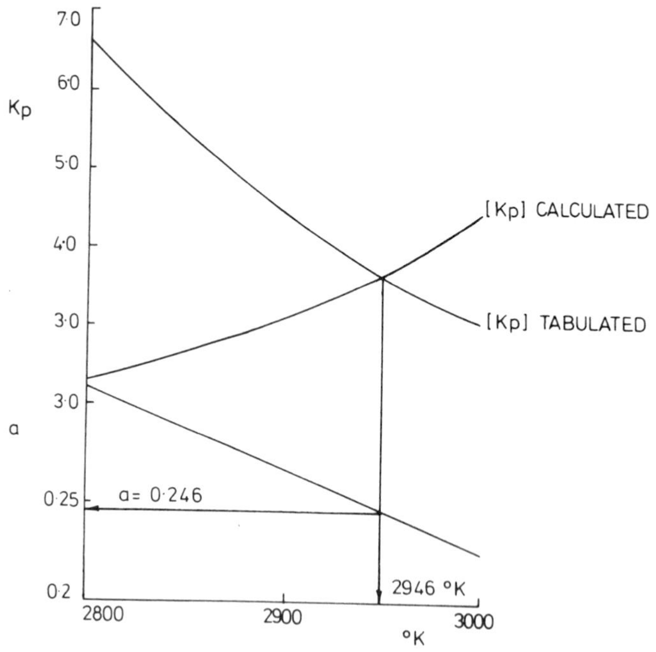

FIG. 4.2.

latter is described by Benson and Baruah† for the Otto cycle with combustion. The analyses can be simplified for certain gases by the method given below. These are called the ideal dissociating gas.

The Lighthill Ideal Dissociating Gas

In studies of air flow over high-speed projectiles or gas flow in hypersonic wind tunnels the thermodynamics of a dissociating gas may be required, in particular the dissociation of the diatomic molecules of oxygen, nitrogen and hydrogen into their atomic form may be significant.

† R.S. Benson and P.C. Baruah, A Generalized Calculation for an Ideal Otto Cycle with Hydrocarbon-Air Mixture Allowing for Dissociation and Variable Specific Heats. IJMEE Vol.4, No.1, 1976, p.49.

Let us consider a diatomic gas A_2 which can dissociate according to the relation†

$$A_2 \rightleftharpoons 2A. \tag{4.114}$$

At some temperature T_2 the gas exists in the form

$$(1-\alpha)A_2 + 2\alpha A. \tag{4.115}$$

The partial pressures will be

$$\left.\begin{aligned} p_A &= \frac{2\alpha}{1+\alpha}\, p \quad \text{for } A \\ p_{A_2} &= \frac{1-\alpha}{1+\alpha}\, p \quad \text{for } A_2 \end{aligned}\right\} \tag{4.116}$$

where p is the total pressure.

The equilibrium equation for the reaction will be

$$K_p = \frac{4\alpha^2}{1-\alpha^2}\, p. \tag{4.117}$$

The state equation is

$$pV = MR_{mol}T$$

and

$$M = 1+\alpha.$$

Hence

$$pV = (1+\alpha)R_{mol}T. \tag{4.118}$$

It is often convenient to express (4.118) in terms of the original substance A_2. Let m_A be the mass of atom A. The total mass of the mixture will be $2m_A$. The density ρ is equal to

$$\frac{\text{mass}}{\text{volume}} = \frac{2m_A}{V}.$$

Substituting into (4.118)

†The symbol A_2 represents any diatomic gas, the symbol A the atomic form of the gas.

$$\frac{p}{\rho} = \frac{(1+\alpha)R_{mol}T}{2m_A}. \qquad (4.119)$$

In terms of the mass of A_2 (equal to $2m_A$) we have

$$\frac{p}{\rho} = (1+\alpha)\frac{R_{mol}}{m_{A_2}}T \qquad (4.120)$$

$$\frac{p}{\rho} = (1+\alpha)R_{A_2}T. \qquad (4.121)$$

R_{A_2} is the characteristic gas constant for the gas A_2. Equation (4.121) is the state equation for the gas A_2. The factor $(1+\alpha)$ is sometimes called the deviation from the perfect gas.

The equilibrium constant K_p is calculated from equation (4.83). In this equation $\nu_P = 2$, $\nu_R = 1$ and

$$\ln K_p = \left(\frac{g(T)}{R_{mol}T}\right)_{A_2} - 2\left(\frac{g(T)}{R_{mol}T}\right)_A - \frac{\Delta H_o}{R_{mol}T}. \qquad (4.122)$$

The functions $g(T)/R_{mol}T$ can be calculated from data in Table A.2 and the expression (4.87).

The stoichiometric equation is

$$A_2 = 2A.$$

If this is an endothermic process ΔH_o has a positive value equal to the heat of dissociation D at absolute zero. Tabulated values for ΔH_o and K_p are given in the tables at the end of the book.

Van't Hoff's equation for the reaction is

$$\frac{d(\ln K_p)}{dT} = \frac{Q_p}{R_{mol}T^2}$$

where

$$Q_p = \sum(\nu h)_P - \sum(\nu h)_R$$

$$= 2h_A - h_{A_2}.$$

The heat of reaction is

$$Q_p = \left(2h(T) + 2h_o\right)_A - \left(h(T) + h_o\right)_{A_2}$$

$$= 2h(T)_A - h(T)_{A_2} + \Delta H_o$$

or

$$Q_p = \Delta H(T) + \Delta H_o \tag{4.123}$$

where

$$\Delta H(T) = 2h(T)_A - h(T)_{A_2}. \tag{4.124}$$

Van't Hoff's equation then becomes

$$d(\ln K_p) = \frac{(\Delta H(T) + \Delta H_o)}{R_{mol}} \frac{dT}{T^2} \tag{4.125}$$

Integrating (4.125) between the limits T_o and T and letting B be the equilibrium constant at T_o we have

$$\ln K_p - \ln B = -\frac{\Delta H_o}{R_{mol}T} + \phi(T) \tag{4.126}$$

where

$$\phi(T) = \int_{T_o}^{T} \frac{\Delta H(T)\, dT}{R_{mol}T^2} + \frac{\Delta H_o}{R_{mol}T_o} \tag{4.127}$$

Equation (4.126) can be rewritten in the form

$$K_p = Be^{-\Delta H_o/R_{mol}T} e^{\phi(T)}.$$

Now from (4.117)

$$K_p = \frac{4\alpha^2}{1-\alpha^2} p \tag{4.128}$$

and

$$\frac{\alpha^2}{1-\alpha} = \frac{1+\alpha}{4p} Be^{\phi(T)} e^{-\Delta H_o/R_{mol}T}. \tag{4.129}$$

From (4.119)

$$\frac{1+\alpha}{p} = \frac{2m_A}{\rho R_{mol} T}.$$

Substituting into (4.129) we have

$$\frac{\alpha^2}{1-\alpha} = \frac{e^{-\Delta H_o/R_{mol}T}}{\rho} \frac{Bm_A}{2R_{mol}T} e^{\phi(T)}. \tag{4.130}$$

Lighthill† has shown that for oxygen and nitrogen over the temperature range 1000°K to 7000°K $\left((Bm_A/2R_{mol}T)e^{\phi(T)}\right)$ may be considered to be a constant. This constant is called a characteristic density ρ_d. Equation (4.130) becomes

$$\frac{\alpha^2}{1-\alpha} = \frac{\rho_d}{\rho} e^{-\Delta H_o/R_{mol}T} \tag{4.131}$$

For the dissociation reaction

$$A_2 \rightleftharpoons 2A.$$

ΔH_o equals the heat of dissociation D at absolute zero. The ratio D/R_{mol} is the characteristic temperature of dissociation θ_D and (4.121) can be written in the form

$$\frac{\alpha^2}{1-\alpha} = \frac{\rho_d}{\rho} e^{-\theta_D/T}. \tag{4.132}$$

Typical values for θ_D and ρ_d for oxygen and nitrogen are given in Table 4.1.

TABLE 4.1

	θ_D (°K)	ρ_d (g/cm³) when T (°K) is						
		1000	2000	3000	4000	5000	6000	7000
Oxygen	59,000	145	170	166	156	144	133	123
Nitrogen	113,000	113	135	136	133	128	123	118

†M.J. Lighthill, Dynamics of a dissociating gas: Part 1, Equilibrium flow, Journal of Fluid Mechanics, vol.2 (1957), p.1.

Let us examine the characteristic density ρ_d again:

$$\rho_d = \frac{Bm_A}{2R_{mol}T} e^{\phi(T)} \tag{4.133}$$

or

$$\left(\frac{2R_{mol}\rho_d}{m_A B}\right) T = e^{\phi(T)}. \tag{4.134}$$

Taking the logarithm of (4.134) we have

$$\ln\left(\frac{2R_{mol}\rho_d}{m_A B}\right) + \ln T = \phi(T) = \int_{T_o}^{T} \frac{\Delta H(T)\,dT}{R_{mol}T^2} + \frac{\Delta H_o}{R_{mol}T_o}. \tag{4.135}$$

Differentiating (4.135) with respect to T,

$$\frac{dT}{T} = \frac{\Delta H(T)\ dT}{R_{mol}T} \tag{4.136}$$

since $2R_{mol}\rho_d/m_A B$ and $\Delta H_o/R_{mol}T_o$ are constants. From (4.136)

$$\Delta H(T) = R_{mol}T. \tag{4.137}$$

Now

$$\Delta H(T) = 2h(T)_A - h(T)_{A_2},$$

therefore

$$2h(T)_A - h(T)_{A_2} = R_{mol}T. \tag{4.138}$$

The specific enthalpy h is

$$h = C_pT = (C_v + R_{mol})T.^{\dagger}$$

Substituting for h in (4.138) we have, after cancellation of the R_{mol} terms,

$$2(C_v)_A - (C_v)_{A_2} = 0$$

or

$$2(C_v)_A = (C_v)_{A_2}. \tag{4.139}$$

---

†C_p is the average specific heat over the temperature range T_o to T_1.

The specific heats of non-ionized monatomic gases are independent of temperature.   The assumption that ρ_d is a constant therefore implies that the relationship between the specific heats of the diatomic molecule and the monatomic molecule is independent of temperature.   For a monatomic gas with no ionization there are only three degrees of freedom; and from (3.15a)

$$(C_v)_A = \frac{3}{2} R_{mol}.$$

Thus from (4.139)

$$(C_v)_{A_2} = 3R_{mol}. \qquad (4.140)$$

For a diatomic gas there are five degrees of freedom for the rigid molecule model (see Chapter 3, p. 74).   One would therefore expect from (3.15a) that, since $f = 5$, the specific heat $(C_v)_A$ would be $\frac{5}{2} R_{mol}$.   The assumption of constant ρ_d implies that there is intra-molecular vibration energy equal to one degree of freedom (i.e. $\frac{1}{2} R_{mol}T$).   It can be shown for a diatomic gas that the maximum intra-molecular vibrational energy is equal to $R_{mol}T$.   Thus the vibrational degrees of freedom are considered to be half-excited over the temperature range 1000-7000°K.   This gas is called the Lighthill Ideal Dissociating Gas.   Calculations of thermodynamic or gas dynamic processes with the Lighthill ideal dissociating gas are less complex than with the ideal dissociating gas[†] as will be seen from the following example.

Example 7

A mol of gas A_2 is heated.   Show that, by considering the gas to be an ideal dissociating gas, the internal energy of the gas at temperature T is given by

$$3R_{mol}T + \alpha\Delta H_o$$

and that the internal energy per unit mass of A_2 is

$$\frac{3p}{(1+\alpha)\rho} + \frac{\alpha}{m_{A_2}} \Delta H_o.$$

[†]In some texts an "ideal" dissociating gas is called a "real" gas.

The gas mixture is

$$A_2 \rightarrow (1-\alpha)A_2 + 2\alpha A.$$

The internal energy per mol is

$$u = u(T) + u_o.$$

The total internal energy of the mixture is

$$E = (1-\alpha)u(T)_{A_2} + 2\alpha u(T)_A + (1-\alpha)u_{o_{A_2}} + 2\alpha u_{o_A}.$$

Now for the reaction $A_2 = 2A$.

$$\Delta u_o = 2u_{o_A} - u_{o_{A_2}}.$$

If the reference temperature is taken as the absolute zero $\Delta u_o = \Delta H_o$. The internal energy of the mixture is, therefore,

$$E = (1-\alpha)u(T)_{A_2} + 2\alpha u(T)_A + \alpha\Delta H_o + u_{o_{A_2}}.$$

If we take $u_{o_{A_2}} = 0$, i.e. the energy of the gas is referred to the zero energy level of the undissociated molecule A_2, then

$$E = (1-\alpha)\ u(T)_{A_2} + 2\alpha u(T)_A + \alpha\ \Delta H_o.$$

Now

$$u(T) = C_v T$$

and for the Lighthill ideal dissociating gas

$$(C_v)_A = \frac{3}{2} R_{mol}, \qquad (C_v)_A = 3R_{mol},$$

therefore

$$E = (1-\alpha)\ 3R_{mol}T + 2\alpha\ \frac{3}{2}\ R_{mol}T + \alpha\ \Delta H_o$$

$$E = 3R_{mol}T + \alpha\ \Delta H_o.$$

The internal energy per unit mass of A_2 is

$$\frac{E}{m_{A_2}} = \frac{3R_{mol}T}{m_{A_2}} + \frac{\alpha\Delta H_o}{m_{A_2}}.$$

The state equation for the gas is

$$\frac{p}{\rho} = \frac{(1+\alpha)R_{mol}T}{m_{A_2}}$$

and the internal energy per unit mass of A_2 is

$$\frac{E}{m_{A_2}} = \frac{3p}{(1+\alpha)\rho} + \frac{\alpha}{m_{A_2}} \Delta H_o .$$

Ionization of Monatomic Gases

Above about 5000 K (depending on the pressure and the gas) monatomic gases will shed electrons and become ionized (that is they can conduct electricity). The gas will be a mixture of neutral atoms A, ionized atoms A^+ and electrons e. Ionized gases are often called Plasmas. If the number of electrons equals the number of positive or ionized atoms the plasma is considered to be neutral. A fairly elementary, but useful, approach to the thermodynamics of real gases may be made if the plasma and its constituents (the neutral atoms, the ionized atoms and the electrons) are considered to be perfect gases. It is also assumed that the ionization process may be represented by the law of mass action

$$A \rightleftharpoons A^+ + e. \tag{4.141}$$

Let us consider a neutral plasma in the equilibrium state. The ionization reaction will be represented by

$$A \rightarrow (1-\alpha)\ A + \alpha A^+ + \alpha e \tag{4.142}$$

where α is the degree of ionization of the atom A.

The partial pressures of the neutral atoms, the positive ions A^+ and the electrons e are given by

$$p_A = \frac{1-\alpha}{V} R_{mol}T$$

$$p_{A^+} = \frac{\alpha}{V} R_{mol}T$$

$$p_e = \frac{\alpha}{V} R_{mol}T$$

where V is the volume of the plasma, R_{mol} is the universal gas constant and T the bulk gas temperature. The total pressure p is the sum of the partial pressures. We then have

$$p = \frac{(1+\alpha)R_{mol}T}{V} \tag{4.143}$$

If we let R be the characteristic gas constant of the plasma and m the molecular weight, then the density ρ is given by

$$\rho = \frac{m}{V}$$

and $R_{mol} = mR.$

Hence on substitution into (4.143) we have

$$p = (1+\alpha)\rho RT. \tag{4.144}$$

This is the gas law or state equation of the plasma (ionized gas). Now

$$R_{mol} = m_A R_A = m_{A+}R_{A+} = m_e R_e = mR$$

where m_A, m_{A+} and m_e are the "molecular weights" of the neutral atoms, the ionized atoms (ions) and the electrons. Now $m_A \simeq m_{A+}$ and $R_A \simeq R_{A+}$. The mean molecular weight m for the mixture is given by

$$m = (1-\alpha)\, m_A + \alpha m_{A+} + \alpha m_e \simeq m_A$$

the "molecular weight" of the electron m_e being negligible compared with the molecular weight of the neutral atom. Thus

$$R_A \simeq R_{A+} \simeq R.$$

We can now derive the state equation or gas law for the gas A by substituting R_A for R in (4.144), thus

$$p = (1+\alpha)\rho R_A T. \tag{4.145}$$

The term $(1+\alpha)$ is called the <u>deviation</u> of the real gas A from the perfect gas.

To determine the degree of ionization α we use the law of mass action

$$A \rightleftharpoons A^+ + e. \tag{4.141}$$

The equilibrium constant K_p is given by

$$K_p = \frac{(p_{A^+})\ (p_e)}{(p_A)}. \tag{4.146}$$

The ionized gas composition is from (4.142)

$$(1-\alpha)A + \alpha A^+ + \alpha e.$$

The partial pressures are therefore

$$p_{A^+} = \frac{\alpha}{1+\alpha}\, p$$

$$p_e = \frac{\alpha}{1+\alpha}\, p$$

$$p_A = \frac{1-\alpha}{1+\alpha}\, p.$$

Substituting for p_{A^+}, p_e, p_A in (4.146)

$$K_p = \frac{\alpha^2}{1-\alpha^2}\, p \tag{4.147}$$

Van't Hoff's equation for the plasma is

$$\frac{d(\ln K_p)}{dT} = \frac{Q_p}{R_{mol}T^2}.$$

Q_p is the ionization energy at temperature T. For the "reaction" $A = A^+ + e$, Q_p is given by

$$Q_p = h_e + h_{A^+} - h_A$$

$$Q_p = h(T)_e + h(T)_{A^+} - h(T)_A + (h_o)_e + (h_o)_{A^+} - (h_o)_A$$

$$Q_p = h(T)_e + h(T)_{A+} - h(T)_A + \Delta H_o. \tag{4.148}$$

ΔH_o is the ionization energy at the zero energy state and is equal to $(h_o)_e + (h_o)_{A+} - (h_o)_A$.

The neutral atoms, the positive ions and the electrons may be considered to be "rigid spheres" with three translational degrees of freedom. The specific heats at constant volume are

$$(C_v)_A = (C_v)_{A+} = (C_v)_e = \frac{3}{2} R_{mol}.$$

Now

$$h(T) = (C_v + R_{mol})T = \frac{5}{2} R_{mol}T.$$

The ionization energy at temperature T therefore becomes

$$Q_p = \frac{5}{2} R_{mol}T + \frac{5}{2} R_{mol}T - \frac{5}{2} R_{mol}T + \Delta H_o$$

$$Q_p = \frac{5}{2} R_{mol}T + \Delta H_o. \tag{4.149}$$

The ionization process is endothermic and ΔH_o is positive.

Substituting for Q_p in Van't Hoff's equation

$$\frac{d(\ln K_p)}{dT} = \frac{\frac{5}{2} R_{mol}T + \Delta H_o}{R_{mol}T^2}$$

Integrating between T_o and T

$$\ln\left(\frac{K_p}{K_{p_o}}\right) = \frac{5}{2}\ln\left(\frac{T}{T_o}\right) - \frac{\Delta H_o}{R_{mol}T} + \frac{\Delta H_o}{R_{mol}T_o}$$

$$\ln\left\{\left(\frac{K_p}{K_{p_o}}\right)\left(\frac{T_o}{T}\right)^{\frac{5}{2}}\right\} = -\frac{\Delta H_o}{R_{mol}T} + \frac{\Delta H_o}{R_{mol}T_o}$$

or

$$\frac{K_p}{K_{p_o}}\left(\frac{T_o}{T}\right) = e^{-\Delta H_o/R_{mol}T}\, e^{\Delta H_o/R_{mol}T_o}$$

or

$$K_p = \text{constant}\ T^{\frac{5}{2}}\, e^{-\Delta H_o/R_{mol}T} \tag{4.150}$$

where the constant is given by

$$\frac{K_{p_o}}{T_o^{5/2}}\, e^{\Delta H_o/R_{mol} T_o}. \tag{4.151}$$

Now from (4.147)

$$K_p = \frac{\alpha^2}{1-\alpha^2}\, p$$

and

$$\frac{\alpha^2}{1-\alpha^2} = \text{constant}\ \frac{T^{5/2}}{p}\, e^{-\Delta H_o/R_{mol} T}. \tag{4.152}$$

The ratio $\Delta H_o/R_{mol}$ is called the characteristic temperature of ionization θ_I. Equation (4.152) can be written in the form

$$\frac{\alpha^2}{1-\alpha^2}\, p = CT^{5/2}\, e^{-\theta_I/T}.$$

Equation (4.152) is the well-known Saha equation for ionized gases (single ionization).† Saha's equation can also be developed by the methods of statistical thermodynamics.††

In order to develop the relations for the ionized gas we have considered the plasma to be a mixture of ideal gases, since it was assumed that the specific heats were <u>independent of temperature</u>. The internal energy of ionized monatomic gas consists of

---

†From the methods of statistical thermodynamics it can be shown that the constant (4.151) equals

$$k^{5/2}\left(\frac{2\pi m_e}{h^2}\right)^{3/2} \frac{2Z_1}{Z_o}$$

where k = Boltzmann constant; h = Planck's constant; m_e = mass of electron; Z_1 = partition function for neutral atom; Z_o = partition function for ion.

††A simple derivation using the methods of statistical thermodynamics is given in Plasma Physics and Magnetofluidmechanics by A.B. Cambel, McGraw-Hill, 1963.

(a) energy due to translation,

(b) ionization energy,

(c) electronic excitation energy.

If we assume the electrons, the ions and the neutral atoms have specific heats independent of temperature we neglect the electronic excitation energy. This will be seen from the following analysis.

The internal energy of the gas mixture is

$$E = (1-\alpha)\, u_A + \alpha u_A + \alpha u_e;$$

now

$$u = \frac{3}{2} R_{mol} T + u_o$$

and

$$E = (1+\alpha)\frac{3}{2} R_{mol} T + \alpha\, \Delta H_o + u_o$$

where

$$\Delta H_o = u_{oA} + u_{oe} - u_{oA}.$$

u_{oA} is the zero state energy of the neutral atom A.

If we consider u_{oA} is zero, that is we measure the energy above the zero state energy level of the neutral atom A, then

$$E = (1+\alpha)\frac{3}{2} R_{mol} T + \alpha\, \Delta H_o.$$

The first term is due to the energy of translation, the second term is due to the ionization energy. A third term due to the electronic excitation should be included. This third term is a function of temperature. Below about 7000°K it may be neglected, above this temperature the electronic energy should be included. The Saha equation neglects the electronic excitation energy and for most problems of engineering interest this is sufficiently accurate. For a modified form of the Saha equation allowing for electronic energy the methods of statistical thermodynamics must be used.† The Saha equation is applicable only to homogeneous systems in thermal equilibrium. In particular problems, in which relaxation process occur, a modified form of Saha's equation must be used. In the present discussion we have only considered ionization processes for

† Plasma Physics and Magnetofluidmechanics by A.B. Cambel, McGraw-Hill, 1963.

single ions, the methods may be readily extended to processes producing more than one ion. It will be seen from the Saha equation that, for a particular gas, the degree of ionization increases with increase in temperature and decrease in pressure.

Non-equilibrium Processes. Equilibrium and Frozen Flows

In engineering problems, in which either dissociation or ionization or both have a significant influence on the thermodynamic properties of systems, it is usual to consider the system to be in thermodynamic equilibrium. If the temperature of the system is raised or lowered it is assumed that the time rate of change of temperature has no effect on the equilibrium relations as represented by the law of mass action. That is, for each temperature the composition of the system can be determined by the methods outlined in this chapter, and from this the thermodynamic properties of the system evaluated. Indeed we assume that the time rate of change has no influence at all on the thermodynamic properties. In some texts these processes are called Thermostatic Processes. For an open system (control volume), where there is a mass exchange, the "thermostatic" process corresponds to the system being in "thermodynamic" equilibrium. For example, if we consider a high-temperature gas flowing through a De Laval nozzle, as the gas expands the velocity increases and the temperature drops. The degree of dissociation of the gas decreases with the drop in temperature - in other words recombination is taking place. For such a process, in "thermodynamic" equilibrium, no matter what the geometry of the nozzle, providing the exit to throat area is fixed for shock-free flow, the pressure, temperature, velocity and gas composition will be the same at the same area ratio (the area of a nozzle in a particular plane divided by the throat area). In practice this is not so. It is found that for nozzles of different shapes, depending on the rate of increase in area downstream of the throat with the length, the pressure, temperature, velocity and composition at equal area ratios may be different. This is due to the finite time taken for the dissociation and the recombination processes. This time is called the relaxation time. If the expansion of the gas is extremely rapid, recombination may not take place to a sufficient degree to satisfy the thermodynamic equilibrium conditions and the composition of the gas may remain constant from the throat to the nozzle exit. These flow conditions are generally called frozen flow, since the gas composition is "frozen". Under certain conditions the flow downstream of the nozzle may start under equilibrium conditions and then suddenly "freeze".

Strictly it is not possible to study these types of flow using equilibrium thermodynamics. However, for most practical purposes it is possible to use the thermodynamic properties of pressure, temperature, density, etc., and assume equilibrium conditions for these properties, but in place of the equilibrium equation for determining the degree of dissociation the chemical reaction-rate equation is used. The reaction equation is so formulated that for zero reaction time the usual equilibrium equation is obtained. Extensive research is currently being carried out throughout the world to evaluate chemical reaction rate data for gases of interest to engineers. Calculations using reaction rates are generally extremely complex and require the use of high-speed digital computers and a knowledge of statistical thermodynamics. It is beyond the scope of this book to describe in detail the methods used.

However, a brief outline of the procedures used in these calculations for two examples of current interest will be given. The first example will refer to the dissociation at high temperatures of a diatomic gas flowing through a De Laval nozzle and the second example will illustrate the methods used to predict the nitric oxide concentration in an internal combustion engine cylinder.

Although the nature of the chemical process is extremely complex, the usual model for describing the dissociation and recombination process is fairly simple.

A dissociation process is considered to be due to the collision of molecules, say A_2, with another body M_i thus producing two atoms of A. This is represented by

$$A_2 + M_i \xrightarrow{k_D^i} A + A + M_i, \tag{4.153}$$

The suffix i is used to specify the species of the body M_i. The symbol k_D^i is called the reaction "velocity" for dissociation. The recombination process is considered to be a three-body collision of the two atoms A and a body M_i and is represented by

$$A + A + M_i \xrightarrow{k_R^i} A \;\; + M_i. \tag{4.154}$$

The reaction "velocity" for recombination is k_R^i. The body M_i can be another molecule of A such as A_2, or a free atom A, or any inert substance. The time rate of formation of $[A]$, where the square brackets indicate the number of mols of A per unit volume, is usually expressed in the form

$$\frac{d[A]}{dt} = [A_2] \sum k_D^i \, [M_i] - [A]^2 \sum k_R^i \, [M_i]. \tag{4.155}$$

The reaction "velocities" k_D^i, k_R^i are related through the equilibrium constant K_c

$$K_c = \frac{k_D^i}{k_R^i}. \tag{4.156}$$

The raised index refers to the constants for the particular collision process.

As an example let us consider the dissociation and recombination reactions for oxygen O_2. If there are no inert substances present the collision body can be either the atomic O or the molecule O_2 or both.

If the collision body is atomic oxygen O then $M_i = M_1 = O$ (collision process 1) and

$$\frac{d(O)}{dt} = (O_2)\ k_D^1(O) - (O)^2\ k_R^1(O). \tag{4.157}$$

If the collision body is molecular oxygen O_2 then $M_i = M_2 = O_2$ (collision process 2) and

$$\frac{d(O)}{dt} = (O_2)\ k_D^2(O_2) - (O)^2\ k_R(O_2). \tag{4.158}$$

If there are two collision bodies O and O_2 (collision processes 1 and 2) then

$$\frac{d(O)}{dt} = k_D^1(O_2)(O) + k_D^2(O_2)^2 - k_R^1(O)^2\ (O) - k_R^2(O_2)(O)^2. \tag{4.159}$$

Although k_R^1, k_R^2, k_D^1, k_D^2 are related through K_c the laws for the reaction velocities are only available in empirical form. In general one can express the dissociation velocity constant k_D as a function of $T^{-r}\ e^{-\theta_D/T}$ where r lies between 0 and 3, whilst the recombination velocity is of the form T^{-s} where s lies between 0 and 2.5.† Despite the lack of precise information of these laws, calculations†† show that the overall results are not greatly affected by the numerical values of r and s for the problems of interest to engineers.

The following analysis is a simple approach to the problem which, although not as rigorous as the work referred to above, will give an insight into the phenomenon. We will consider an

---

†K.N.C. Bray, Atomic recombination in a hypersonic wind-tunnel nozzle, Journal of Fluid Mechanics, vol.6 (1959) p.1.

††J.G. Hall and A.L. Russo, Studies of Chemical Non-equilibrium in Hypersonic Nozzle Flows. Kinetic Equilibria and Performance of High Temperature Systems, Butterworth, p.219.

ideal dissociating gas A_2. We will assume that the thermodynamic equilibrium properties pressure p, temperature T and density ρ are related by equation (4.132) for the Lighthill dissociating gas namely

$$\frac{\alpha^2}{1-\alpha} = \frac{\rho_d}{\rho} e^{-\theta_D/T} \tag{4.132}$$

where α is the degree of dissociation of A_2 given by

$$A_2 \rightarrow (1-\alpha)\, A_2 + 2\alpha A, \tag{4.160}$$

For a <u>finite</u> reaction rate we <u>assume</u> that the rate equation is of the form

$$\frac{d(A)}{dt} = k_d(A_2) - k_r(A)^2 \tag{4.161}$$

where k_d, k_r are the velocity constants for dissociation and recombination and are related through the equilibrium constant K_c by

$$K_c = \frac{k_d}{k_r}. \tag{4.162}$$

For the ideal dissociating gas given by (4.132) the equilibrium constant is†

$$K_c = \frac{2}{m_A} \rho_d e^{-\theta_D/T}. \tag{4.163}$$

---

† This expression is derived as follows:

From (4.128)

$$K_p = Be^{-\theta_D/T} e^{\phi(T)}.$$

Multiplying both sides by $m_A/2R_{mol}T$, then

$$\frac{K_p}{2} \frac{m_A}{R_{mol}T} = \frac{Bm_A}{2R_{mol}T} e^{\phi(T)} e^{-\theta_D/T} = \rho_d e^{-\theta_D/T}$$

since

$$\rho_d = \frac{Bm_A}{2R_{mol}T} e^{\phi(T)}.$$

Hence

$$\frac{K_p}{R_{mol}T} = \frac{2}{m_A} \rho_d e^{-\theta_D/T}.$$

Now from (4.68b) K_p and K_c are related. By substituting for the stoichiometric coefficients $\nu_a = 1$, $\nu_b = 0$, $\nu_c = 2$, $\nu_d = 0$, we have

$$K_c = K_p(R_{mol}T)^{-1} = \frac{K_p}{R_{mol}T}.$$

and

$$K_c = \frac{2}{m_A} \rho_d e^{-\theta_D/T}.$$

Now the mol fractions are

$$(A) = \frac{2\alpha}{V} = \alpha\frac{\rho}{m_A} \tag{4.163a}$$

$$(A_2) = \frac{1-\alpha}{V} = \frac{1-\alpha}{2}\frac{\rho}{m_A} \tag{4.163b}$$

since

$$\rho = \frac{2m_A}{V}.$$

From (4.163a)

$$\frac{d(A)}{dt} = \frac{\rho}{m_A}\frac{d\alpha}{dt}. \tag{4.163c}$$

Substituting for $d(A)/dt$, (A), (A_2) and k_d in (4.161) we obtain after simplification

$$\frac{d\alpha}{dt} = k_r\left[\frac{K_c}{2}(1-\alpha) - \alpha^2\frac{\rho}{m_A}\right].$$

Substituting for K_c from (4.163) and rearranging

$$\frac{d\alpha}{dt} = k_r\frac{\rho_d}{m_A}\left[(1-\alpha)\,e^{-\theta_D/T} - \alpha^2\frac{\rho}{\rho_d}\right]. \tag{4.164}$$

Comparison of equations (4.161) and (4.155) shows that the two velocity coefficients k_r and k_R are related:

$$k_r = \frac{k_R}{V} = \frac{\rho}{2m_A}k_R.$$

Furthermore, since the problems of frozen and equilibrium flow are of interest in ducts, we can rearrange (4.164) to give the rate of change of the degree of dissociation of A_2 along the length by noting

$$\frac{d\alpha}{dt} = \frac{d\alpha}{dx}\frac{dx}{dt} = c\frac{d\alpha}{dx}$$

where x is the unit of length and c is the axial velocity.

The variation in the degree of dissociation along the duct is then

$$\frac{d\alpha}{dx} = \frac{k_R \rho_d}{2m_A^2}\frac{1}{c}\left[\rho(1-\alpha)\,e^{-\theta_D/T} - \alpha^2\frac{\rho^2}{\rho_d}\right]. \tag{4.165}$$

If the reaction rate is fast we have equilibrium flow. For if k_R approaches infinity, then either $d\alpha/dx$ approaches infinity, or the term in the bracket approaches zero. But $d\alpha/dx$ has a finite value, therefore, the term in the bracket approaches zero. In the limit the term in the bracket is the equilibrium equation (4.132).

For a finite reaction rate the first term in the bracket,

$$\rho(1-\alpha)\,e^{-\theta_D/T} = D$$

is of similar form to the expression for the rate of dissociation given by statistical thermodynamics, whilst the second term

$$\alpha^2\frac{\rho^2}{\rho_d} = R$$

is of similar form to the expression for the rate of recombination. The net rate of dissociation along the duct will therefore be proportional to (D-R), i.e.

$$\frac{d\alpha}{dx} \propto (D-R).$$

The results of some recent calculations† of the flow of a dissociating gas in a hypersonic nozzle show that for a <u>finite</u> reaction rate R is initially slightly greater than D, but D is very much greater than

$$\frac{d\alpha}{dt}\left(= c\,\frac{d\alpha}{dx}\right),$$

hence $d\alpha/dx$ is negative. Recombination is occurring and the conditions <u>approximate to equilibrium flow</u>. As the gas temperature drops along the nozzle we encounter a transition region where both ρ and T have fallen slightly so that D, R and $d\alpha/dt$ are of the same order. At this point it is observed that there is a departure from equilibrium. The temperature is below the corresponding equilibrium temperature. Since D is strongly dependent on the temperature it drops very rapidly

---

†K.N.C. Bray, Atomic recombination in a hypersonic wind-tunnel nozzle, <u>Journal of Fluid Mechanics</u>, vol.6 (1959), p.1.

to a negligible value. At this point the composition of the gas remains constant and "freezing" commences. We now have c(dα/dx) proportional to -R. The drop in temperature causes a fall in density and since R varies as the square of the density it rapidly becomes negligible and dα/dx is zero and "freezing" has set in completely.

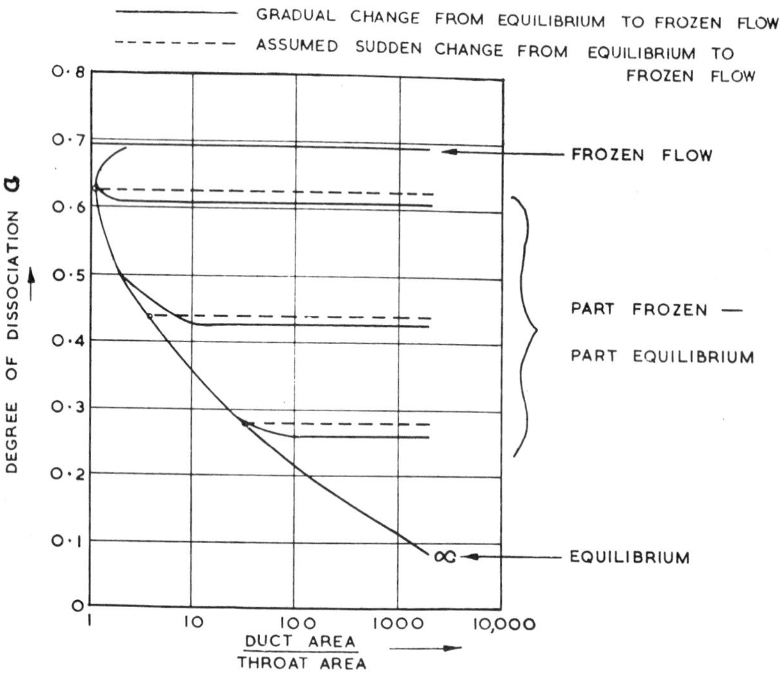

FIG. 4.3. Dissociation fraction for hypersonic flows. (After Bray, *J. Fluid Mechanics*, vol. 6, 1959.)

A full discussion of the phenomenon is given by Bray.† In Fig. 4.3 the results of some of Bray's calculations for isentropic flow of an ideal gas are given. In Fig. 4.4 test results by Nagamatsu et al.†† are shown for air flowing in a hypersonic nozzle. The effect of "frozen" flow is to reduce the gas velocity, static pressure and temperature below the equilibrium values. In the case of hypersonic ram jets or rocket motors this will reduce the engine performance.

---

†See footnote on p.185.

††H.T. Nagamatsu, J.B. Workman and R.E. Sheer, Jr., Hypersonic nozzle expansion of air with atom recombination present, Journal of the Aerospace Sciences, vol.28, No.11, Nov. 1961.

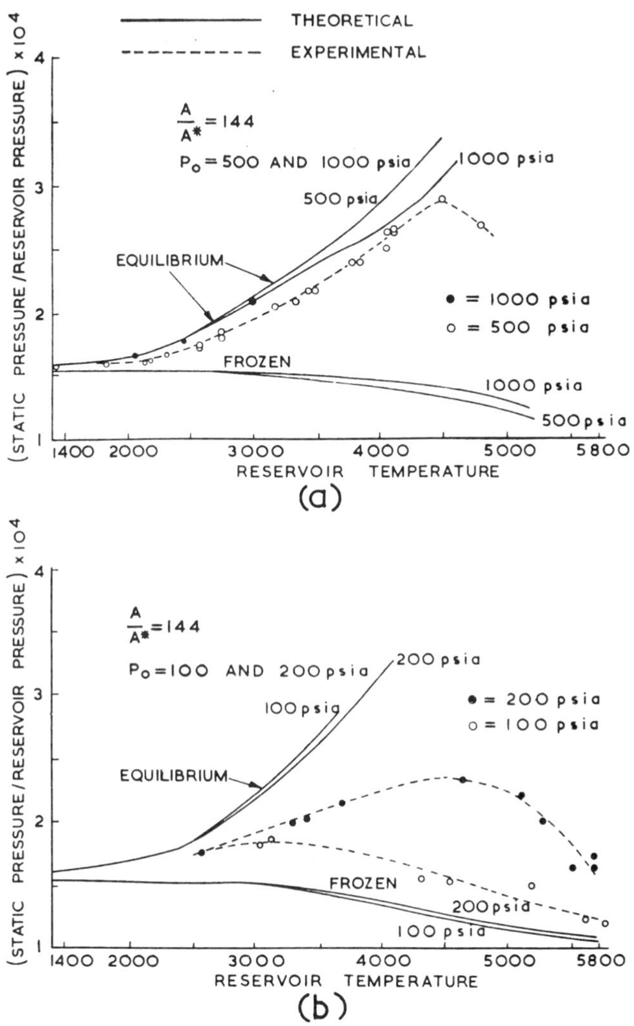

FIG. 4.4. Variation of static pressure ratio with reservoir temperature. (After Nagamatsu *et al.*, *J. Aerospace Sciences*, vol. 28, 1961.)

In practice it is usual to prepare design data for the two limiting cases of equilibrium flow and frozen flow.

The second example to be outlined is the procedure used to determine the concentration of nitric oxide in the cylinder of an internal combustion engine. During combustion the nitrogen and oxygen in the air combine in a sequence of chemical reactions to form nitric oxide. If the reactions were in equilibrium the nitric oxide at the end of the expansion stroke would be insignificant. In practice, however, the nitric oxide concentrations exceed the equilibrium concentrations. At the time of writing there is still an incomplete understanding of all the reactions which effect the formation of nitric oxide (NO). A sequence of reactions have been given by Lavoie et al†; there will be referred to later. A method for calculating the rate of production of NO developed by Lavoie†, described by Benson et al††, is given below.

We will consider a chemical reaction between A and B to form C and D. We can write (4.153) and (4.154) in the form

$$A + B \underset{k_b}{\overset{k_f}{\rightleftharpoons}} C + D \tag{4.166}$$

where k_f is the forward reaction velocity constant, corresponding to k_D^i in (4.153) and k_b is the backward reaction velocity constant corresponding to k_R^i.

The rate of formation of A in molar units per unit volume will be

$$\frac{1}{V}\frac{d((A)V)}{dt} = -k_f\,(A)(B) + k_b(C)(D) \tag{4.167}$$

the rate of formation of B

$$\frac{1}{V}\frac{d((B)V)}{dt} = -k_f\,(A)(B) + k_b(C)(D) \tag{4.168}$$

the rate of formation of C

$$\frac{1}{V}\frac{d((C)V)}{dt} = k_f\,(A)(B) - k_b(C)(D) \tag{4.169}$$

---

†G.A. Lavoie, J.B. Heywood, J.C. Keck, Combustion Sc. Tech. Vol.1, p.313 (1970).

††R.S. Benson, W.J.D. Annand, P.C. Baruah, Int. J. Mech. Sci. Vol.17, p.97-124 (1975).

and the rate of formation of D

$$\frac{1}{V}\frac{d((D)V)}{dt} = k_f (A)(B) - k_b (C)(D) \tag{4.170}$$

Equilibrium is reached when the concentrations of A, B, C and D are constant, or,

$$\frac{d(A)}{dt} = 0, \quad \frac{d(B)}{dt} = 0, \quad \frac{d(C)}{dt} = 0, \quad \frac{d(D)}{dt} = 0 \tag{4.171}$$

If we call the equilibrium concentrations $(A)_e$, $(B)_e$, $(C)_e$ and $(D)_e$; then at equilibrium,

$$k_f (A)_e (B)_e = k_b (C)_e (D)_e = R_f \tag{4.172}$$

R_f is called the <u>one-way equilibrium rate</u>.

We can relate the time dependent concentrations to the equilibrium concentrations if we define,

$$\alpha = \frac{(A)}{(A)_e}, \quad \beta = \frac{(B)}{(B)_e}, \quad \gamma = \frac{(C)}{(C)_e}, \quad \delta = \frac{(D)}{(D)_e} \tag{4.173}$$

Each of the parameters, α, β, γ, δ, is the ratio of the time dependent concentration to the equilibrium concentration. In some reactions some of the time dependent concentrations may equal the equilibrium concentrations in which case the appropriate parameters will be unity.

The rate equations (4.167) to (4.170) may be written in terms of the parameters α, β, γ, δ and the one way equilibrium constant R_f. For example, if we substitute (4.173) into (4.167) we obtain

$$\frac{1}{V}\frac{d((A)V)}{dt} = -k_f \alpha\beta (A)_e (B)_e + k_b \gamma\delta (C)_e (D)_e \tag{4.174}$$

and substituting (4.172),

$$\frac{1}{V}\frac{d((A)V)}{dt} = -\alpha\beta R_f + \gamma\delta R_f = (-\alpha\beta + \gamma\delta) R_f \tag{4.175}$$

All four rate equations can be expressed in the general form

$$(\mp \alpha\beta \pm \gamma\delta) R_f \tag{4.176}$$

The upper signs are for the formation rate of A and B and the lower signs are for the formation rate of C and D.

Let us consider the reaction

$$NO + N \rightleftharpoons N_2 + O$$

In this reaction atomic nitrogen, N, combines with nitric oxide, NO, to form molecular nitrogen, N_2, and atomic oxygen, O. Experiments show that N_2 and O are always present in equilibrium concentrations, thus $(N_2) = (N_2)_e$ and $(O) = (O)_e$. The NO and N concentrations are rate controlled. Using the notation in (4.166); then if we set $A \equiv NO$, $B \equiv N$, $C \equiv N_2$, $D \equiv O$, it follows that

$$\alpha = \frac{(NO)}{(NO)_e}, \quad \beta = \frac{(N)}{(N)_e}, \quad \gamma = \frac{(N_2)}{(N_2)_e} = 1$$

$$\delta = \frac{(O)}{(O_2)_e} = 1$$

and from (4.176) the rate of formation of nitric oxide is,

$$\frac{1}{V}\frac{d((NO)V)}{dt} = (-\alpha\beta + 1)\, R_f$$

where R_f is given by

$$R_f = k_f(N)_e(NO)_e = k_b(N_2)_e(O)_e$$

The one-way equilibrium rate R_f is obtained from the equilibrium constant K_c and the forward reaction rate constant k_f. The equilibrium constant K_c in the familiar form of (4.68(a)) and (4.156), is given by,

$$\frac{(C)_e(D)_e}{(A)_e(B)_e} = \frac{k_f}{k_b} = K_c \qquad (4.177)$$

The forward reaction rate constant, k_f, is normally expressed in the form

$$k_f = a\, T^b\, e^{-c/T}$$

where a, b and c are constants and T is the temperature. Hence, at a given temperature T both k_f and K_c are determined, whence the equilibrium concentrations and R_f.

In an internal combustion engine the following reactions have been suggested*† for NO formation

(1)	$NO + N$	$\rightleftharpoons$	$N_2 + O$	k_{f1}, K_{c1}, R_{f1}
(2)	$N + O_2$	$\rightleftharpoons$	$NO + O$	k_{f2}, K_{c2}, R_{f2}
(3)	$N + OH$	$\rightleftharpoons$	$NO + H$	k_{f3}, K_{c3}, R_{f3}
(4)	$H + N_2O$	$\rightleftharpoons$	$N_2 + OH$	k_{f4}, K_{c4}, R_{f4}
(5)	$O + N_2O$	$\rightleftharpoons$	$N_2 + O_2$	k_{f5}, K_{c5}, R_{f5}
(6)	$O + N_2O$	$\rightleftharpoons$	$NO + NO$	k_{f6}, K_{c6}, R_{f6}
(7)	$N_2O + M$	$\rightleftharpoons$	$N_2 + O + M$	k_{f7}, K_{c7}, R_{f7}

The rates of energy producing reactions in a flame are sufficiently fast so that the burned gases are close to thermal equilibrium and the pressure and temperature may be calculated from equilibrium data. The hydrogen, H, hydroxyl, OH, oxygen, O, and O_2, and nitrogen, N_2, are present in equilibrium concentrations. The constituents which are rate controlled are NO, N and N_2O. Using the methods outlined above it can be shown*† that the <u>total</u> production rate of NO per unit volume is

$$\frac{1}{V}\frac{d((NO)V)}{dt} = -\alpha(\beta R_{f1} + R_{f2} + R_{f3} + 2\alpha R_{f6}) + R_{f1} + \beta(R_{f2} + R_{f3}) + 2\gamma R_{f6} \qquad (4.178)$$

where $\alpha = \dfrac{(NO)}{(NO)_e}$, $\beta = \dfrac{(N)}{(N)_e}$, $\gamma = \dfrac{(N_2O)}{(N_2O)_e}$

Similar type of expressions for N and N_2O (see Benson et al*). when combined with (4.178), give a single expression for the production of NO in terms of α and the one-way equilibrium rates R_{fn}, n = 1 to 7.

$$\frac{1}{V}\frac{d((NO)V)}{dt} = 2(1-\alpha^2)\left[\frac{R_{f1}}{1+\alpha\left[\dfrac{R_{f1}}{(R_{f2}+R_{f3})}\right]} + \frac{R_{f6}}{1+\left[\dfrac{R_{f6}}{(R_{f4}+R_{f5}+R_{f7})}\right]}\right] \qquad (4.179)$$

† G.A. Lavoie, J.B. Heywood, J.C. Keck, Combustion Sc. Tech. Vol.1, p.313 (1970).

* R.S. Benson, W.J.D. Annand, P.C. Baruah, Int. J. Mech. Sci. Vol.17, p.97-124, (1975).

Cycle calculations allowing for rate equations are complex.† Typical results for one such calculation are shown in Fig. 4.5 where the nitric oxide concentrations in parts per million are shown for a typical cycle calculation. The differences between the rate controlled values and the equilibrium values are quite significant. Notice the 'freezing' of NO during the expansion stroke, similar to the frozen flow in Fig. 4.3. The overall effect of the rate controlled reactions is to increase the concentration of nitric oxide in the exhaust gases of engines.

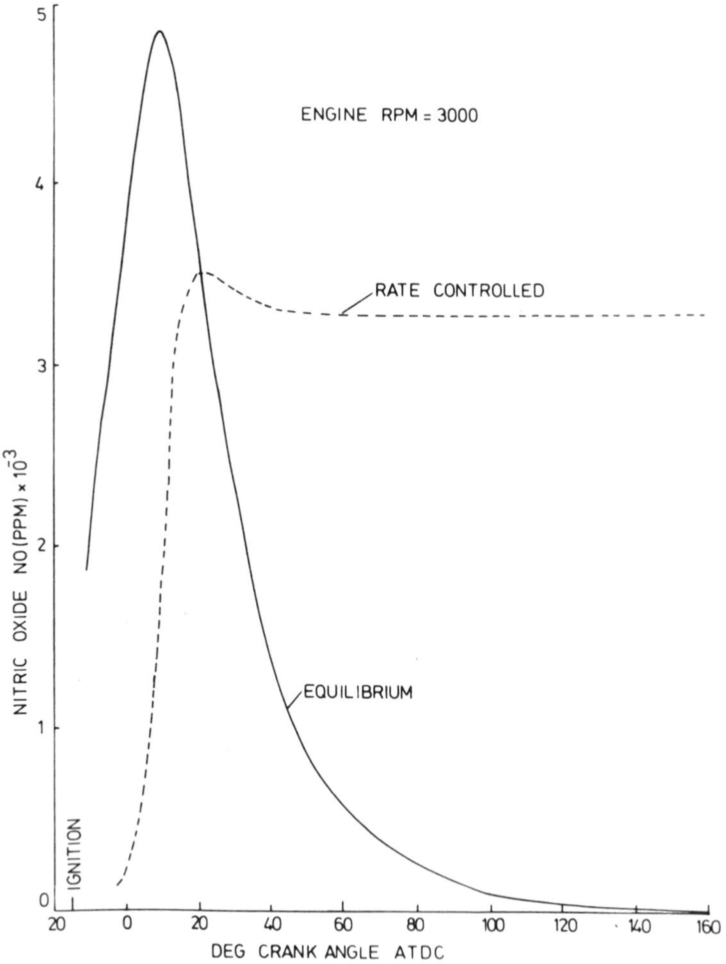

FIG 4·5

† R.S. Benson, W.J.D. Annand, P.C. Baruah, Int. J. Mech. Sci. Vol.17, p.97-124, (1975).

Exercises

1. Show that for the reaction

$$aA + bB \rightleftharpoons cC + dD$$

the equilibrium constant is

$$K_p = \frac{p_C^c p_D^d}{p_A^a p_B^b} p_o^{(c+d)-(a+b)}$$

where p_o is the standard reference pressure.

For the reaction

$$CO + H_2O \rightarrow CO_2 + H_2$$

calculate the equilibrium constant at 2000°K for the standard reference pressures at one atmosphere and one bar respectively.

2.(a) A correct mixture of hydrogen and oxygen is exploded in a vessel. Calculate the percentage by volume of the gases present at 4040°F if the initial temperature and pressure were 140°F and 1 atm respectively. Determine the pressure at 4040°F.

(b) A mixture of 50 per cent CO_2 and 50 per cent H_2 by volume is ignited in a closed vessel at an initial temperature and pressure of 140°F and 1 atm respectively. Calculate the percentage by volume of the gases present at 4040°F. Also determine the pressure at 4040°F. The equilibrium constants are

$$\text{(a)} \quad \frac{p_{H_2O}}{p_{H_2}\sqrt{p_{O_2}}} = 169.0. \qquad \text{(b)} \quad \frac{p_{H_2O}\, p_{CO}}{p_{H_2}\, p_{CO_2}} = 6.422$$

3. A reaction chamber contains a mixture of water vapour and carbon monoxide in the proportions of 2 to 1 by volume. The temperature of the chamber is raised to 1860°F, and maintained at this temperature so that equilibrium is completed. If the value of the equilibrium constant $\dfrac{(CO_2)(H_2)}{(CO)(H_2O)}$ is 0.6 at 1860°F find the percentage composition by volume of the gases present.

4. A 10 per cent rich mixture of heptane, C_7H_{16} and air initially at 1 atm and 100°C, is compressed to one-sixth of its volume and ignited at constant volume. If the resultant gases be CO_2, CO, H_2O, H_2, O_2 and N_2 and if the maximum temperature be

2900°C abs, at which temperature the partial pressure relationships are

$$\frac{p_{CO}p_{H_2O}}{p_{CO_2}p_{H_2}} = 6.717 \quad \text{and} \quad \frac{(p_{CO})^2 p_{O_2}}{(p_{CO_2})^2} = 0.0539 \text{ atm},$$

show that 30.2 per cent of the carbon has been burned to CO only. Air contains 21 per cent by volume of oxygen.

(Univ. Liv.)

5. A 15 per cent rich propane (C_3H_8)- air mixture is supplied to a gas engine. Show that in addition to CO, CO_2, H_2O, H_2 and N_2 in the products of combustion at 4200°R there is a small quantity of oxygen present. Determine the concentration of O_2 in parts per million of O_2 present in the original mixture. The compression ratio is 7:1 and the pressure and temperature at the beginning of the compression stroke are 1 atm abs and 140°F. The combustion is at constant volume. The relationships between the partial pressures at 4200°R are

$$\frac{p_{H_2O} \cdot p_{CO}}{p_{H_2} \cdot p_{CO_2}} = 5.918$$

$$\frac{p_{CO_2}}{p_{CO}\sqrt{p_{O_2}}} = 67.92.$$

92 per cent of the hydrogen in the fuel forms H_2O at 4200°R. Air contains 21 per cent O_2 by volume.

(Univ. Liv.)

6. A gas containing 3 parts by volume methane (CH_4) and one part by volume carbon monoxide (CO) is mixed with 8 parts by volume of oxygen and ignited. Determine the composition of the resultant gases at 5000°R. The relationships between the partial pressures at 5000°R are

$$\frac{p_{CO_2}}{p_{CO}\sqrt{p_{O_2}}} = 7.0259 \qquad \frac{p_{H_2O}}{p_{H_2}\sqrt{p_{O_2}}} = 50.23$$

Initial temperature and pressure 140 F at 1 atm. Concentration of hydrogen in products of combustion at 5000°R = 0.035 mols/mol methane in original mixture.

(Univ. Liv.)

7. A mixture of one volume of hydrogen and two volumes of carbon-dioxide at a pressure of 1 atm abs and temperature 140°F is heated to 2540°F. If the final volume of the mixture is twice the initial volume calculate the final pressure and the volumetric composition of the gases at 2540°F.

At 2540°F the relationship between the partial pressures is

$$\frac{p_{H_2O}\ p_{CO}}{p_{H_2}\ p_{CO_2}} = 3.417$$

(Univ. Liv.)

8. A vessel is filled with hydrogen and carbon dioxide in equal parts by volume, and the mixture ignited. If the initial pressure and temperature are 2 atm abs and 140 F respectively and the maximum pressure 11.8 atm abs estimate (a) the maximum temperature and (b) the equilibrium constant, K_p, and the volumetric analysis of the products of combustion at the maximum temperature.

$$\log_{10}K_p = 1.3573 - \frac{2.437 \times 10^3}{T}$$

where

$$K_p = \frac{p_{H_2O} \cdot p_{CO}}{p_{H_2} \cdot p_{CO_2}}$$

and T = Temperature °R.

(Univ. Liv.)

9. A vessel contains a correct mixture of carbon monoxide and oxygen at a pressure of 2 atm abs and temperature of 600°F, abs. The mixture is ignited. Calculate the volumetric composition of the products of combustion when the pressure is 10 atm abs. Calculate the temperature corresponding to this pressure

$$\log_{10}K_p = -4.442 + \frac{26.332}{T} \times 10^3, \quad \text{(A)}$$

where

$$K_p = \frac{p_{CO_2}}{p_{CO}\sqrt{p_{O_2}}} \quad \text{(B)}$$

T = temperature °F abs.

(Univ. Liv.)

10. Explain what is meant by "chemical equilibrium" and "frozen-equilibrium" of a rapidly expanding mixture of gases. Give one engineering example of each type.

A cylinder containing 1 lb mass of carbon dioxide gas is compressed in an ideal manner. Show that the pressure, volume and temperature are related by the equations

$$\frac{1-a}{a}\sqrt{\left(\frac{2+a}{a}\right)} = K_p\sqrt{\left(\frac{p}{2117}\right)}$$

and

$$pv = \frac{G(2+a)T}{88}.$$

If the compression process is adiabatic, show that

$$\left(\frac{\partial p}{\partial h}\right) = 44.32\,\frac{p}{T}\left(\frac{1}{2+a}\right).$$

Is the process isentropic?

a = degree of dissociation of CO_2.

p = pressure of gas mixture lbf/ft^2.

$K_p = \dfrac{p_{CO_2}}{p_{CO}p_{O_2}^{\frac{1}{2}}}$ equilibrium constant for the reaction.

$$CO + \tfrac{1}{2}\,O_2 \rightleftharpoons CO_2.$$

p_{CO_2} = partial pressure CO_2 atmospheres.
p_{CO} = partial pressure CO atmospheres.

p_{O_2} = partial pressure O_2 atmospheres.

v = specific volume of gas mixture ft^3/lb.

h = specific enthalpy of mixture Btu/lb.

T = absolute temperature $^{\circ}$R.

G = universal gas constant 1545.4 ft lbf/lb mol$^{\circ}$R.

(Univ. Liv.)

11. Consider the dissociation of 1 mol of nitrogen tetroxide

$$N_2O_4 \rightleftharpoons 2NO_2.$$

Derive a relation between p, V, T, R_{mol} and α (where α is the degree of dissociation and R_{mol} is the Universal Gas Constant). Derive also an expression for the equilibrium constant

$$K_p = \frac{p^2_{NO_2}}{p_{N_2O_4}},$$

as a function of p and α.

Obtain an expression for the product of the Joule-Kelvin coefficient (μ) and the specific heat at constant pressure (C_p) as a function of p, T, R_{mol}, K_p and dK_p/dT.

(Univ. Liv.)

12. Oxygen dissociates according to the relation

$$O_2 \rightarrow 2O.$$

Show that the equilibrium constant is given by

$$K_p = \frac{4\alpha^2 p}{(1-\alpha^2)}$$

where α is the degree of dissociation, and obtain an expression for the rate of variation of α with temperature, $(\partial\alpha/\partial T)_p$, in terms of α, dK_p/dT and the total pressure p.

At a temperature of $3800^\circ K$, $K_p = 1$ atm, and $\Delta H/R_{mol} = 59.000^\circ C$ abs. Calculate the value of α, and $(\partial\alpha/\partial T)_p$ at a pressure p of 1 atm.

(Univ. Liv.)

13. For the reaction $CO_2 \rightleftharpoons CO + \frac{1}{2}O_2$, show that the equilibrium constant K_p is given by

$$K_p = \frac{\alpha^{\frac{3}{2}} p^{\frac{1}{2}}}{(2+\alpha)^{\frac{1}{2}}(1-\alpha)}$$

where α is the degree of dissociation per mol of CO_2.

1 mol of CO and $\frac{1}{2}$ mol of O_2 initially at 500 F abs are burnt adiabatically at constant atmospheric pressure.

Determine the temperature after combustion using the following data.

T °F abs	500	5000	5500	6000
Enthalpy of CO_2 above 0°F abs. Btu per mol	3,729	64,132	71,630	79,165
Heat of reaction at constant pressure $(h_{CO}+\frac{1}{2}h_{O_2}-h_{CO_2})$ Btu per mol of CO	121,643	117,545	116,850	116,173
$K_p\,(At)^{\frac{1}{2}}$	-	0.142	0.426	1.0

(Univ. Liv.)

14. A mixture of 2 parts carbon monoxide, 1 part oxygen and 6 parts nitrogen by volume is ignited in a closed vessel. If the initial pressure and temperature are 4.5 atm absolute and 140 F respectively estimate the adiabatic temperature rise allowing for dissociation.

Thermodynamic data:

Temperature °R	600	4300	4400	4500
Internal energy $(E)_{CO_2}$ Btu/lb mol	3434	45,187	46,465	47,746
Internal energy $(E)_{N_2}$ Btu/lb mol	2979	25,441	26,116	26,793
K_p		48.75	35.56	26.30
Q_v Btu/lb mol CO	-121,198	-114,234	-113,999	-113,763

where the equilibrium constant K_p is given by

$$K_p = \frac{p_{CO_2}}{p_{CO}\sqrt{p_{O_2}}}$$

and the internal energy is measured above 0°R.

(Univ. Liv.)

15. A spark ignition engine operates with a 10 per cent rich mixture of carbon monoxide and air. At the end of the compression stroke the cylinder pressure and temperature are 8.5 atm abs and 1100°F abs respectively. If combustion takes place at constant volume with no heat loss, calculate the maximum temperature and pressure. The enthalpy of reaction at absolute zero ΔH_o = -120,163 Btu/lb mol CO.

Internal energy u(T) above $0^{o}R$ in Btu/lb mol:

Gas	Temperature ^{o}R			
	1100	5200	5300	5400
N_2	5518	31,560	32,245	32,931
O_2	5669	33,680	34,429	35,179
CO	5537	31,891	32,581	33,271
CO_2	7638	56,796	58,097	59,400

Equilibrium constant $K_p = \dfrac{p_{CO_2}}{p_{CO}\sqrt{p_{O_2}}}$ for reaction

$$CO + \tfrac{1}{2}\,O_2 \rightleftharpoons CO_2;$$

Temperature ^{o}R	5200	5300	5400
K_p	4.458	3.598	2.929

(Univ. Liv.)

16. Starting from Van't Hoff's equation derive the following expression for the "Lighthill Ideal Dissociating Gas" ($A_2 \rightleftharpoons 2A$)

$$\frac{\alpha^2}{1-\alpha} = \frac{\rho_d}{\rho}\, e^{-\theta_D/T}$$

where α is the degree of dissociation of A_2, ρ is the density of the gas, T is the gas temperature, ρ_d is the characteristic density, θ_D is the characteristic temperature of dissociation.

Hence show that the specific heat at constant volume for the molecular form of A is twice the corresponding specific heat of the atomic form. Discuss the methods used in your analysis, the assumptions made and the implications of the numerical value of the specific heats at constant volume of A_2 in terms of the degrees of freedom for the molecule model.

(Univ. Manch.)

17. Consider the partial ionization of 1 mol of argon,

$$A \rightleftharpoons A^+ + e^-.$$

(i) Show that for the ionized mixture

$$pV = (1+\alpha)R_{mol}T$$

$$H = \frac{5}{2}\,R_{mol}T(1+\alpha) - \alpha\Delta H_o + h_{A_o}$$

and

$$K_p = \frac{\alpha^2 p}{(1-\alpha^2)},$$

where p is the pressure, V is the volume, T is the temperature, α is the degree of ionization, H is the enthalpy, R_{mol} is the universal gas constant, $\Delta H_o = (h_A - h_{A+} - h_{e-})$ at absolute zero, and h_{A_o} is the enthalpy per mol of argon at absolute zero.

It may be assumed that the argon, the ions and the electrons behave as perfect monatomic gases.

(ii) Obtain the change in entropy of the mixture in isothermal expansion at temperature T, as a function of the change in degree of ionization from α_1 to α_2.

(Univ. Liv.)

18. Show for a partially ionized monatomic gas, A, that

$$\left(\frac{d\alpha}{dT}\right)_p = \frac{\alpha}{2T}(1-\alpha^2)\left(\frac{5}{2} + \frac{\theta_I}{T}\right)$$

$$\left(\frac{d\alpha}{dT}\right)_V = \frac{\alpha}{T}\frac{(1-\alpha)}{(2-\alpha)}\left(\frac{3}{2} + \frac{\theta_I}{T}\right)$$

where α is the degree of ionization at pressure p and temperature T, θ_I is the characteristic temperature of ionization.

Obtain an expression for the specific heats at constant pressure and constant volume for the plasma.

19. A vessel contains 0.04 lb of ionized argon at a temperature of 18,000°C abs and a pressure of 14.7 p.s.i.a. Determine the degree of ionization and the volume of the vessel.

Molecular weight of argon = 40, characteristic ionization temperature for argon = 182,900 K, constant C in Saha's equation = 4.85×10^{-6} p.s.i.$(°K)^{5/2}$.

20. Outline a procedure for calculating the equilibrium composition of the products of an ideal gas reaction which takes place at constant pressure (use approximately 100 words). It may be assumed that the initial state of the reactants is known, that there is a known heat exchange with the surroundings and that information is available on the properties of all the gases taking part in the reaction.

A reaction chamber is supplied with a steady flow of carbon monoxide and steam at 400°K and two atmospheres absolute pressure. The steam flow rate is 20% greater than the stoichiometric value. If the reaction takes place adiabatically and at constant pressure, estimate the final temperature and the percentage by volume of carbon dioxide in the exhaust gases.

For the reaction $CO + H_2O = CO_2 + H_2$ the heat of reaction at constant pressure, $Q_p = -42,000$ kJ/kg mol of CO at 400°K.

Temperature T°K	Enthalpy at temperature T°K- Enthalpy at 400°K, kJ/kg mol				Equilibrium Constant $K_p = \frac{p_{CO_2}p_{H_2}}{p_{CO}p_{H_2O}}$
	CO	H_2O	CO_2	H_2	
750	10,700	12,800	17,500	10,300	5.84
800	12,300	14,700	20,000	11,800	4.13

(Univ. Manch)

21. A rigid vessel contains a 20% rich mixture of carbon monoxide and air at a temperature of 700°K and a pressure of 8.6 atmospheres absolute. The mixture is ignited and burns without heat loss to a temperature of 3,000°K.

Calculate:

(a) The degree of dissociation of the carbon dioxide at 3,000°K.

(b) The equilibrium constant $K_p = \frac{p_{CO_2}}{p_{CO}\sqrt{p_{O_2}}}$ at 3,000°K.

If the dissociated products are now cooled to 700°K, at which temperature the equilibrium constant as defined above is very large, what will be the composition of the products at 700°K and how much heat must be transferred per kilogram mol of the original mixture?

For the reaction $CO + \frac{1}{2}O_2 \rightarrow CO_2$ the heat of reaction at 0°K is -280,000 kJ per kg mol of CO.

Temperature °K	Internal Energy u(T) above 0°K kJ/kg mol			
	CO	O_2	N_2	CO_2
700	14,900	15,400	14,800	21,300
3000	77,300	81,800	76,500	138,000

(Univ. Manch.)

22. A mixture consists of two parts by volume of steam and three parts by volume of carbon monoxide. The mixture is contained within a cylinder at a pressure of one bar absolute and a temperature of 300°K. After isothermal compression to one tenth of the original volume, the mixture reacts at constant volume.

Calculate the pressure within the cylinder at equilibrium and the composition of the products of combustion, assuming that these consist only of steam, carbon monoxide, hydrogen and carbon dioxide.

The following data applies:

Temperature T°K	(Internal energy at temperature T)-(Internal energy at 300°K kJ/kg mol				Equilibrium Constant
	H_2O	CO	H_2	CO_2	$K_p = \frac{p_{H_2}\, p_{CO_2}}{p_{H_2O}\, p_{CO}}$
750	12,400	9,900	9,500	17,700	6.03
800	14,000	11,100	10,600	19,800	4.13

For the reaction $H_2O + CO = H_2 + CO_2$ the heat of reaction at constant pressure at 300°K is -42,500 kJ/kg mol of H_2O.

How would the final cylinder pressure have been affected if the compression stroke preceding the reaction had been

(a) adiatabic and frictionless, or
(b) adiabatic but frictional.

Indicate the direction of the pressure change in each case, giving the reason.

(Univ. Manch.)

23. A mixture of propane gas C_3H_8 and air burns at a constant pressure of 2 bars, reaching equilibrium at a temperature of 2800°K. The quantity of air in the reactants is 20% greater than stoichiometric. If the products of combustion contain only CO_2, CO, O_2, H_2O, H_2 and N_2, show that the degree of dissociation of CO_2 is approximately 0.32 and calculate the percentage by volume of H_2 in the products.

When partial pressures are expressed in bars, the equilibrium constants are given by

$$\frac{p_{CO_2}}{p_{CO}\sqrt{p_{O_2}}} = 6.6 \text{ bar}^{-1/2} \quad \text{and} \quad \frac{p_{CO_2}p_{H_2}}{p_{H_2O}p_{CO}} = 0.15$$

(Univ. Manch.)

24. The volumetric composition of a gaseous fuel is 80% methane (CH_4) and 20% inert gases. The fuel is mixed with the stoichiometric quantity of air and enters an engine cylinder at atmospheric pressure and 300°K. After ignition the products reach a temperature of 2,700°K at a point in the stroke where the volume is one sixth of the initial volume. Assuming that the products are in equilibrium and contain CO_2, CO, H_2O, H_2, O_2 and N_2, together with the inert gas present in the fuel show that the degree of dissociation of the water vapour is approximately 2.18% and calculate the degree of dissociation of the carbon dioxide.

$$\frac{p_{CO_2}}{p_{CO}\sqrt{p_{O_2}}} = 10.2\ \text{atm}^{-\frac{1}{2}} \qquad \frac{p_{CO_2}\cdot p_{H_2}}{p_{CO}\cdot p_{H_2O}} = 0.15$$

Assume that air contains 21% O_2 by volume.

(Univ. Manch.)

25. A gaseous fuel is a mixture of equal parts by volume of carbon monoxide and methane (CH_4). The gas burns with a quantity of air which is 20% greater than the stoichiometric amount. The products of combustion, which contain only CO_2, CO, H_2O, H_2, O_2 and N_2, are in equilibrium when they are discharged at atmospheric pressure. If the carbon monoxide content of the exhaust is not to exceed 1% by volume, show that the temperature of the exhaust must not be greater than 2,360°K approximately. Calculate also the hydrogen content of the exhaust.

At 2,360°K the equilibrium constants, based on a reference pressure p_o of one atmosphere, have the values

$$\frac{p_{CO_2}/p_o}{(p_{CO}/p_o)(p_{O_2}/p_o)^{\frac{1}{2}}} = 60$$

and

$$\frac{(p_{CO_2}/p_o)(p_{H_2}/p_o)}{(p_{CO}/p_o)(p_{H_2O}/p_o)} = \frac{1}{6}$$

Both equilibrium constants decrease with increasing temperature. Air contains 21% oxygen by volume.

(Univ. Manch.)

26. Say what is meant by the terms "equilibrium flow" and "frozen flow" as applied to the flow of gases along a duct.

A steady stream of nitrogen gas is initially at low temperature and in the undissociated molecular form. The gas is then heated to 7,500°K at a pressure of 10 bars, at which state equilibrium is reached between the molecular and the atomic forms of the gas. Calculate the degree of dissociation of the molecular nitrogen.

The stream then enters a nozzle which expands the gas to a static pressure of 1 bar and a static temperature of 6,200°K. Assuming equilibrium flow, calculate the degree of dissociation at the nozzle exit.

The nozzle is then replaced by a second, shorter nozzle, capable of expanding the gas stream from the same equilibrium inlet state to the same outlet static pressure, but with frozen flow within the nozzle. If the static temperature at the outlet from the short nozzle is again found to be 6,200°K, estimate the difference in kinetic energy produced by the two nozzles, per mol of the original undissociated nitrogen.

Neglect ionisation and consider only the reaction $N_2 \rightleftharpoons 2N$. The equilibrium constant K_p, defined as

$$K_p = \frac{(p_N/p_o)^2}{p_{N_2}/p_o}$$

has a value of 3.0 at 7,500°K, taking the reference pressure p_o as one bar. The variation of K_p with temperature is given by Van't Hoff's equation

$$\frac{d}{dT}(\ln K_p) = \frac{Q_p}{R_{mol}T^2}$$

Over the appropriate temperature range, the heat of reaction at constant pressure Q_p remains constant at the value $Q_p = 10^6$ kJ per kg mol of N_2.

(Univ. Manch.)

27. Considering the dissociation reaction $N_2 \rightleftharpoons 2N$ define the term "characteristic temperature of dissociation". If nitrogen behaves as an ideal dissociating gas, comment on the relative magnitudes of the specific heats of the atomic and molecular forms of the gas. Show that

$$\ln K_{p_1} - \ln K_{p_2} = \theta_D\left(\frac{T_1 - T_2}{T_1 T_2}\right)$$

where θ_D is the characteristic temperature of dissociation, K_{p_1} and K_{p_2} are the values of the equilibrium constant for the dissociation reaction at temperatures T_1 and T_2. If the heat of dissociation of nitrogen 996,000 kJ/kg mol of N_2 and K_p = 0.25 atmospheres at 6,500°K, estimate the value of the equilibrium constant at 7,000°K.

A mass 28 grammes of nitrogen is in equilibrium at 7,000°K and 0.5 atmospheres. Calculate the degree of dissociation.

(Univ. Manch.)

28. Show that the specific enthalpy h and the specific Gibbs free energy g are related by

$$h = -T^2\left(\frac{\partial}{\partial T}\left(\frac{g}{T}\right)\right)_p$$

Hence prove that the variation of equilibrium constant K_p with temperature T is given by

$$\frac{d}{dT}(\ln K_p) = \frac{Q_p}{R_{mol}T^2}$$

where Q_p is the heat of reaction at constant pressure.

For the reaction $\frac{1}{2}N_2 + \frac{1}{2}O_2 = NO$
$Q_p = 10^5$ kJ/kg mol of NO and $K_p = 1.5 \times 10^{-3}$ at a temperature of 1400°K. K_p is defined as

$$K_p = \frac{p_{NO}}{(p_{N_2})^{\frac{1}{2}}(p_{O_2})^{\frac{1}{2}}}$$

A stream of exhaust gases from a combustion process is in equilibrium at 1400°K and a pressure of one atmosphere. By volume, 75% of the gas is N_2 and 8% is O_2, the remainder being H_2O and CO_2 together with lesser amounts of H_2, CO and NO. Calculate the NO content of the stream, in parts per million by volume. If, in one application, the maximum permissible NO content is to be 10 parts per million, estimate the limiting permissible value of the exhaust gas temperature.

Mention any assumptions made in the calculation and comment briefly on their validity.

(Univ. Manch.)

29. For the reaction $N_2 + O_2 = 2NO$ the equilibrium constant K_p at a temperature of 1100°C has a value of 2.25×10^{-6}, where K_p is defined as

$$K_p = \frac{(p_{NO})^2}{(p_{N_2})(p_{O_2})}$$

A stream of exhaust gases is at a temperature of 1100°C. The gases contain 75% nitrogen and 8% oxygen by volume. Calculate the NO content at equilibrium. The NO content is to be reduced to ten parts per million by volume, and this is to be achieved by cooling the gas stream. Estimate the temperature to which the exhaust gas must be lowered. Assume that, compared with the NO, the changes of concentration of all other gases are small. The variation of equilibrium constant K_p with temperature is given by Van't Hoff's equation:

$$\frac{d}{dT}(\ln(K_p)) = \frac{Q_p}{R_{mol}T^2}$$

where $Q_p = 2 \times 10^5$ kJ per kg mol of N_2, and is constant over the temperature range in question.

(Univ. Manch.)

CHAPTER 5

APPLICATION OF THERMODYNAMICS TO SPECIAL SYSTEMS

Thermodynamics of elastic systems, surface tension, magnetic systems, reversible electrical cell and fuel cell. Introduction to irreversible thermodynamics, Onsager reciprocal relation, thermoelectricity.

Notation

A	cross sectional area
B	induced magnetism
C_p	specific heat at constant pressure
C_v	specific heat at constant volume
C_F	specific heat at constant tension
C_H	specific heat at constant magnetic field
C_J	specific heat at constant magnetic intensity
C_L	specific heat at constant length
C_H^*	specific heat at constant magnetic field at magnetic temperature T^*
C_1, C_2	constants in Curie's Law
E	isothermal Young's modulus
f	specific Helmholtz free energy function
F	tensile force
F	Farad
G	Gibbs free energy function
h	specific enthalpy
$\bar{h}$	specific magnetic enthalpy $u-H\bar{j}$
	specific elastic enthalpy, $u-Fl$
H	enthalpy
H	magnetic field strength
I	electric current
j	valency of ion, intensity of magnetization
$\bar{j}$	specific magnetization
J	thermodynamic velocity or flow
J_I	electrical current flux
J_Q	heat energy flux
J_S	entropy flux
k	thermal conductivity
$\bar{k}$	magnetic susceptibility
l	length per unit mass, coordinate

L	length
L	transport property in irreversible thermodynamics
L_{ii}, L_{kk}	primary coefficients
L_{ik}, L_{ki}	cross coupling coefficients
M	number of mols
N	number of turns in toroid
p	pressure
p_i	partial pressure
q	heat transfer per unit mass
Q	heat transfer
Q_p	heat of reaction at constant pressure
Q^*	heat of transport
R_{mol}	universal gas constant
s	specific entropy
S	entropy
S^*	entropy of transport
T	temperature absolute
T_c	critical temperature
u	specific internal energy in the absence of motion, gravity, etc.
U	internal energy in the absence of motion, gravity, etc.
V	volume
W	work
W_ε	electrical work
x	coordinate
X	coordinate, thermodynamic force
Z	electrical charge
α	coefficient of linear expansion
ε	electrical potential difference
ε_R	reversible e.m.f.
η_I	ideal efficiency of fuel cell

θ	rate of entropy production per unit volume
λ	heat addition to a surface film at constant temperature, electrical conductivity
μ	chemical potential
$\bar{\mu}$	permeability
$\bar{\mu}_o$	permeability of free space
ν	stoichiometric coefficient
π	Peltier coefficient
σ	surface tension, Thomson coefficient
ϕ	magnetic flux

In the previous chapters we have discussed the application of thermodynamics to gaseous systems. In this chapter we will consider the application of the same methods to elastic systems, systems with surface tension, electrical cells and magnetic systems; finally, we will examine thermoelectricity. The main problem is the formulation of the thermodynamic properties of the system and the equation of state. These must be determined by experiment.

Elastic Systems

Consider an elastic system comprising a wire length L under a tensile force F at temperature T. Experiments show that this system can be completely described by these three thermodynamic properties. The equation of state is therefore

$$L = L(F,T) \tag{5.1}$$

and in differential form

$$dL = \left(\frac{\partial L}{\partial T}\right)_F dT + \left(\frac{\partial L}{\partial F}\right)_T dF. \tag{5.2}$$

The differential coefficients will be determined by experiment.

If a bar of uniform cross-sectional area is heated uniformly with the ends non-constrained the ratio of the increase in length per unit length to the temperature rise is called the coefficient of linear expansion (α)

$$\alpha = \left(\frac{dL/L}{dT}\right)_{F=\text{constant}} = \left(\frac{\partial L}{\partial T}\right)_F \frac{1}{L}. \tag{5.3}$$

In a tensile or compressive test on a bar at constant temperature, the ratio of the change in stress to the change in strain is called the isothermal Young's modulus (E)

$$E = \left(\frac{dF/A}{dL/L}\right)_{T=\text{constant}} = \left(\frac{\partial F}{\partial L}\right)_T \frac{L}{A}. \tag{5.4}$$

Substituting the differential coefficients $(\partial L/\partial T)_F$ and $(\partial L/\partial F)_T$ into (5.2) we obtain

$$\frac{dL}{L} = \alpha \, dT + \frac{1}{EA} \, dF. \tag{5.5}$$

If the system is initially at temperature T_o and length L_o with zero force F then, if α is independent of temperature, integration of (5.5) gives the relationship between the new length L at temperature T with a force F applied to the ends of the wire. Thus

$$L = L_o \exp \left\{ \alpha(T-T_o) + \frac{F}{EA} \right\}. \tag{5.6}$$

For small temperature differences

$$L = L_o \left\{ 1 + \alpha(T-T_o) + \frac{F}{EA} \right\}. \tag{5.7}$$

Equations (5.6) and (5.7) are the equations of state of the system.

The intensive properties of the system are the temperature T and the force F, the extensive property is the length L.

The work done by an elastic system is

$$đW = -FdL. \tag{5.8}$$ †

Since an applied force F produces an increase in length dL work is done on the system. As the force F is reduced, the change in length dL will decrease and work will be done by the system. It will be seen that the work is the product of the intensive property and the change in the extensive property.

The first law of thermodynamics for the system is

$$dU = đQ - đW = đQ + F \, dL$$

or

$$đQ = dU - F \, dL. \tag{5.9}$$

The internal energy U is a function of any two of the three variables F, T, L.

The specific heat at constant length (analogous to C_v) is defined as

---

† The derivative dW (with the cross đ) indicates an inexact differential.

$$C_L = \left(\frac{\partial q}{\partial T}\right)_1 \tag{5.10}$$

where 1 is the length per unit quantity of mass.

The specific heat at constant tension (analogous to C_p) is defined as

$$C_F = \left(\frac{\partial q}{\partial T}\right)_F. \tag{5.11}$$

Substituting Q = q, L = 1 and U = u, in equation (5.9) for 1 mol of substance in the system we have

$$C_L = \left(\frac{\partial q}{\partial T}\right)_1 = \left(\frac{\partial u}{\partial T}\right)_1 \tag{5.10}$$

$$C_F = \left(\frac{\partial q}{\partial T}\right)_F = \left(\frac{\partial u}{\partial T}\right)_F - F\left(\frac{\partial l}{\partial T}\right)_F. \tag{5.11}$$

From (5.3) for L = 1

$$\left(\frac{\partial l}{\partial T}\right)_F = \alpha l$$

and

$$C_F = \left(\frac{\partial u}{\partial T}\right)_F - \alpha l F. \tag{5.12}$$

The relationship between the specific heats can be obtained using the methods described in Chapter 2.

For the <u>elastic</u> systems we can write for the second law of thermodynamics

$$dS = \frac{dQ}{T}$$

and for reversible changes

$$dQ = T\,dS = dU - F\,dL.$$

Hence

$$dU = T\,dS + F\,dL. \tag{5.13}$$

In specific units

$$du = T\,ds + F\,dl. \tag{5.14}$$

We shall define the "elastic enthalpy" $\bar{h}$ as

$$\bar{h} = u - Fl. \tag{5.15}$$

The corresponding Helmholtz and Gibbs free energy functions are

$$f = u - Ts$$
$$g = \bar{h} - Ts.$$

In differential form we have the following expressions

$$du = T\,ds + F\,dl \quad \text{or} \quad u = u(s,v) \tag{5.14}$$
$$d\bar{h} = T\,ds - l\,dF \quad \text{or} \quad \bar{h} = \bar{h}(s,F) \tag{5.16}$$
$$df = F\,dl - s\,dT \quad \text{or} \quad f = f(l,T) \tag{5.17}$$
$$dg = -l\,dF - s\,dT \quad \text{or} \quad g = g(p,T). \tag{5.18}$$

These equations correspond to I(a), II(a), III(a), IV(a) (p.35). If we compare these expressions we see that

$$F \equiv -p, \quad l \equiv v. \tag{5.19}$$

Now for a reversible process

$$d\bar{h} = T\,ds - l\,dF = dq - l\,dF. \tag{5.20}$$

For a constant force process $d\bar{h} = dq$, and

$$C_F = \left(\frac{\partial q}{\partial T}\right)_F = \left(\frac{\partial \bar{h}}{\partial T}\right)_F. \tag{5.21}$$

The analogous relations to (2.51) and (2.52) for the specific heats are obtained as follows:

$$C_L = \left(\frac{\partial u}{\partial T}\right)_l = \left(\frac{\partial u}{\partial s}\right)_l \left(\frac{\partial s}{\partial T}\right)_l. \tag{5.22}$$

From (5.14)

$$\left(\frac{\partial u}{\partial s}\right)_l = T \tag{5.23}$$

and

$$C_L = T\left(\frac{\partial s}{\partial T}\right)_l \tag{5.24}$$

$$C_F = T\left(\frac{\partial s}{\partial T}\right)_F. \tag{5.25}$$

The equivalent relationship to (2.62),

$$\frac{C_p - C_v}{T} = \left(\frac{\partial p}{\partial T}\right)_v \left(\frac{\partial v}{\partial T}\right)_p, \tag{2.62}$$

can be obtained using the analogous relations noted in (5.19) and

$$C_F \equiv C_p, \quad C_L \equiv C_v. \tag{5.26}$$

From (5.19):

$$\left(\frac{\partial p}{\partial T}\right)_v \equiv -\left(\frac{\partial F}{\partial T}\right)_l \tag{5.27}$$

$$\left(\frac{\partial v}{\partial T}\right)_p \equiv \left(\frac{\partial l}{\partial T}\right)_F. \tag{5.28}$$

Therefore

$$\frac{C_F - C_L}{T} = -\left(\frac{\partial F}{\partial T}\right)_l \left(\frac{\partial l}{\partial T}\right)_F. \tag{5.29}$$

The state equation is

$$\frac{l}{l_o} = \frac{L}{L_o} = \exp\left(\alpha(T - T_o) + \frac{F}{EA}\right).$$

Therefore

$$\left(\frac{\partial F}{\partial T}\right)_l = -\alpha EA, \quad \left(\frac{\partial l}{\partial T}\right)_F = \alpha l$$

and

$$\frac{C_F - C_L}{T} = \alpha^2 EAl \quad \text{in specific units.} \tag{5.30}$$

For materials in which α is small (say the order of 10^{-6}) the difference between the specific heats is very small. In these cases C_F is usually determined by direct experiment.

The technique used to develop expression (5.30) from (5.26) is an indirect method of determining the thermodynamic relations for special systems. The procedure is to use the "conventional" Maxwell relations for a p, v, T system and to replace these "coordinates" by the equivalent thermodynamic coordinates for the systems under examination. This technique will be used in the other systems to be described in this chapter.

Systems with Surface Tension

Surface tension is a property of the interface between a system and its surroundings. In thermodynamic studies of systems in which surface tension is present, the properties of the system and surroundings must be taken into account when the forces on the interface are considered. The phenomenon of surface tension arises because the particles of an interface are subject to forces differing from those in the interior of the system. For practical purposes the interface may be considered to be a thin film (surface film). The properties of the film and its effect on the system it encloses and the surrounding is of importance in, for example, the studies of free liquid surfaces in which the liquid may be in equilibrium with its vapour or with other gaseous substances, or in the studies of droplet formation, e.g. soap bubbles, fuel droplets, boiling, etc. A simple example to illustrate the application of the methods of thermodynamics is shown in Fig. 5.1.

A soap film BCD is stretched over a wire frame ABCDE. The soap film has two surfaces and encloses a space containing some vapour. By definition the force F due to the surface tension of a film length L is given by $F = \sigma L$ where σ is the surface tension (force/unit length).

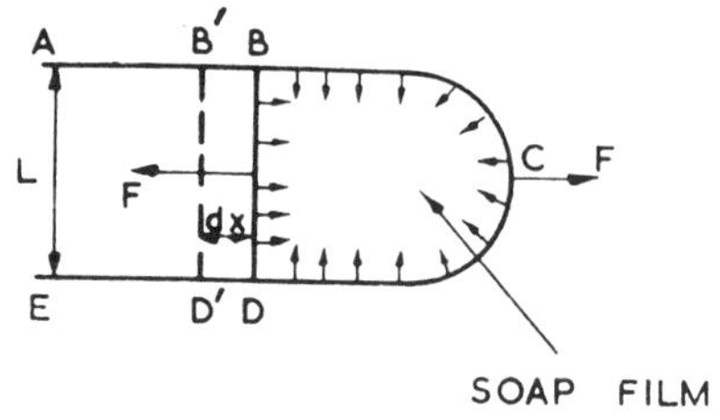

FIG. 5.1.

For the system shown in Fig. 5.1 the thermodynamic properties are p, T, V, σ and film area A. If the wire BD is moved to the left, a distance dX, the surface area of the film is increased 2LdX (since there are two sides of the film). The force causing the increase in the film size is

$$F = 2\sigma L. \tag{5.31}$$

The work done in increasing the area of the film dA is

$$đW = -F\,dX = -2\sigma L\,dX$$

or

$$đW = -\sigma\,dA. \tag{5.32}$$

The minus sign indicates that work is done on the film when the surface area is increased. The increased film area dA equals 2L dX. The surface tension σ is an intensive property. The work done is equal to the product of the intensive variable (σ) and the change in the extensive variable (A). Equation (5.32) is the general equation for the work done on a system due to surface tension. For a change in pressure p and volume V the total work would be

$$đW = -\sigma \, dA + p \, dV. \tag{5.33}$$

The state equation relating the thermodynamic properties of the system is determined experimentally. For a pure liquid in equilibrium with its own vapour the following empirical expression may be used

$$\sigma = \sigma_o \, (1 - T/T_c)^n. \tag{5.34}$$

σ_o is the surface tension at 0^oC, T_c is the critical temperature and n is an experimentally determined index. At the critical temperature, T_c, the surface tension is zero.

For systems in which the volume changes are infinitesimal and can be neglected, the first law is

$$dU = đQ - đW$$

or

$$đQ = dU - \sigma \, dA. \tag{5.35}$$

The internal energy, U, is a function of any two of the variables T, σ, A.

Experiments show that effect of the increase in the surface film is to cause a mass transfer from the vapour to the liquid phase. This is accompanied by heat addition if the temperature of the film is to remain constant. For a surface film enclosing a volume V, the heat added to keep the temperature T constant, when the area of the film increases dA, is given by

$$đQ = \lambda \, dA \tag{5.36}$$

where λ is in energy units per unit area. The numerical value of λ can be determined from experimental data by the use of the thermodynamic relations discussed in Chapter 2.

The first law for the system is

$$đQ = \lambda \, dA = dU - \sigma \, dA$$

whence

$$dU = (\lambda + \sigma) \, dA. \tag{5.37}$$

For a reversible heat exchange process

$$dQ = T \, dS = dU - \sigma \, dA$$

or

$$dU = T\,dS + \sigma\,dA. \tag{5.38}$$

The "surface tension enthalpy" H can be defined by

$$H = U - \sigma A \tag{5.39}$$

and therefore

$$dH = dU - \sigma\,dA - A\,d\sigma = T\,dS - A\,d\sigma. \tag{5.40}$$

For a heating process at <u>constant temperature</u>, the surface tension is constant, and

$$dQ = \lambda\,dA = T\,dS = dH$$

or

$$\left(\frac{\partial H}{\partial A}\right)_\sigma = \lambda. \tag{5.41}$$

Now

$$H = H(S,\sigma) \tag{5.42}$$

and

$$dH = \left(\frac{\partial H}{\partial S}\right)_\sigma dS + \left(\frac{\partial H}{\partial \sigma}\right)_S d\sigma. \tag{5.43}$$

Also from (5.40)

$$dH = T\,dS - A\,d\sigma \tag{5.40}$$

and

$$\left.\begin{aligned} T &= \left(\frac{\partial H}{\partial S}\right)_\sigma \\ A &= -\left(\frac{\partial H}{\partial \sigma}\right)_S \\ \left(\frac{\partial T}{\partial \sigma}\right)_S &= -\left(\frac{\partial A}{\partial S}\right)_\sigma. \end{aligned}\right\} \tag{5.44}$$

Substituting (5.44) into (5.41) we have

$$\lambda = \left(\frac{\partial H}{\partial A}\right)_\sigma = \left(\frac{\partial H}{\partial S}\right)_\sigma \left(\frac{\partial S}{\partial A}\right)_\sigma = T\left(\frac{\partial S}{\partial A}\right)_\sigma. \tag{5.45a}$$

Finally, substituting for $(\partial S/\partial A)_\sigma$ from (5.44) into (5.45a)

$$\lambda = -T\left(\frac{\partial \sigma}{\partial T}\right)_S. \tag{5.45b}$$

If the variation of the surface tension σ with temperature T is known the numerical value of λ can be obtained. Since the surface tension always decreases with temperature, λ is always positive; thus heat is supplied when the surface area of a film is increased at constant temperature. In general the surface tension is a monotonic function of temperature only and equation (5.45b) may be expressed in the form

$$\lambda = -T\frac{d\sigma}{dT}. \tag{5.46}$$

Consider a droplet with surface area A at temperature T. If the surface film is in equilibrium with its own vapour then

$$\sigma = \sigma_o\left(1 - \frac{T}{T_c}\right)^n,$$

therefore

$$\frac{d\sigma}{dT} = -\frac{n\sigma_o}{T_c}\left(1 - \left(\frac{T}{T_c}\right)\right)^{n-1} = -\frac{n\sigma}{T_c - T}$$

and

$$\lambda = n\sigma_o\left(\frac{T}{T_c}\right)\left(1 - \frac{T}{T_c}\right)^{n-1} = \frac{n\sigma T}{T_c - T}. \tag{5.47}$$

At the critical temperature T_c, λ is zero. The maximum value of λ will occur at some temperature between zero and T_c. If we differentiate λ with respect to T we have

$$\frac{d\lambda}{dT} = \frac{n\sigma}{(T_c - T)^2}(T_c - nT). \tag{5.48}$$

Thus λ increases with temperature when T is less than T_c/n and decreases with temperature when T is greater than T_c/n. The maximum value of λ corresponds to $T = T_c/n$ when

$$\lambda = \frac{n\sigma}{n-1}. \tag{5.49}$$

The surface energy of the film is defined as the increase in internal energy per unit area of the film during the production of the film.

From the first law (5.37) and (5.46)

$$dU = (\lambda + \sigma)\,dA = \left(\sigma - T\frac{d\sigma}{dT}\right)dA. \tag{5.50}$$

If U_o is the internal energy at A = 0 and U the internal energy when the film has a surface area A, then the surface energy is $(U-U_o)/A$.

Integrating between U_0, U and A = 0, A = A the surface energy of the film is

$$\frac{U-U_0}{A} = \left(\sigma - T\frac{d\sigma}{dT}\right). \qquad (5.51)$$

Now
$$\frac{d\sigma}{dT} = \frac{n\sigma}{T_c - T}$$

for a film in contact with its own vapour and the surface energy of the film is

$$\frac{U-U_0}{A} = \sigma\left(\frac{T_c + (n-1)T}{T_c - T}\right) \qquad (5.52)$$

This expression shows that the surface energy is a function of temperature only.

For a more detailed study of the application of thermodynamics to surface films the student is recommended to study more advanced texts.

Reversible Cell

A reversible cell consists of two electrodes each immersed in a different electrolyte. When the electrodes are coupled to an external electric circuit, current flows from the positive terminal to the negative terminal in the external circuit. Within the cell, "ions" flow from the negative terminal to the positive terminal thus completing the circuit. The electromotive force of the cell (e.m.f.) depends on the nature of the materials of the electrodes and electrolytes, the concentration of the electrolytes and the temperature.

The important feature of a reversible cell is that chemical changes, accompanying the transfer of electricity in one direction, take place to the same extent in the reverse direction when the same quantity of electricity is transferred in the reverse direction. The chemical reaction takes place reversibly in the thermodynamic sense (i.e. the reversible operation will return the chemical state of the cell to the same condition as at the start).

A typical reversible cell is the Daniell Cell diagrammatically illustrated in Fig. 5.2.

The cell in Fig. 5.2 consists of a copper electrode immersed in a saturated copper sulphate solution and a zinc electrode immersed in a saturated zinc solution. The two solutions are separated by a porous wall. The conventional method of representing the cell is

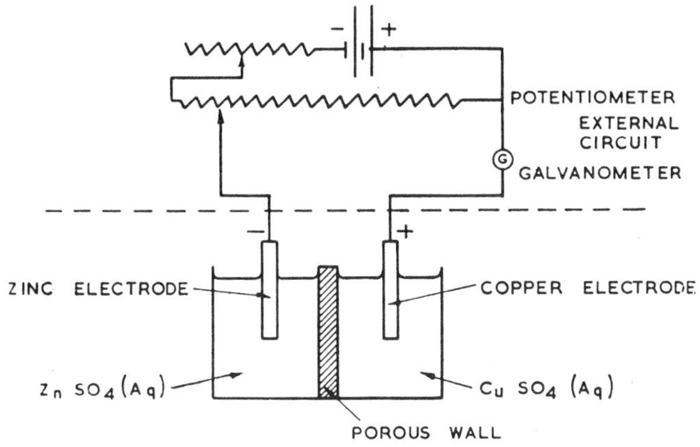

FIG. 5.2. Reversible Daniell cell

$$Zn\,|\,ZnSO_4\,|\,CuSO_4\,|\,Cu$$

each vertical line representing an interface between the phases. The copper electrode is, by convention, the positive electrode, the zinc, the negative electrode. For a spontaneous change the sequence of events takes place from left to right.

Let the cell be connected to a potentiometer and the potential difference arranged to be slightly smaller than the e.m.f. of the cell. The following reactions take place within the cell:

$$Zn \rightarrow Zn^{++} + 2e \quad \text{at zinc electrode}$$

$$Zn^{++} + CuSO_4 \rightarrow ZnSO_4 + Cu^{++} \quad \text{in solutions}$$

$$Cu^{++} + 2e \rightarrow Cu \quad \text{at copper electrode}$$

The zinc electrode will have an excess of electrons, whilst the copper electrode will have a deficit of electrons. The latter is therefore positively charged.

The sequence of events is as follows:

The zinc goes into solution and zinc sulphate is formed, copper is deposited and copper sulphate used up. The electricity in the external circuit "flows" from the copper electrode to the zinc electrode. In the internal circuit it is considered that charged ions "flow" in the electrolyte from the zinc to the copper electrode. The overall chemical reaction is therefore:

$$Zn + CuSO_4 \rightarrow ZnSO_4 + Cu.$$

The quantity of electricity flowing follows Faraday's laws of electrolysis. The first law of electrolysis states that the mass of ions moving to an electrode is proportional to the current strength. The second law of electrolysis states that 96,500 coulombs of electricity deposit one gram-mol equivalent of any ion in electrolysis. Thus, if j is the valency of an ion and dM the number of mols of ions deposited at an electrode, the quantity of electricity flowing is

$$dM \cdot j \cdot F$$

where F is 96,500 coulombs or one farad.

In the above example, since the valency of copper is 2, the quantity of electricity flowing per mol of copper deposited is

2 farads or 193,000 coulombs.

The change in charge, dZ, of the cell is equal to

$$dZ = -jF\ dM \text{ coulombs.} \tag{5.53}$$

The negative sign is used since for a flow of electricity from the cell there will be a drop in the charge.

The work done by an electrical system when a quantity of charge dZ flows between two points on the boundary of the system is

$$đW = -\varepsilon\, dZ \tag{5.54}$$

where ε is the potential difference between the two points on the boundary.

Thus the work done by the cell whilst discharging will be equal to

$$đW = -\varepsilon_R\, dZ \tag{5.55}$$

where ε_R is the reversible e.m.f. of the cell. If the cell is discharging, then dZ is negative, $đW$ is positive and work is done by the cell.

If the potential difference of the external circuit is arranged to be slightly higher than the e.m.f. of the cell, the following reactions take place:

$$Cu \rightarrow Cu^{++} + 2e \qquad \text{at copper electrode}$$

$$Cu^{++} + ZnSO \rightarrow Zn^{++} + CuSO \qquad \text{in solutions}$$

$$Zn^{++} + 2e \rightarrow Zn \qquad \text{at zinc electrode.}$$

If the quantity of electricity flowing into the cell is exactly the same as the decrease in charge when the cell was discharging, then the chemical state of the cell will be restored (neglecting polarization currents).

For an ideal reversible cell, therefore, the thermodynamic properties are the e.m.f., ε_R, the charge, Z, and the usual properties of pressure, p, volume, V, and temperature, T.

The Daniell cell is a battery in which the reactants are permanently stored. If some of the reactants continuously flow into and out of the cell the system is called a fuel cell.

Fuel Cell

A fuel cell is a battery in which some of the reactants (usually the active electrodes) are continuously flowing into and out of the cell. For a given size a conventional reversible cell or battery will yield a given quantity of electricity between charging, but the fuel cell only produces electricity whilst the active reactants flow through the cell. A fuel cell may be a component of a heat-engine in which electrical energy is produced directly from chemical energy.

A diagrammatic drawing of a carbon-oxygen fuel cell is shown in Fig. 5.3.

Coal (carbon) is supplied at the anode where it interacts with the oxide ions in the electrolyte to form carbon dioxide (CO_2) and releases electrons to the external circuit. Electrons do work on the way to the cathode where they combine with oxygen from the air passing through the cathode. The oxygen ions thus formed "flow" through the electrolyte to the anode completing the circuit. In an efficient cell the oxygen ions carry all the current and the electrolyte composition remains constant. Using the same convention as for the reversible Daniell cell the electrical current "flows" from the cathode to the anode in the external circuit. (The convention for the Daniell cell states that the current flows from a high potential to a low potential, that is in the opposite direction to the electron drift.) In the ideal situation the reactions in the fuel cell will be

$$C+2O^{--} = CO_2+4e \qquad \text{at anode}$$

$$O_2+4e = 2O^{--} \qquad \text{at cathode.}$$

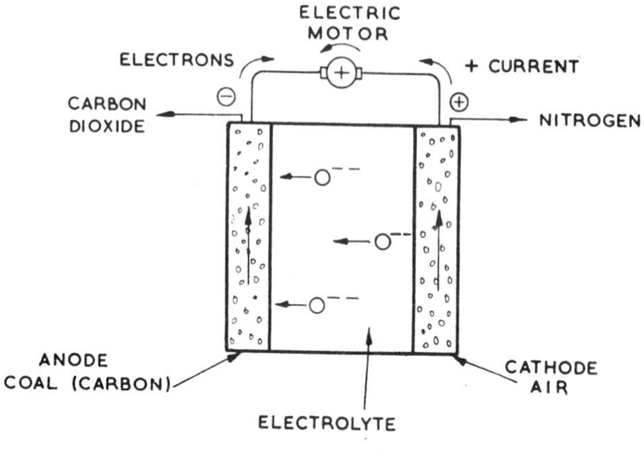

FIG. 5.3. Carbon–oxygen fuel cell.

The fuel (carbon) enters and the products CO_2 leave the anode, air enters and nitrogen leaves the cathode.

Many practical problems must still be overcome to make the simple fuel cell a viable economic product and intensive research is being carried out to produce a low cost efficient fuel cell. One of the promising types is a hydrogen-oxygen fuel cell. In Fig. 5.4 a concentric hydrogen-air fuel cell developed by Union Carbide is shown.

The electrolyte is a 30 per cent potassium hydroxide solution. The cell produces electricity as soon as hydrogen is fed into the inner porous carbon tube. The outer carbon tube is exposed to air. Oxygen from the air diffuses through the carbon tube. The hydrogen reacts on catalytically active sites on the electrode surface (a platinum group catalyst is used). The catalyst converts the diatomic hydrogen gas, H_2, to the atomic form, 2H, which is absorbed on the catalyst surface. The absorbed H reacts with the OH^- radical in the electrolyte to produce electrons. The electrons "flow" from the hydrogen electrode through the external circuit. In the electrolyte the OH^- ions migrate from the oxygen electrode carrying the current. The oxygen absorbed on the carbon element combines with the electrons and water to produce the OH^- ions† at the anode. A full description of various types of fuel cell is given in _Fuel Cells_††.

†Experiments show that both OH^- and HO_2^- radicals are produced with carbon electrodes.

††_Fuel Cells_, edited by George J. Young, Reinhold Publishing Corp., 1960.

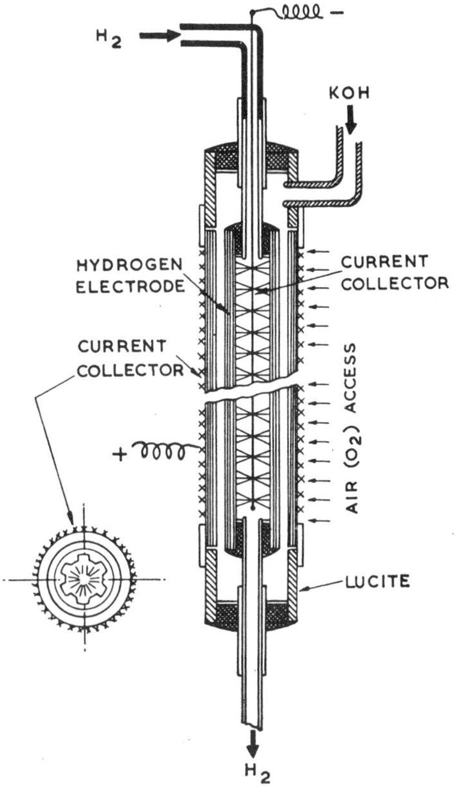

FIG. 5.4. Concentric hydrogen–air fuel cell. (After Young, *Fuel Cells* Reinhold, 1960.)

The thermodynamics of reversible cells are equally applicable to the conventional battery (Daniell cell) and the fuel cell.

The work done by a reversible cell due to a flow of electricity is

$$đW = p\,dV - \varepsilon_R dZ.$$

The first and second laws of thermodynamics give

$$dU = đQ - đW = đQ - p\,dV + \varepsilon_R dZ$$

$$\therefore \quad dU = T\,dS - p\,dV + \varepsilon_R dZ$$

or

$$\varepsilon_R dZ = dU - T\,dS + p\,dV. \tag{5.56}$$

For most applications, cells operate at constant pressure and temperature. If the initial and final states of a cell are 1 and 2 then for a change in charge ΔZ, we have on integration of (5.56)

$$\varepsilon_R \Delta Z = (U_2 - U_1) - T(S_2 - S_1) + p(V_2 - V_1)$$

or

$$\varepsilon_R \Delta Z = (U_2 - TS_2 + pV_2) - (U_1 - TS_1 + pV_1)$$

or

$$\varepsilon_R \Delta Z = G_2 - G_1 \qquad (5.57)$$

where G is the Gibbs function (H-TS).

For a cell which is discharging ΔZ is negative and G_1 is greater than G_2. If the cell is charging G_2 is greater than G_1. The product $-\varepsilon_R \Delta Z$ is the electrical work dW_ε (for positive work by the cell).

Hence

$$dW_\varepsilon = -dG. \qquad (5.58)$$

Now the electrical work dW_ε is related to the e.m.f., ε_R, and the change in charge, dZ, by equation (5.54); and the change in charge is related to the chemical reaction by equation (5.53). We thus have on simplification

$$-dG = Fj\varepsilon_R \, dM. \qquad (5.59)$$

The magnitude of F in thermodynamic units is 23,052 calories per volt† if the system of units is gram-mol, calories, °K.

Although the chemical reaction in the cell is by electrolysis, we can use the thermodynamic relations for systems of variable composition developed in Chapter 4 to calculate the e.m.f. of the cell.

The process is represented by equation (4.29) (p.124):

---

†Electrical energy = Joules = Coulombs x Volts.

4.186 Joules = 1 calorie.

Hence the energy in calories $= \dfrac{\text{coulombs}}{4.186}$ x volts.

$$\frac{96{,}500 \text{ coulombs}}{4.186 \text{ Joules/cal}}$$

$$= 23{,}052 \text{ cals/volt.}$$

$$dG = V\,dp - S\,dT + \sum_{i=1}^{i=k} \mu_i\, dM_i. \qquad (4.29)$$

For a reaction at constant pressure and temperature:

$$dG = \sum_{i=1}^{i=k} \mu_i\, dM_i.$$

Hence from (5.59)

$$Fj\varepsilon_R dM = -\sum_{i=1}^{i=k} \mu_i\, dM_i \qquad (5.60)$$

or

$$\varepsilon_R = \frac{\sum_{i=1}^{i=k} \mu_i\, dM_i}{Fj dM}. \qquad (5.61)$$

In calculating the e.m.f. ε_R care must be taken to specify the direction of the chemical reaction. An example will be given to illustrate the method.

A simple fuel cell consists of hydrogen and oxygen electrodes and potassium hydroxide electrolyte. The cell can be represented by

$$H_2|KOH|O_2.$$

The basic electrolysis reactions are

$H_2 + 2OH^- = 2H_2O + 2e$ †at hydrogen electrode (cathode)

$\frac{1}{2}O_2 + H_2O + 2e = 2OH^-$ †at oxygen electrode (anode)

The overall chemical reaction is

$$H_2 + \tfrac{1}{2}O_2 = H_2O.$$

The KOH does not change its composition. The valency of hydrogen is 2.

The change in Gibbs function for the above reaction, for dM mols of H_2, is

---

†The cathode is the positive electrode using the conventional method, the anode is the negative electrode using the conventional method.

$$\sum_{i=1}^{i=k} \mu_i \, dM_i = \left(\mu_{H_2O} - \mu_{H_2} - 0.5\mu_{O_2}\right) dM.$$

Note the "nominal" reactants have "negative" stoichiometric coefficients. Hence the e.m.f. of the cell will be

$$\varepsilon_R = \frac{\left(\mu_{H_2} + 0.5\mu_{O_2} - \mu_{H_2O}\right)}{2F} \tag{5.62}$$

The numerator of equation (5.62) is equal to $(-\Delta G_T)$ for the reaction $H_2 + 0.5O_2 = H_2O$ and

$$\varepsilon_R = \frac{-\Delta G_T}{2F} \tag{5.63}$$

At temperature of 298°K, ΔG_T equals -54,635 cal/g mol when the partial pressures of H_2, O_2 and H_2O are all equal to 1 atm. Substituting into (5.63) the e.m.f. of the cell is

$$\varepsilon_R = \frac{54,635}{2 \times 23,052} = 1.17 \text{ volts}$$

The general expression for the e.m.f. of any type of reversible cell is

$$\varepsilon_R = \frac{-\Delta G_T}{jF}. \tag{5.64}$$

For gas cells the e.m.f. developed is dependent on both the partial pressures of the reactants and the products and the cell temperature. This will be readily seen by using the methods discussed in Chapter 4:

$$\mu = \mu^o + R_{mol} T \ln p \tag{4.36}$$

and $$\Delta G_T = (\nu\mu)_{products} - (\nu\mu)_{reactants}.$$

Hence $$\Delta G_T = \left[(\nu\mu^o)_P - (\nu\mu^o)_R\right] + R_{mol} T \ln \left[\frac{\Pi(p^\nu)P}{\Pi(p^\nu)R}\right]$$

or $$\Delta G_T = \Delta G_T^o + R_{mol} T \ln \left[\frac{\Pi(p^\nu)_{products}}{\Pi(p^\nu)_{reactants}}\right] \tag{5.65}$$

where $-\Delta G_T^o$ is the free energy of formation for the chemical reaction. This corresponds to the difference in the standard Gibbs functions for the active reactants and products at temperature T, the partial pressures of each of the reactants and products being 1 atm abs. (1.01325 bar).

The e.m.f. of the cell is therefore

$$\varepsilon_R = \frac{\Delta G_T^o}{jF} - \frac{R_{mol}T}{jF} \ln \left[\frac{\Pi(p^\nu)_{products}}{\Pi(p^\nu)_{reactants}}\right]. \tag{5.66}$$

For the fuel cell $H_2|KOH|O_2$ the e.m.f. will be

$$\varepsilon_R = -\frac{\Delta G_T^o}{jF} + \frac{R_{mol}T}{jF} \ln \frac{p_{H_2}\sqrt{p_{O_2}}}{p_{H_2O}} \tag{5.67}$$

The first term of the above expression is related to the equilibrium constant K_p

$$-\frac{\Delta G_T^o}{jF} = \frac{R_{mol}T}{jF} \ln K_p.$$

Hence (5.66) may be written in the form

$$\varepsilon_R = \frac{R_{mol}T}{jF}\left[\ln K_p + \ln \frac{\Pi(p^\nu)_{reactants}}{\Pi(p^\nu)_{products}}\right]. \tag{5.68}$$

If the partial pressures of the reactants and products correspond to the equilibrium conditions, $dG)_{p,T} = 0$, then $\varepsilon_R = 0$.

For a <u>fuel</u> cell, operating as a <u>direct conversion device</u>, the maximum external work performed will be

$$dW_\varepsilon = -dG$$

or

$$W_\varepsilon = G_1 - G_2 = -(\Delta H - T\Delta S) = -\Delta G_T. \tag{5.69}$$

If we define the ideal efficiency η_I of a fuel cell as

$$\eta_I = -\frac{\text{Maximum work}}{\text{Heat of reaction}} \times 100^\dagger$$

$$\eta_I = -\left(\frac{W_\varepsilon}{Q_p}\right) \times 100 = \left(\frac{\Delta G_T}{Q_p}\right) \times 100$$

†The negative sign is used since Q_p will be negative.

$$\eta_I = \left(\frac{\Delta H - T\Delta S}{Q_p}\right) \times 100 = \left(1 - \frac{T\Delta S}{Q_p}\right) \times 100 \qquad (5.70)$$

since $\Delta H = Q_p$.

The heat of reaction, Q_p, will be negative. For most reactions Q_p is nearly constant. If the $T\Delta S$ term is positive an ideal efficiency, greater than 100 per cent may be obtained: in this case heat will be received from the surroundings. If $T\Delta S$ is negative the fuel cell efficiency will decrease with increase in operating temperature and heat will be rejected to the surroundings†.

In practice, fuel cells do not operate under isothermal conditions and there are heat losses in addition to those required for an isothermal ideal reaction. The actual e.m.f. on load is less than the reversible e.m.f. due to unwanted reactions in the cell, polarization effects at the anode and cathode, concentration gradients in the electrolyte and ohmic heating in the electrolyte.

Magnetic Systems

If a magnetic material is placed in a magnetic field of strength H the total induced magnetism B in the material is given by

$$B = \mu_o H + j \quad \text{weber/metre} \qquad (5.71)$$

where j is the intensity of magnetism (weber/m²), μ_o is permeability of free space (mks system of units).

The permeability (μ) is defined as

$$\mu = \frac{B}{\mu_o H} \qquad (5.72)$$

and the magnetic susceptibility $\bar{k}$ as

---

† The fuel cell does not operate on a complete thermodynamic cycle as, for example, closed cycle gas turbines, refrigerators or steam plants. The fuel cell is equivalent to some component in the cycle, its efficiency therefore is a component efficiency. For example, in a steam or gas turbine the efficiency of the expansion process is measured by the turbine efficiency, in the case of the fuel cell the ideal efficiency quoted above is equivalent to the turbine efficiency.

$$\bar{k} = \frac{j}{\mu_o H}. \qquad (5.73)$$

Hence

$$\mu = 1 + \bar{k}.$$

If μ is very much greater than one for a material, then the material is said to be ferromagnetic (e.g. steel, iron, nickel). If μ is of the order of unity, the material is weakly magnetic. If μ is slightly greater than unity the material is paramagnetic; if it is slightly less than unity the material is diamagnetic.

In general, paramagnetic materials become magnetic when located in a magnetic field. When the field is removed they are non-magnetic. Hence the intensity of magnetization j is not a constant but depends on the field strength H.

Consider a toroid (Fig. 5.5) of length L consisting of a coil of N turns wound round a magnetic material. If the current I is flowing in the wire the field strength is given by

$$H = \frac{NI}{L} \text{ ampere turns/metre.} \qquad (5.75)$$

The back e.m.f. due to the change in flux of the magnetic field is given by Faraday's Law

$$\varepsilon = -N \frac{d\phi}{dt} \qquad (5.76)$$

where ϕ = total flux = BA.

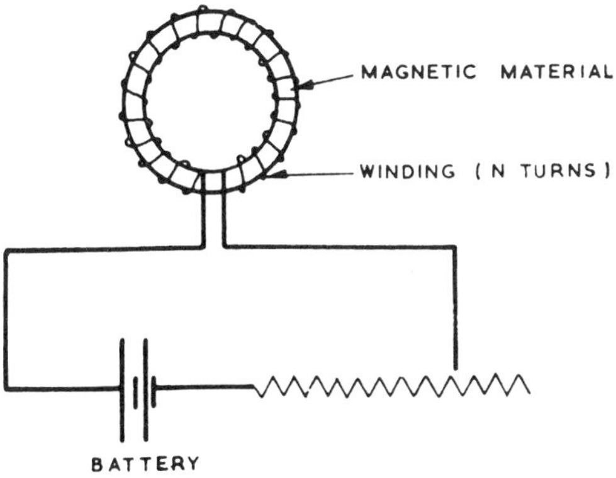

FIG. 5.5. Toroid.

The rate of doing work by the system, against the back e.m.f., to maintain a current I is

$$\frac{đW}{dt} = \varepsilon I = -IN\frac{d\phi}{dt} \tag{5.77}$$

or

$$đW = -INA\ dB. \tag{5.78}$$

If the current I flowing in the wire is changed and causes an <u>increase</u> in the magnetic induction B, work will be done <u>on</u> the system by the surroundings.

For a thermally insulated system the first law of thermodynamics gives

$$đQ = dU + đW = 0$$

$$dU = -đW = INA\ dB. \tag{5.79}$$

Hence the internal energy of the system will increase with increase in the magnetic induction B.

Now from (5.71)

$$dB = \mu_o\ dH + dj \tag{5.80}$$

and from (5.75)

$$INA = ALH = VH. \tag{5.81}$$

Substituting in (5.78) the work done by the system is

$$đW = -VH(\mu_o\ dH + dj). \tag{5.82}$$

We observe that for an increase in H or j work will be done <u>on</u> the system.

The magnetization of the material J (or the magnetic moment) is equal to the product of the intensity of magnetization j and the volume V, assuming uniform magnetization, and

$$J = Vj \tag{5.83}$$

$$đW = -(V\mu_o H\ dH + H\ dJ). \tag{5.84}$$

The first term is the work necessary to increase the magnetic field in a vacuum of volume V, the second term is the work required to increase the magnetization J of the material. The property J is an extensive property, whilst H, the applied field, is an intensive property. For paramagnetic materials, therefore, in order to increase the magnetization, the work required is

$$đW = -H\ dJ. \tag{5.85}$$

The properties of paramagnetic salts are particularly useful in cryogenics. Experiments show that at atmospheric pressure there are only minute volume changes in these salts when magnetized. At room temperatures an approximate equation of state is Curie's Law.

$$T = \frac{\mu_o C_1 H}{\bar{j}} = \frac{\mu_o C_2 H}{j} = \frac{C_2}{\bar{k}} \tag{5.86}$$

where either C_1 or C_2 is Curie's constant in the appropriate units, $\bar{j}$ is the specific magnetization (J/M), j is the intensity of magnetization, k is the susceptibility, Curie's Law is satisfactory down to about 1^oK, but below this temperature more complex equations must be used. The ratio $C_2/\bar{k}$ is called the magnetic temperature T^*. Above 1^oK the magnetic and thermodynamic temperatures are equal, below 1^oK the two temperatures are not equal, but the thermodynamic temperature T can be derived from measured values of the magnetic temperature T^*.

For reversible magnetization of paramagnetic solids the first and second laws of thermodynamics give

$$dQ = T\,dS = dU + đW = dU - H\,dJ. \tag{5.87}$$

In specific quantities

$$du = T\,ds + H\,d\bar{j}, \quad u = u(s,\bar{j}). \tag{5.88}$$

The "magnetic enthalpy" $\bar{h}$ is defined as

$$\bar{h} = u - H\bar{j} \tag{5.89}$$

$$dh = du - Hd\bar{j} - \bar{j}dH = T\,ds - \bar{j}dH, \quad \bar{h} = \bar{h}(s,H). \tag{5.90}$$

If we compare equations I(a) and II(a) (p.35) with (5.88) and (5.90) respectively, we see that

$$H \equiv -p, \quad \bar{j} \equiv v. \tag{5.91}$$

Now for a reversible magnetization dq = T ds and

$$du = dq + Hd\bar{j}. \tag{5.92}$$

$$d\bar{h} = dq - \bar{j}dH. \tag{5.93}$$

The specific heat at constant $\bar{j}$ is given by

$$C_J = \left(\frac{\partial q}{\partial T}\right) = \left(\frac{\partial u}{\partial T}\right) \text{ from (5.92)} \tag{5.94}$$

The specific heat at constant H is given by

$$C_H = \left(\frac{\partial q}{\partial T}\right)_H = \left(\frac{\partial \bar{h}}{\partial T}\right)_H \quad \text{from (5.93).} \tag{5.95}$$

The corresponding specific heats for the magnetic system and the gaseous system are

$$C_H \equiv C_p; \quad C_J \equiv C_v. \tag{5.96}$$

A complete set of thermodynamic relations can be developed for magnetic systems in the same manner as described for p, v, T systems in Chapter 2. The second Tds equation ((2.57) p.40) for a gaseous system is

$$T\,ds = C_p\,dT - T\left(\frac{\partial v}{\partial T}\right)_p dp. \tag{2.57}$$

Using (5.91) and (5.96) the second T ds equation for a paramagnetic solid is

$$T\,ds = C_H dT + T\left(\frac{\partial \bar{J}}{\partial T}\right)_H dH. \tag{5.97}$$

For a reversible isothermal process, dT = 0, the heat exchanged per mol of magnetic material is

$$q = \int T\,ds = T\int_{H_1}^{H_2}\left(\frac{\partial \bar{J}}{\partial T}\right)_H dH. \tag{5.98}$$

From Curie's Law (5.86)

$$\left(\frac{\partial \bar{J}}{\partial T}\right)_H = -\frac{\mu_o C_1 H}{T^2} \tag{5.99}$$

and

$$q = -\frac{\mu_o C_1}{T}\int_{H_1}^{H_2} H dH$$

$$q = -\frac{\mu_o C_1}{2T}(H_2^2 - H_1^2). \tag{5.100}$$

Thus, heat is rejected if the magnetic field is increased from H_1 to H_2 and vice versa if the magnetic field is decreased. Although Curie's Law does not apply for temperatures below 1°K, experiments show that the differential coefficient $(\partial\bar{J}/\partial T)_H$ is negative in this region, and the above remarks are applicable over the range about absolute zero to 1°K.

For a reversible adiabatic change in field, (the process is called adiabatic magnetization in the literature but more strictly isentropic magnetization), ds = 0 and equation (5.97) becomes

$$C_H\, dT = -T \left(\frac{\partial \bar{J}}{\partial T}\right)_H dH. \qquad (5.101)$$

Above $1^{\circ}K$ we can substitute (5.99) for the differential coefficient $(\partial\bar{J}/\partial T)_H$ and,

$$C_H\, dT = \frac{\mu_o C_1 H\, dH}{T}$$

or

$$\int_{T_1}^{T_2} T\, dT = \mu_o C_1 \int_{H_1}^{H_2} \frac{H dH}{C_H}. \qquad (5.102)$$

The specific heat C_H is always positive. Therefore for an <u>increase</u> in field from H_1 to H_2 the right-hand side of expression (5.102) is positive and T_2 is greater than T_1. If the field is <u>decreased</u> the temperature T_2 is less than T_1. The same effects are present below $1^{\circ}K$. Summarizing, therefore, for adiabatic <u>magnetization</u> the temperature of the material <u>increases</u> and for adiabatic <u>demagnetization</u> the temperature <u>falls</u>. This phenomenon is known as the <u>magnetocaloric</u> effect.

By a combination of reversible isothermal magnetization and adiabatic demagnetization processes it is possible to produce very low temperatures using paramagnetic salts. In practice this technique is used as the last stage of a cryogenic plant for temperatures below $1^{\circ}K$. The first two stages incorporate liquefaction by the Joule-Thomson effect (p.48) and the rapid vaporization of liquid helium. The final stage is a combination of isothermal magnetization and adiabatic demagnetization of a paramagnetic salt.

A diagrammatic arrangement of an apparatus used in the Royal Society Mond Laboratory is shown in Fig. 5.6.

The paramagnetic salt is suspended in a vessel containing some helium gas to establish thermal contact between the salt and the walls. The vessel is surrounded by liquid helium boiling under reduced pressure. When the magnetic field is switched on the temperature of the salt rises, but the heat is rapidly conducted away by the helium gas so that the salt is left in the magnetic field at the temperature of the helium bath. The helium gas is then pumped away and the salt is permanently isolated. The magnetic field is then removed. The temperature of the salt falls to T_3. If the substance obeys Curie's Law before magnetization at temperature T_1, then the magnetic temperature T_3^* can be calculated, if the susceptibility $(\bar{k})$ of the salt is measured at states 1 and 3. (For an excellent description of methods of producing very low

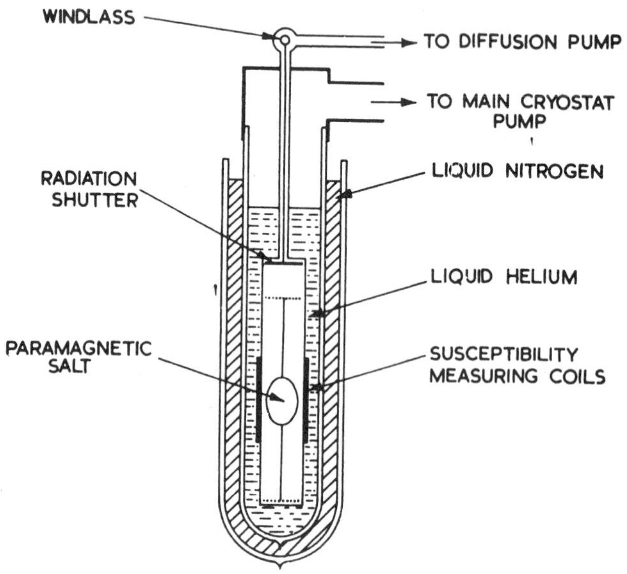
WINDLASS
TO DIFFUSION PUMP
TO MAIN CRYOSTAT PUMP
RADIATION SHUTTER
LIQUID NITROGEN
LIQUID HELIUM
PARAMAGNETIC SALT
SUSCEPTIBILITY MEASURING COILS

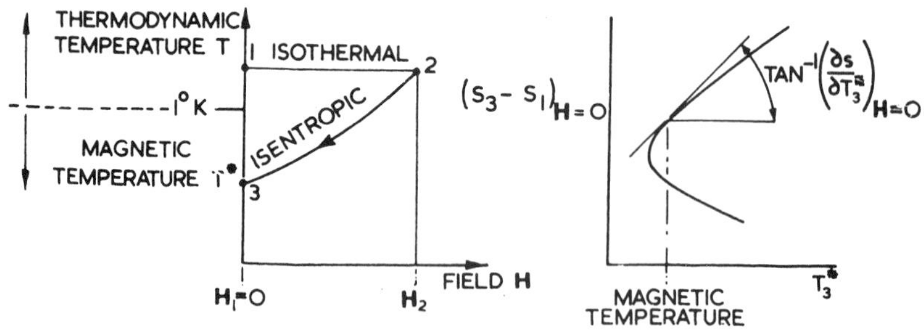
THERMODYNAMIC TEMPERATURE T
1 ISOTHERMAL
2
1° K
ISENTROPIC
MAGNETIC TEMPERATURE T*
3
FIELD H
H1=O
H2
(S3 - S1)H=O
TAN-1(∂s/∂T3*)H=O
MAGNETIC TEMPERATURE
T3*

FIG. 5.6.

temperature see Roberts and Miller.)†

From Curie's Law and the definition of magnetic temperature

$$T_3^* = \frac{\overline{k}_1}{\overline{k}_3} T_1. \tag{5.103}$$

To obtain the thermodynamic temperature T_3 from the magnetic temperature T_3^* a series of experiments over a range of temperatures is required. A brief outline follows.

At zero field, (H=0), equation (5.88) gives

$$T = \left(\frac{\partial u}{\partial s}\right)_{H=0}. \tag{5.104}$$

Now

$$\left(\frac{\partial u}{\partial s}\right)_{H=0} = \left(\frac{\partial u}{\partial T^*}\right)_{H=0} \left(\frac{\partial T^*}{\partial s}\right)_{H=0}.$$

Hence

$$T = \frac{\left(\frac{\partial u}{\partial T^*}\right)_{H=0}}{\left(\frac{\partial s}{\partial T^*}\right)_{H=0}}. \tag{5.105}$$

We can define the specific heats for paramagnetic materials in terms of the thermodynamic temperature or the magnetic temperature. For a paramagnetic salt the specific heat in terms of the magnetic temperature is $C_H^* = (\partial q/\partial T^*)_H$. The numerator in expression (5.105) is therefore the specific heat C_H^* at $H = 0$.†† To obtain C_H^* it is necessary to supply energy to the salt with zero field at various values of T_3^*. A number of methods have been used including induction heating, resistance heating and gamma rays. By using these techniques, the variation of $(C_H^*)_{H=0}$ with T^* can be obtained.

The denominator of equation (5.105) is obtained directly from an adiabatic demagnetization experiment. Referring to Fig. 5.6, the process 1 to 2 corresponds to the first stage of isothermal magnetization and 2 to 3 to the second stage of adiabatic demagnetization.

For 1 mol of paramagnetic salt

$$s_3 - s_1 = s_2 - s_1. \tag{5.106}$$

---

†<u>Heat and Thermodynamics</u>, by Roberts and Miller - 6th edition published by Blackie & Sons, 1960.

††At zero field $\overline{h} = u$ and $q = u$.

From (5.97) for a reversible isothermal process

$$T\,ds = T\left(\frac{\partial \bar{J}}{\partial T}\right)_H dH \tag{5.107}$$

and

$$s_2 - s_1 = \int\left(\frac{\partial \bar{J}}{\partial T}\right)_H dH. \tag{5.108}$$

For a given field strength H the coefficient $(\partial\bar{J}/\partial T)_H$ can be evaluated from the equation of state for the salt, and the entropy difference can be computed from (5.108). From (5.106) the entropy difference is equal to the entropy difference $s_3 - s_1$ at H = 0 for a given magnetic temperature T_3^*. If the temperature T_1 is fixed then s_1 is fixed, and s_3 will vary with T^* if the field H is varied. A series of experiments can be performed with fixed T_1 but with variable H and the results plotted in the form $s_3 - s_1$ against T_3^*, as shown in Fig. 5.6. Since the entropy s_1 is fixed and the state 3 is variable, the slope of the graph at temperature $T^* = T_3^*$ will be

$$\left(\frac{\Delta(s_3 - s_1)}{\Delta T_3^*}\right)_{\substack{H=0\\ T^*=T_3}} = \left(\frac{\partial s}{\partial T^*}\right)_{H=0}. \tag{5.109}$$

At any given magnetic temperature T_3^* the corresponding thermodynamic temperature T_3 is then

$$T_3 = \left(\frac{(C_H^*)_{H=0}}{\left(\dfrac{\partial s}{\partial T^*}\right)_{H=0}}\right)_{T^*=T^*}. \tag{5.110}$$

Steady State or Irreversible Thermodynamics

We commenced our study of thermodynamics by examining the equilibrium of thermodynamic systems. It was stated that a system was in equilibrium when no spontaneous process took place and all the thermodynamic properties remained unchanged. Furthermore, we stated that the properties were uniform throughout the system. At equilibrium, the derivative dS/dt is zero. These concepts correspond to a system in which the macroscopic properties are spatial and time invariant.

Let us examine the system shown in Fig. 5.7.

A rod is connected to two large reservoirs at temperatures T_2 and T_1 respectively. The walls of the rod are thermally insulated so that heat flow is by conduction in the longitudinal direction. If the reservoirs are large compared with the

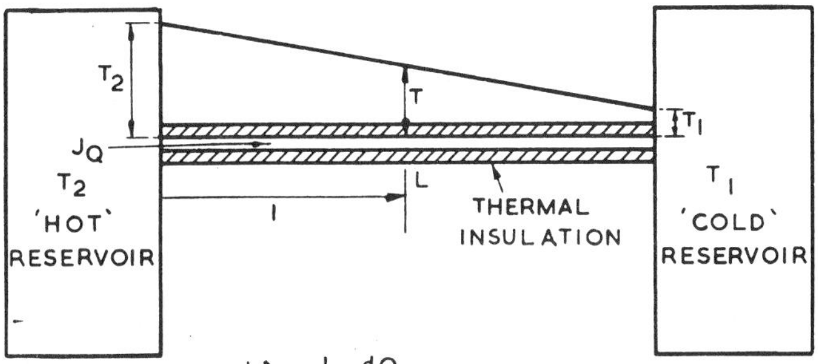

$$J_Q^{\bullet} = \frac{1}{A}\frac{dQ}{dt}$$

FIG. 5.7.

dimensions of the rod, the rate of heat flow dQ/dt entering the rod will equal the rate of heat flow leaving the rod - dQ/dt. If we were to insert a thermometer at any point in the rod the thermometer reading will not change with time, but on the other hand for each point in the bar the thermometer reading will be different. For a bar of uniform cross section the thermometer reading will vary linearly with the distance from one reservoir, the highest reading at the end adjacent to the reservoir at temperature T_2.

If the heat flow into the bar at the left-hand end is with an infinitesimal temperature difference, the left-hand reservoir loses entropy at the rate

$$\frac{dS_2}{dt} = -\frac{dQ}{dt}\cdot\frac{1}{T_2}.$$

At the right-hand end, if the heat flow is with an infinitesimal temperature difference, the right-hand reservoir gains entropy at the rate

$$\frac{dS_1}{dt} = \frac{dQ}{dt}\cdot\frac{1}{T_1}.$$

The total change of entropy is

$$\frac{dS}{dt} = \frac{dS_1}{dt} + \frac{dS_2}{dt} = \frac{dQ}{dt}\left(\frac{1}{T_1} - \frac{1}{T_2}\right) = \frac{dQ}{dt}\left(\frac{T_2-T_1}{T_1T_2}\right) \qquad (5.111)$$

The experiment has shown that thermometer readings in the bar are constant with time, but differ from point to point. Yet there is a net increase of entropy with time since T_2 is greater than T_1.

Before proceeding further, we must examine the meaning of the thermometer readings. It is evident from the experiment that these readings are constant with time at each point yet variable with distance from either end. At each point the thermometer will be in equilibrium with the volume of the rod in contact with the thermometer. Hence, the thermometer will read the "temperature". On the other hand, it is clear that the reading will depend on the size of the sensing element since the thermometer readings vary with distance. Under these conditions we define the temperature at a point as the final equilibrium temperature of an isolated small element, enclosing the point in contact with the recording device. The volume element is small compared with the dimensions of the system but large enough to avoid molecular fluctuations. The other thermodynamic properties such as pressure, density and entropy are defined in the same way.

Now reverting to the system we see that at each point in the rod the entropy is invariant with time; however, there is a net transfer of entropy from the left hand reservoir to the right-hand reservoir, i.e. entropy is "flowing" along the rod. The thermodynamic properties of this system are time independent, but the system is not in thermodynamic equilibrium since entropy is produced. To distinguish the thermodynamics of these systems from those in which there is both spatial and temporal invariance of the macroscopic properties, the study of thermodynamics is usually subdivided into non equilibrium thermodynamics and thermostatics, the former subdivision referring to temporal invariant systems and the latter to spatial and temporal invariant systems. An alternative heading for non-equilibrium thermodynamics is Irreversible Thermodynamics whilst Denbigh has suggested "The Thermodynamics of the Steady State".†

"Steady state" systems are associated with transport processes. Typical processes are heat conduction, electrical current flow, diffusion. The distinguishing feature of the thermodynamics of these systems is that, if two or more of these processes can occur simultaneously, they may influence each other. It is not necessary for the simultaneous processes to be imposed separately; it may be possible for the initiation of one process to cause the other to occur spontaneously. For example, in a thermocouple the difference in the temperatures at the junctions at each end of a pair of wires causes electrical current to flow in the wires. In this case we have the coupled transport process of heat conduction and electrical current flow. The selection of the coupling processes depends on experimental experience and the laws of these processes are at present axiomatic - that is they cannot be proved except by experiment.

It has been suggested that in the development of the relations used in the thermodynamics of the steady state, the

---

†Thermodynamics of the Steady State, by K.G. Denbigh, published by Methuen, 1951.

the study should proceed on analogous lines to the study of the dynamics of particles. For this purpose the concepts of thermodynamic velocities or flows and thermodynamic forces have been introduced. The thermodynamic forces are represented by X and the velocities by J. The relationship between J and X is of the form

$$J = LX \tag{5.112}$$

where L is a scalar quantity and is a function of the state of the system. It is sometimes called a transport property of the system.

The form of (5.112) is based on experiment. The selection of the appropriate parameters for J, L and X depend on simple rules to be discussed shortly. Equation (5.112) is limited to systems not far removed from "thermodynamic equilibrium", i.e. the conventional macroscopic thermodynamic properties (pressure, temperature, density, etc.) must be meaningful.

Simple types of relationships which illustrate (5.112) are Fourier's equation for heat flow,

$$\frac{dq}{dt} = -kA\,\frac{dT}{dl} \tag{5.113}$$

where k is the thermal conductivity of the material and Ohm's law,†

$$I = -\lambda A\,\frac{d\varepsilon}{dl} \tag{5.114}$$

where λ is the electrical conductivity of the material.

The selection of the thermodynamic velocity J is generally straightforward; thus in equations (5.113) and (5.114), respectively,

$$J = J_Q = \frac{1}{A}\frac{dq}{dt}$$

$$J = J_I = \frac{I}{A}.$$

The selection of X and L will be dependent on the nature of the problem. For example, we can make $X = -dT/dl$ in (5.113) and L will be k, or we can make $X = -(1/T)\,(dT/dl)$ and L will be kT and so on. In all cases, however, the forces, X, are associated with the gradient of one of the properties, e.g. the temperature gradient in (5.113) or the potential gradient in (5.114).

---

†This form of Ohm's Law may appear to be unusual. For current I to flow along a wire $d\varepsilon/dl$ is negative, hence the negative sign in equation (5.114).

The rules for the selection of L and X have been formulated by Onsager† and are associated with the coupling of the transport processes. If two transport processes are coupled (for example, heat conduction and electricity in thermoelectricity) then Onsager states that the laws of the two processes can be expressed in the form

$$J_1 = L_{11}X_1 + L_{12}X_2 \tag{5.115}$$

$$J_2 = L_{21}X_1 + L_{22}X_2. \tag{5.116}$$

Thus, if we have two primary processes (say heat conduction and electricity) the basic or primary laws will be of the form

$J_1 = L_{11}X_1$ for process 1 alone (say heat conduction)

$J_2 = L_{22}X_2$ for process 2 alone (say electrical flow)

where X_1 and X_2 are the primary forces and L_{11} and L_{22} the primary coefficients.

If process 2 influences process 1, then the effect of 2 on 1 will be given by

$J_1 = L_{12}X_2$ (say the quantity of heat flow due to electrical potential).

Similarly, if process 1 influences process 2, the effect of 1 on 2 will be given by

$J_2 = L_{21}X_1$ (say the electrical current flow due to temperature gradient).

The latter two processes are called the coupled processes. The coefficients L_{12} and L_{21} are called the coupling coefficients. Onsager's equations (5.115) and (5.116) say that the primary and coupled process can be linearly superimposed to give the overall magnitude of the thermodynamic velocity or flow J.

The primary coefficients can be generally obtained by experiments in uncoupled processes. Onsager has suggested the coupling coefficients are related by the equation

$$L_{12} = L_{21}. \tag{5.117}$$

This is called Onsager's Reciprocal Relation.†† An

†L. Onsager, Reciprocal relations in irreversible processes, Phys. Rev., vol.37, p.405; vol.38, p.2265 (1931).

††Strictly speaking, since we are concerned with flows and forces, the equations we have developed relating J and X should be in vector form. We are using the cartesian form of these equations.

interpretation of the above relations in terms of microscopic reversibility has been given by Denbigh.† Expression (5.117) in terms of macroscopic thermodynamics is axiomatic and its application to problems is dependent on experience. It is, however, a powerful tool in the thermodynamics of the steady state.

If the coupling coefficients L_{12}, L_{21} are zero, the flows are dependent only on the primary forces and are not coupled.

For k transport phenomena the most general set of relations is

$$J_i = \Sigma_k L_{ik} X_k \tag{5.118}$$

and Onsager's reciprocal relation is

$$L_{ik} = L_{ki}. \tag{5.119}$$

It was pointed out that, although the selection of the thermodynamic velocity or flow J is fairly straightforward, there may be some latitude in the selection of the thermodynamic force X and hence the transport property L. Onsager has suggested that the force X_i should be selected in such a way that the sum of the products of each flow (J_i) and the appropriate (X_i) is equal to the rate of entropy production per unit volume of the system (θ) multiplied by the temperature (T) of the volume element in which the flows and forces have their particular values.

Thus

$$T\theta = J_1X_1 + J_2X_2 + ,\ldots, = \Sigma_i J_i X_i \tag{5.120}$$

or

$$\theta = \Sigma_i \frac{J_i X_i}{T}. \tag{5.121}$$ ††

It should be emphasized that in applying Onsager's equations (5.112) to (5.121) it has to be ascertained in every case whether these equations are a good approximation and in particular, it has to be shown that the forces X are proportional to the flows J for each transport process.

Before applying the above principles we will examine the concept of entropy flow rate per unit volume.

† Thermodynamics of the Steady State, by K.G. Denbigh, published by Methuen, 1951.

†† See footnote ††, page 242.

In the example illustrated in Fig. 5.7, at any point in the bar L, the entropy flux J_S is defined as the entropy flow rate per unit cross section†.

At any point L the entropy flow rate will be

$$\frac{dS}{dt} = \frac{d(Q/T)}{dt} = \frac{dQ}{dt}\frac{1}{T}.$$

The entropy flux is

$$J_S = \frac{1}{A}\frac{dS}{dt} = \frac{1}{A}\frac{dQ}{dt}\frac{1}{T}. \tag{5.122}$$

The heat flow rate J_Q is defined by

$$J_Q = \frac{dQ/dt}{A} \tag{5.123}$$

Hence

$$J_S = \frac{J_Q}{T}. \tag{5.124}$$

The rate of production of entropy per unit volume is

$$\theta = \frac{d(dS/dt)}{dV} = \frac{d(dS/dt)}{Adl} = \frac{d(1/A)(dS/dt)}{dl}$$

$$\theta = \frac{dJ_S}{dl}. \tag{5.125}$$

Substitute for J_S

$$\theta = \frac{d(J_Q/T)}{dl} = -\frac{J_Q}{T^2}\frac{dT}{dl}. \tag{5.126}$$

Expression (5.126) gives the rate of production of entropy per unit volume at the point L where the temperature is T, if there is heat conduction only. If the rod is of uniform temperature then $dT/dl = 0$ and $\theta = 0$. Since for heat conduction dT/dl is negative, θ is positive, as would be expected.

Another important case is the flow of electricity along a wire. Consider a wire in contact along its length with a reservoir at temperature T. Let an electric current $J_I (=I/A)$

---

†In time Δt the entropy "created" or produced by the bar will from (5.111) be

$$\Delta S = \frac{dS}{dt}\Delta t = \frac{dQ}{dt}\left(\frac{1}{T_1} - \frac{1}{T_2}\right)\Delta t.$$

Since the quantity of entropy entering the bar ΔS_2 was less than that leaving the bar ΔS_1 by the above value, ΔS is "created" or produced.

flow with a potential difference $d\varepsilon$, then since the wire is at constant temperature T the electrical work will be equal to the heat produced. The rate at which electrical work is done is equal to $-J_I A.d\varepsilon$ and the total heat produced is $\dot{Q}$ per second. Hence

$$\dot{Q} = -J_I \, A d\varepsilon. \tag{5.127}$$

The total entropy production in a volume dV is

$$\theta.dV = \frac{\dot{Q}}{T} = -\frac{J_I A \, d\varepsilon}{T} \tag{5.128}$$

The rate of entropy production per unit volume is therefore

$$\theta = -\frac{J_I}{T}\frac{d\varepsilon}{dl} \tag{5.129}$$

since

$$dV = A \, dl.$$

As before, since $d\varepsilon/dl$ is negative for electrical current flow, θ is positive and entropy is created. Equation (5.129) applies for electrical flow <u>only</u> with no heat conduction.

We are now in a position to apply the thermodynamics of the steady state to practical problems. We will look at thermo-electricity in the next section.

Thermoelectricity

Experiments show that if the junctions of two wires of dissimilar materials are at different temperatures, then an electrical current will flow in the wires, the heat flow and the electrical flow being coupled. This phenomenon and its reverse are of practical interest in the measurement of temperature and the direct conversion of heat to electricity. We will first develop the thermodynamic equations for a single wire and then proceed to apply these equations to two wires of dissimilar materials joined at the ends.

The heat flow along a wire is given by $J_Q = (dQ/dt)/A$, the electrical flow by $J_I = I/A$. We will use suffix 1 to refer to the heat flow processes and suffix 2, to the electrical flow processes. We will first assume that the two types of flow are uncoupled, that is, we will examine the heat flow without electrical flow and vice versa.

Heat Flow

For zero electrical flow $X_2 = 0$, $J_2 = J_I = 0$, and the rate of entropy production per unit volume is from (5.121) and (5.126)

$$\theta = \frac{J_1 X_1}{T} = \frac{J_Q X_1}{T} = -\frac{J_Q}{T^2}\frac{dT}{dl} \tag{5.130}$$

and

$$X_1 = -\frac{1}{T}\frac{dT}{dl}. \tag{5.131}$$

X_1 is the primary force for heat conduction.

Electrical Flow

For zero heat flow $X_1 = 0$, $J_1 = J_Q = 0$. Hence the rate of entropy production per unit volume is from (5.121) and (5.129)

$$\theta = \frac{J_2 X_2}{T} = \frac{J_I X_2}{T} = -\frac{J_I}{T}\frac{d\varepsilon}{dl} \tag{5.132}$$

and

$$X_2 = -\frac{d\varepsilon}{dl}. \tag{5.133}$$

X_2 is the primary force for electrical conduction.

Heat and Electrical Flow

For the coupled flow we use the Onsager relations (5.115) and (5.116). Substituting for J and X we obtain

Heat Flow

$$J_Q = -\frac{L_{11}}{T}\frac{dT}{dl} - L_{12}\frac{d\varepsilon}{dl}. \tag{5.134}$$

Electrical Flow

$$J_I = -\frac{L_{21}}{T}\frac{dT}{dl} - L_{22}\frac{d\varepsilon}{dl}. \tag{5.135}$$

The entropy flux is $J_S = J_Q/T$

$$J_S = -\frac{L_{11}}{T^2}\frac{dT}{dl} - \frac{L_{12}}{T}\frac{d\varepsilon}{dl}. \tag{5.136}$$

We have assumed that both the electrical and heat flow phenomena can be represented by empirical laws of the type $J = LX$. From experiment we know that at constant temperature Ohm's Law gives

$$I = -\lambda A \frac{d\varepsilon}{dl}$$

or

$$J_I = \frac{I}{A} = -\lambda \frac{d\varepsilon}{dl} \tag{5.137}$$

where λ is the conductivity of the wire at constant temperature. If in (5.135) dT is set to zero, then equating (5.135) and (5.137) for this case we have

$$(J_I)_T = -L_{22}\frac{d\varepsilon}{dl} = -\lambda\frac{d\varepsilon}{dl}$$

and

$$L_{22} = \lambda. \tag{5.138}$$

The ratio of the entropy flux to the electrical current flowing at constant temperature is called the entropy of transport. This can be considered as the entropy transported by the current and is given the symbol S^*.

From (5.135) and (5.136) for $dT = 0$

$$\left(\frac{J_S}{J_I}\right)_T = \frac{L_{12}}{TL_{22}} = S^*. \tag{5.139}$$

Hence

$$L_{12} = TS^*L_{22} = \lambda TS^* \tag{5.140}$$

from (5.138).

Using Onsager's reciprocal relation (5.117) or (5.119) the coupling coefficients are

$$L_{21} = L_{12} = \lambda TS^* \tag{5.141}$$

Substituting for L_{12}, L_{22}, L_{21} in (5.134) and (5.135) we have

$$J_Q = -\frac{L_{11}}{T}\frac{dT}{dl} - \lambda TS^*\frac{d\varepsilon}{dl} \tag{5.142}$$

$$J_I = -\lambda S^*\frac{dT}{dl} - \lambda\frac{d\varepsilon}{dl}. \tag{5.143}$$

To obtain L_{11} we use the experimental law for heat conduction. From Fourier's Law at zero electrical current flow ($J_I = 0$)

$$J_Q = \frac{dQ/dt}{A} = -k\frac{dT}{dl} \tag{5.144}$$

and from (5.143) for zero electrical current flow $J_I = 0$,

$$\frac{d\varepsilon}{dl} = -S^*\frac{dT}{dl}. \tag{5.145}$$

Substituting (5.144) and (5.145) into (5.142) we have

$$-\frac{k\,dT}{dl} = -\frac{L_{11}}{T}\frac{dT}{dl} + TS^{*2}\frac{dT}{dl}$$

whence

$$L_{11} = (k + \lambda TS^{*2})T. \qquad (5.146)$$

The coupling equations are

<u>Heat Flow</u>

$$J_Q = -(k + \lambda TS^{*2})\frac{dT}{dl} - \lambda TS^{*}\frac{d\varepsilon}{dl}. \qquad (5.147)$$

<u>Electrical Flow</u>

$$J_I = -\lambda S^{*}\frac{dT}{dl} - \lambda\frac{d\varepsilon}{dl}. \qquad (5.148)$$

<u>Entropy Flux</u>

$$J_S = -\frac{(k + \lambda TS^{*2})}{T}\frac{dT}{dl} - \lambda S^{*}\frac{d\varepsilon}{dl}. \qquad (5.149)$$

Equations (5.147) to (5.149) are the governing thermodynamic equations for thermoelectricity.

The heat of transport Q^* is defined as

$$Q^* = \left(\frac{J_Q}{J_I}\right)_{T=\text{constant}}.$$

From (5.148) and (5.147) it will be seen that for $dT = 0$

$$\left(\frac{J_Q}{J_I}\right)_{T=\text{constant}} = TS^*.$$

Hence

$$Q^* = \left(\frac{J_Q}{J_I}\right)_{T=\text{constant}} = TS^*. \qquad (5.150)$$

(Compare this expression with the thermodynamic equation $Q = T\Delta S$ for an isothermal reversible process.)

The <u>heat of transport</u> is the heat transported by the electrical current at a constant wire temperature.

For zero current flow ($J_I = 0$) we have from (5.148)

$$\frac{d\varepsilon}{dl} = -S^* \frac{dT}{dl} \tag{5.151}$$

and

$$\left(\frac{d\varepsilon}{dT}\right)_{J_I=0} = -S^*. \tag{5.152}$$

For most materials the entropy of transport, S^*, has a non-zero value. It will be seen that for these materials, when there is a temperature difference along the wire, there will be an e.m.f. generated even if there is no electrical current flow. This is called the Seebeck effect and is the principle upon which the thermocouple is based. If electrical current J_I is flowing in the wire it will be seen, from (5.148), that the potential difference will depend on the quantity of electricity flowing and the temperature difference along the wire. The Seebeck effect is also associated with this condition. If the wire temperature is constant there will be heat flow due to the e.m.f. This heat flow is greater than the normal ohmic heating† and is called the Thomson heat (or effect).

The Thermocouple

A thermocouple is a device for recording the temperature at a point. A diagrammatic arrangement of a thermocouple is shown in Fig. 5.8.

---

†The Ohmic or Joulean heating is equal to the electrical energy dissipated in the wire. Consider a wire of length Δl and cross-sectional area A. With current I flowing along the wire, let the drop in e.m.f. be $\Delta\varepsilon$; then the heat dissipated will be:

$$\frac{dQ}{dt} = I\Delta\varepsilon;$$

now from (5.137)

$$I = -\lambda A \frac{d\varepsilon}{d}$$

$$\frac{d\varepsilon}{dl} = -\frac{\Delta\varepsilon}{\Delta l}.$$

Therefore

$$\frac{dQ}{dt} = \frac{I^2}{\lambda A}\Delta l$$

$$J_I = \frac{I}{A}.$$

Hence

$$J_Q = \frac{1}{A}\frac{dQ}{dt} = \frac{J_I^2}{\lambda}\Delta l$$

is the magnitude of the Ohmic heating.

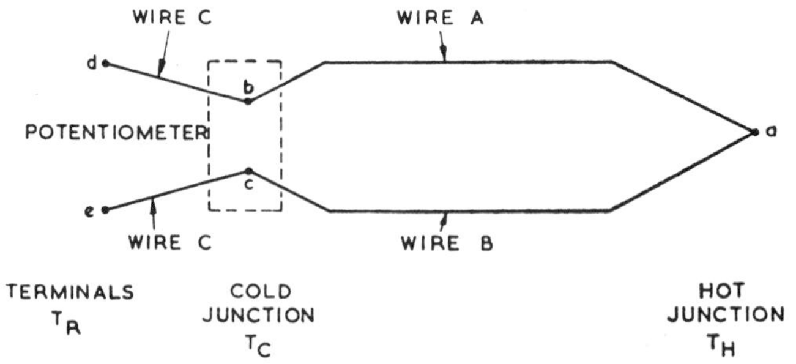

FIG. 5.8.

Two wires of dissimilar materials A and B are joined at the hot junction a. The ends b and c of the two wires are connected to the leads of material C. The joints b and c are immersed in an ice bath to form the cold junction. The leads are connected to a potentiometer at d and e. When the temperature at the hot junction T_H is different from the temperature at the cold junction T_C an electric current will flow until the potentiometer is adjusted to produce zero current ($J_I = 0$). From a knowledge of the materials and the temperature T_C the temperature T_H can be evaluated from the e.m.f. measured with the potentiometer. (At zero current flow the potentiometer e.m.f. equals the "thermal" e.m.f. of the thermocouple.)

The difference in potential at each end of the wire and the temperature difference across the same points are related by (5.152) for zero current flow ($J_I = 0$).

$$d\varepsilon)_{J_I=0} = -S^* \, dT)_{J_I=0}. \tag{5.153}$$

Applying equation (5.153) to each wire in the circuit shown in Fig. 5.8 we have

<u>Wire ec</u>

$$\varepsilon_c - \varepsilon_e = - \int_{T_R}^{T_C} S_C^* \, dT.$$

<u>Wire ca</u>

$$\varepsilon_a - \varepsilon_c = - \int_{T_C}^{T_H} S_B^* \, dT.$$

Wire ab

$$\varepsilon_b - \varepsilon_a = -\int_{T_H}^{T_C} S_A^* \, dT.$$

Wire bd

$$\varepsilon_d - \varepsilon_b = -\int_{T_C}^{T_R} S_C^* \, dT.$$

Adding these equations gives

$$\varepsilon_e - \varepsilon_d = \int_{T_R}^{T_C} S_C^* \, dT + \int_{T_C}^{T_H} S_B^* \, dT + \int_{T_H}^{T_C} S_A^* \, dT + \int_{T_C}^{T_R} S_C^* \, dT$$

$$\varepsilon_e - \varepsilon_d = \int_{T_C}^{T_H} S_B^* \, dT - \int_{T_C}^{T_H} S_A^* \, dT = \int_{T_C}^{T_H} (S_B^* - S_A^*) dT$$

which can be simplified to

$$\varepsilon_{A,B} = \varepsilon_d - \varepsilon_e = \int_{T_C}^{T_H} (S_A^* - S_B^*) \, dT. \qquad (5.154)$$

The thermocouple e.m.f. at any particular temperature T is therefore dependent on the materials between the <u>hot</u> and <u>cold</u> junction, provided that the leads bd and ce are made <u>of</u> the <u>same</u> material. The e.m.f. is independent of the length of the leads.

It will be seen from (5.154) that the e.m.f. for a particular temperature range will depend on the functional relationship between S^* and T for the two wires. If this function is known for a range of wires a suitable combination may be achieved for any particular purpose. In practice the relationship between ε and T at zero current for each wire will be given as a polynomial function and suitable combinations made for standard thermocouples. <u>Pairs</u> of wires are normally stocked with calibrations <u>according</u> to experimental tests. For low temperatures, up to 150°C, copper-nickel may be used, for high temperatures platinum-rhodium may be specified.

If the junctions of a thermocouple are initially at the same temperature and a battery is connected to produce a current in the thermocouple, then the temperature at the junctions will change by an amount greater than the ohmic heat. This additional change in temperature is called the <u>Peltier Effect</u>

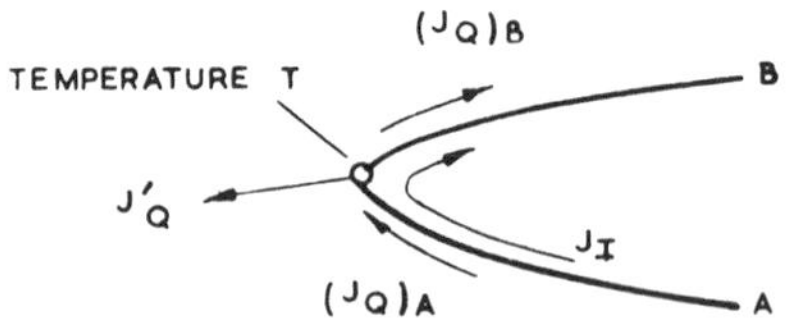

(a) Peltier heating

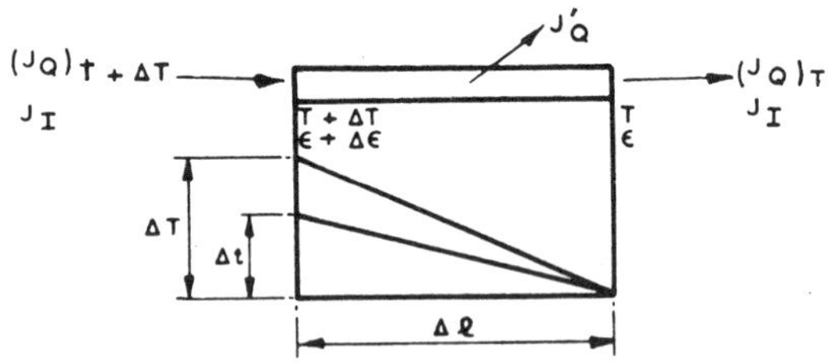

(b) Thomson heating

FIG. 5.9. Peltier and Thomson heating.

and the heating Peltier Heating.

Experiments show that the Peltier heating is linearly proportional to the electric current J_I, the proportionality constant being called the Peltier Coefficient π.

The Peltier heating flux is then

$$J_{Q_{Peltier}} = \pi J_I. \tag{5.155}$$

If we neglect ohmic heating at the junction, shown in Fig. 5.9, we can evaluate the Peltier coefficient, π, in the following manner.

At the junction of the wires A and B an electric current J_I flows in the direction shown. The heat flows $(J_Q)_B$ and $(J_Q)_A$ in the wires and the Peltier heat J'_Q at the junction to maintain the temperature T are given by

$$J'_Q = (J_Q)_A - (J_Q)_B \quad \text{for} \quad T = 0 \tag{5.156}$$

and

$$J'_Q = \pi_{A,B} J_I.$$

Hence

$$\pi_{A,B} = \left(\frac{J_Q}{J_I}\right)_A - \left(\frac{J_Q}{J_I}\right)_B \quad \text{for } \Delta T = 0. \tag{5.157}$$

Now from (5.150)

$$\left(\frac{J_Q}{J_I}\right)_{T=\text{constant}} = Q^* = TS^* \tag{5.150}$$

and at the junction the Peltier coefficient is

$$\pi_{A,B} = T(S_A^* - S_B^*). \tag{5.157}$$

The Peltier heat is reversible; if the current is reversed the heat flow is in the opposite direction.

We have referred earlier to the Thomson heating. The rate at which Thomson heat is transferred into a small region of wire having a temperature difference dT is given by

$$\sigma J_I dT \tag{5.158}$$

where J_I is the current flowing and σ is the Thomson coefficient.

Consider a small element Δl (Fig. 5.9). Initially there is no current but the heat flow J_Q produces a temperature drop ΔT across the element. The current J_I is switched on. Let J_Q' be the heat to be withdrawn from the wire to maintain the same temperatures as before the current was switched on. Under these conditions the heat flows due to heat conduction at each end of the wire are at the same rate as in the initial conditions.

If the wire dimensions are not altered the energy balance is

$$J_Q' = (J_Q)_{T+\Delta T} - (J_Q)_T + J_I\Delta\varepsilon \tag{5.159}$$

the last term being the heat flow generated by the electrical energy $J_I\Delta\varepsilon$. At a given temperature T, the heat of transport is

$$Q^* = \left(\frac{J_Q}{J_I}\right)_T = TS^* \tag{5.150}$$

and

$$(J_Q)_T = TS^*J_I. \tag{5.160}$$

At a temperature $(T + \Delta T)$

$$(J_Q)_{T+\Delta T} = J_I(T+\Delta T)\left\{S^* + \frac{dS^*}{dT}\Delta T\right\}$$

$$(J_Q)_{T+\Delta T} = J_I\left\{TS^* + T\frac{dS^*}{dT}\Delta T + S^*\Delta T\right\} \quad (5.161)$$

to the first order of small quantities.

Now $$\Delta\varepsilon = -\frac{d\varepsilon}{dl}\Delta l \quad (5.162)$$

and $$\Delta T = -\frac{dT}{dl}\Delta l \quad (5.163)$$

The governing equation for electrical flow (5.148) gives

$$\frac{d\varepsilon}{dl} = -\frac{J_I}{\lambda} - S^*\frac{dT}{dl}. \quad (5.164)$$

Substituting (5.160) to (5.164) into (5.159) we have

$$J_Q' = J_I\left\{TS^* + T\frac{dS^*}{dT}\Delta T + S^*\Delta T - TS^*\right\} + J_I\left\{J_I\frac{\Delta l}{\lambda} - S^*\Delta T\right\}$$

$$J_Q' = TJ_I\frac{dS^*}{dT}\Delta T + \frac{J_I^2\,\Delta l}{\lambda}. \quad (5.165)$$

The last term is the ohmic heat generated (see footnote p.249). The first term is the Thomson heat abstracted to maintain the wire temperature. This is equal to $-\sigma J_I\Delta T$. Equating the two expressions for the Thomson heat we have

$$-\sigma J_I\Delta T = TJ_I\frac{dS^*}{dT}\Delta T,$$

hence

$$\sigma = -T\frac{dS^*}{dT}. \quad (5.166)$$

For the thermocouple in Fig. 5.8 the difference in the Thomson coefficients for the two wires is

$$\sigma_A - \sigma_B = -T\frac{d}{dT}(S_A^* - S_B^*). \quad (5.167)$$

We can now group together the basic equations for the thermocouple shown in Fig. 5.8 for the thermocouple A, B.

Seebeck Effect

$$\varepsilon_{A,B} = \int_{T_C}^{T_H} (S_A^* - S_B^*)\, dT. \tag{5.154}$$

Peltier Effect

$$\pi_{A,B} = T(S_A^* - S_B^*). \tag{5.157}$$

Thomson Effect

$$\sigma_A - \sigma_B = -T \frac{d}{dT} (S_A^* - S_B^*). \tag{5.167}$$

Thomson (Lord Kelvin) by a different procedure produced two equations for a thermocouple, these can be obtained from (5.154), (5.157) and (5.167) by eliminating the entropy of transport S^* in the following manner:

The cold junction temperature, T_C, is fixed and the hot junction temperature, T_H, is replaced by the variable temperature T. Equation (5.154) is differentiated with respect to T to give

$$\frac{d\varepsilon_{A,B}}{dT} = S_A^* - S_B^* \tag{5.168}$$

and the result (equation (5.168)) substituted into (5.157) to give the First Thomson equation

$$\pi_{A,B} = T \frac{d\varepsilon_{A,B}}{dT} \tag{5.169}$$

Equation (5.168) is differentiated again with respect to T and the results substituted into (5.167) to obtain the Second Thomson equation

$$\sigma_A - \sigma_B = -T \frac{d^2\varepsilon_{A,B}}{dT^2} \tag{5.170}$$

If, for a given thermocouple, the relationship between the e.m.f. and the temperature is known, then both the Peltier coefficient (5.169) and the difference in the Thomson coefficients (5.170) can be computed.

Exercises

1. The equation of state for a wire is

$$F = \frac{KL}{L_o} - \alpha KT + b.$$

The initial length is L_o, K and b are constants and α is the coefficient of linear expression.

The wire is stretched a length ΔL by a force F_1.

Show that if the process is isothermal and reversible and the force at the original length L_o is zero the heat transfer is

$$Q = \alpha L_o \, TF_1.$$

If the change in length ΔL is small compared with the original length show that the change in internal energy is

$$\Delta U = F_1 L_o \, (L + \alpha T) - K \, \Delta L.$$

2. If the specific heat C_F is positive, show that if the temperature of a certain material increases when stretched isentropically the coefficient of linear expansion is negative.

3. Show that the elongation of a taut wire per unit increase in temperature at constant tention is equal to its isothermal increase of entropy per unit increase in tension.

(Univ. Manch.)

4. The surface tension of a liquid in equilibrium with its own vapour is given by

$$\sigma = A + BT + CT^2 \quad \text{Newtons/cm}$$

where T is the temperature in ^{o}K. Show that

$$\sigma = 2\lambda_{max}\left(1 - \left(\frac{T}{T_c}\right)\right)^2;$$

λ_{max} is the maximum heat addition per unit surface area and T_c is the critical temperature.

5. Derive an expression for the change in Helmholtz and Gibbs functions for a liquid surface film area A and surface tension σ. If the surface tension is a function of temperature only determine the percentage change in the two functions at a given temperature if the surface area of the film is increased by ten percent. Comment on your results.

Show that in general

$$\left(\frac{\partial H}{\partial \sigma}\right)_S = \left(\frac{\partial G}{\partial \sigma}\right)_T$$

$$\left(\frac{\partial F}{\partial T}\right)_A = \left(\frac{\partial G}{\partial T}\right)_\sigma .$$

6. Show that for a reversible cell

$$\frac{d(\ln K_p)}{dT} = \frac{jF}{R_{mol}T^2}\left\{T\left(\frac{\partial \varepsilon_R}{\partial T}\right)_p - \varepsilon_R\right\}$$

$$Q_p = jF\left\{T\left(\frac{\partial \varepsilon_R}{\partial T}\right)_p - \varepsilon_R\right\}.$$

7. A gas cell is of the form

$$H_2(gas)|H_2SO_4(liq)|H_2(gas).$$

The pressure of the hydrogen at the positive terminal is 0.1 atm and at the negative terminal 1 atm. Calculate the e.m.f. of the cell at 300°K if the electrolyte solution is not changed in quantity or concentration.

Is the cell charging or discharging?

8. Calculate the e.m.f. and the ideal efficiency of a hydrogen-oxygen fuel cell operating at the following conditions:

Partial pressures atm H_2	O_2	H_2O	Cell temperature °K
1	1	1	300
2	2	2	600
5	5	5	600

9. Show that the ideal efficiency of a fuel cell is given by

$$\frac{\varepsilon_R}{\varepsilon_R - T\left(\frac{\partial \varepsilon_R}{\partial T}\right)_p}$$

If the heat of reaction for the reactants in a fuel cell is independent of temperature show that for constant pressure operation the ideal efficiency (η_I) is

$$\eta_I = 1 - kT$$

where

$$k = \frac{1}{T_o}\left(\frac{jF}{Q_p}\varepsilon_o + 1\right) = \text{constant.}$$

ε_o is the cell e.m.f. at temperature T_o.

10. For a dielectric material the thermodynamic coordinates are the electric field intensity E, the polarization p', and the temperature T. The work done by the material in the electric field is -E dp'.

Show that

$$T\,ds = C'_p\,dT - T\left(\frac{\partial E}{\partial T}\right)_{p'} dp'$$

$$T\,ds = C_E\,dT + T\left(\frac{\partial p'}{\partial T}\right)_E dE.$$

If the equation of state for the dielectric material is

$$\frac{p'}{V} = \left(a + \frac{b}{T}\right)E,$$

show that

$$\text{(I)}\quad \left(\frac{\partial U}{\partial p'}\right)_T = \left(\frac{aT}{aT + b}\right)E$$

$$\text{(II)}\quad \left(\frac{\partial U}{\partial E}\right)_T = VE_a.$$

11. Obtain the following relations for a paramagnetic salt

$$\left(\frac{\partial u}{\partial \bar{j}}\right)_T = -T\left(\frac{\partial H}{\partial T}\right)_{\bar{j}} + H$$

$$\left(\frac{\partial u}{\partial H}\right)_T = T\left(\frac{\partial \bar{j}}{\partial T}\right)_H + H\left(\frac{\partial \bar{j}}{\partial H}\right)_T$$

$$C_H - C_j = T\left(\frac{\partial \bar{j}}{\partial T}\right)_H^2\left(\frac{\partial H}{\partial \bar{j}}\right)_T$$

$$\left(\frac{\partial C_H}{\partial H}\right)_T = T\left(\frac{\partial^2 j}{\partial T^2}\right)_H.$$

Show that for a salt obeying Curie's Law the internal energy u and the specific heat C_j are independent of j and H and

$$T^2 \left(\frac{\partial C_H}{\partial H}\right)_T = 4\bar{\mu}_o C \, (C_H - C_j).$$

12. The e.m.f. of a copper-iron thermocouple is given by

$$\varepsilon = -13.4t + 0.014t^2 + 0.00013t^3 \ \mu V$$

where t is the temperature in °C. Calculate the Peltier coefficient at 100°C and show that the difference in the Thomson coefficients for the two wires is

$$7.644 + 0.241t + 0.00078t^2 \ \mu V/\text{degree}.$$

13. A system consists of a single fluid substance contained in two vessels which are held at different temperatures and are connected through a porous plate. Show that the difference in pressure between the vessels is given by

$$\frac{dp}{dT} = \frac{h-q}{vT}.$$

h = specific enthalpy of substance, q = energy transported per mol of substance when there is no heat flow, v = specific volume.

[The primary processes are:

Energy (or Heat) Flow J_1,

$$\text{Force} \quad X_1 = -\frac{1}{T}\frac{dT}{dl}$$

Diffusion (or mass) Flow J_2,

$$\text{Force} \quad X_2 = -T\,\frac{d\mu/T}{dl}$$

μ = chemical potential.]

14. Show that, for an elastic rod of length L per unit mass at temperature T and under a tensile force F.

$$Tds = c_L dT - T\left(\frac{\partial F}{\partial T}\right)_L dL$$

where c_L denotes the specific heat at constant length.

The equation of state for a particular rod is

$$F = kT\left(\frac{L}{L_o} - \frac{L_o^2}{L^2}\right)$$

where k is a constant and L_o, the length at zero tension, depends on temperature only. Show

(a) that the work which must be done on the rod to stretch it isothermally and reversibly from the length L_o to $2L_o$ is kTL_o and

(b) that the change in internal energy of the rod is

$$\frac{5}{2} k\, T^2\, L_o\, \alpha_o$$

where α_o is the coefficient of expansion at zero tension, i.e.

$$\alpha_o = \frac{1}{L_o}\frac{dL_o}{dT}$$

(Univ. Manch.)

15. A strip of rubber has the equation of state

$$F = kT\left(\frac{l}{l_o} - \left(\frac{l_o}{l}\right)^2\right)$$

where F is the tension, Newtons
T is the temperature, ^oK
l is the length per unit mass, m/kg
l_o is the unstretched length per unit mass, and is a function of the temperature
k is a constant

By considering the specific entropy s of the rubber to be a function of T and l, or otherwise, show that

$$ds = C_l \frac{dT}{T} + \left(\frac{\partial s}{\partial l}\right)_T dl$$

where C_l is the specific heat at constant length. Hence show that if the rubber is stretched adiabatically and reversibly, the change of temperature is given by

$$\left(\frac{\partial T}{\partial l}\right)_s = \frac{kT}{C_l}\left(\frac{l}{l_o} - \left(\frac{l}{l_o}\right)^2 - \frac{T}{l_o}\frac{dl_o}{dT}\left(\frac{l}{l_o} + 2\frac{l_o^2}{l^2}\right)\right)$$

(Univ. Manch.)

16. A torsion bar is made from a material of uniform properties. The angular twist θ of one end of the shaft relative to the other is a function of the torque L applied to the bar and the temperature T. Show that, if the angular twist of the shaft is held constant as the shaft is heated, the decrease in torque per unit increase in temperature has a value which is exactly equal to the isothermal variation of the entropy S of the shaft with θ.

The torsional stiffness of the bar is defined as $\left(\frac{\partial L}{\partial \theta}\right)_T$ and is a function of T and θ. Show that, at constant θ, the change of torsional stiffness with temperature is equal to $-\left(\frac{\partial^2 S}{\partial \theta^2}\right)_T$.

(Univ. Manch.)

17. Show that, for a reversible cell of electromotive force ε,

$$\left(\frac{\partial \varepsilon}{\partial T}\right)_p = \frac{Q_p}{jFT} + \frac{\varepsilon}{T}$$

where F is the Faraday constant and j is the valency.

For the reaction

$$Pb + Hg_2Cl_2 = PbCl_2 + 2Hg$$

the value of Q_p at 20°C is -95 200 kJ per kg mol of Pb. A cell, in which the above reaction takes place reversibly, takes in heat from the surroundings at the rate of 8 300 kJ per kg mol of Pb. Calculate the electromotive force of the cell and its variation with temperature for constant pressure operation. Estimate the number of Joules of electrical work produced by the cell per kilogram of reactants.

$j = 2$

$F = 96.5 \times 10^6$ coulombs per kg mol equivalent

Atomic weights:

PB = 207
Hg = 200
Cl = 35

(Univ. Manch.)

18. Describe the construction of <u>either</u> a hydrogen-oxygen fuel cell <u>or</u> a carbon-oxygen fuel cell, indicating the principal chemical reactions which occur within the cell. Enumerate the main irreversibilities which occur, and their effect on the electromotive force, as the current taken from the cell is increased.

In a hydrogen-oxygen cell operating at atmospheric temperature and pressure, the overall reduction of Gibbs Free Energy is 119 MJ per kg of hydrogen. The corresponding enthalpy of reaction is -143 MJ per kg of hydrogen. Calculate the e.m.f. of the cell operating reversibly. If, on load, the cell operates with only 75% of the reversible e.m.f., calculate the magnitude and direction of the heat exchange with the surroundings.

Faraday's constant f = 96.5×10^6 Coulombs per kg mol equivalent.

(Univ. Manch.)

19. Give an expression for the e.m.f. of a reversible cell, in terms of the Gibbs free energy of the reactants and products, the valency j of one of the reactants and Faraday's constant F. Explain why an actual cell on load will give an e.m.f. appreciably less than the reversible value.

A hydrogen-oxygen cell operates with reactants and products all at 300 K. 2 bars pressure. From the data given below, calculate the ideal efficiency and the reversible e.m.f. of the cell. If the actual e.m.f. is 65% of the ideal value, determine the rate at which hydrogen must be consumed per kilowatt of electrical power output, and calculate the heat output from the cell to the surroundings under these circumstances.

The properties of hydrogen, oxygen and water at 300 K. 1 bar pressure are as follows:

	Enthalpy kJ/(kg mol)	Entropy kJ/(kg mol.K)
hydrogen	8390	130.6
oxygen	8620	205.0
water	-273000	42.5

Neglect the effect of pressure on the specific entropy of water. F = 96,500 coulombs per gram mol equivalent and j (for H_2) = 2.

(Univ. Manch.)

20. In a hydrogen-oxygen cell the pressures of the H_2, O_2 and H_2O are all equal, but the environment of the cell allows these pressures to fluctuate over the range 4 atmospheres absolute to 0.1 atmosphere absolute. The e.m.f. of the cell is to be maintained constant by control of the temperature. At the highest pressure, the temperature must be 350°K to achieve the required e.m.f. 1.16 volts. Calculate the range of temperature over which the cell must be controlled.

Q_p = -243,000 kJ/kg mol of H_2, and is constant over the temperature range.

Valency, j = 2

(Univ. Manch.)

21. A battery of Hydrogen-Oxygen fuel cells runs at a temperature of 450°K, the reactants and products being at a pressure of 3 atmospheres. The e.m.f. of the battery is 70% of the reversible value. Calculate the rate of consumption of hydrogen (gm. per hour) per kilowatt output. Obtain a comparative figure for the hydrogen consumption rate of a 30% efficient heat engine supplied with the same reactants and driving a generator of 85% efficiency.

1 calorie = 4.186 Joules

Faraday's constant = 96,500 coulombs per gm mol equivalent

Valency = 2

	Properties of gas, cal/gm mol	
	Enthalpy h At 450°K	Gibbs Free Energy g^o At 450 K and 1 atmosphere
O_2	3,140	-20,240
H_2	3,060	-12,280
H_2O	-53,510	-75,310

(Univ. Manch.)

22. A thermocouple is connected to a battery. When the hot junction temperature is 100°C the Peltier heat flux is 2.68 milliwatts/amp and when the hot junction temperature is 200°C the Peltier heat flux is 4.11 milliwatts/amp. In both cases the cold junction is maintained at 0°C. If the e.m.f. of the thermocouple can be represented by the equation

$$\varepsilon = at. + bt^2,$$

calculate the constants a and b, and the e.m.f. at 100°C and 200°C if there is no current flow.

(Univ. Manch.)

APPENDIX

THERMODYNAMIC PROPERTIES

In the tables which follow the thermodynamic properties are calculated from spectroscopic data. The data which form the basis of the calculations are based on the enthalpy $h(T)$ and Gibbs function $g(T)$.

The enthalpy $h(T)$ is defined as

$$h(T) = R_{mol}(a_1T + a_2T^2 + a_3T^3 + a_4T^4 + a_5T^5) \qquad (A.1)$$

This can be expressed in the form

$$h(T) = R_{mol} \sum_{j=1}^{j=5} a_j \, T^j \qquad (A.2)$$

The polynomial coefficients a_1, a_2, a_3, a_4, a_5, are given in Table A.1. Since a number of published papers have in the past referred to these coefficients, the first edition symbols a, b, c, d and e are also given.

The coefficients will give $h(T)$ in the units of R_{mol} multiplied by T deg. K. Thus

if R_{mol} is in J/kg mol °K then $h(T)$ is in J/kg mol.

or if R_{mol} is in kJ/kg mol°K then $h(T)$ is in kJ/kg mol.

The Gibbs function $g(T)$ is defined from (4.87)

$$g(T) = R_{mol}\left\{T(a_1 - \ln T) - \sum_{j=1}^{j=5} \frac{a_j}{j-1} T^j - a_6T\right\} \qquad (A.3)$$

The numerical value of $g(T)$ is dependent on the reference pressure p_o. Thus the coefficient a_6 is <u>reference pressure</u> dependent. In Table A.1, the coefficient a corresponds to a reference pressure p_o = 101325 N/m^2 or 1.01325 bar.

The internal energy u(T) is

$$u(T) = h(T) - R_{mol}T \tag{A.4}$$

The entropy s(T) is

$$s(T) = \frac{1}{T}(h(T) - g(T)) \tag{A.5}$$

Notice that the numerical value of entropy s(T) will depend on the reference pressure p_o since the numerical value of g(T) depends on the reference pressure. Furthermore in calculating the absolute entropy (3.29) the same reference pressure must be used. In the case of the data in Table A.2, the absolute entropy will be

$$s = s(T) - R_{mol} \ln\left(\frac{p}{101325}\right) + s_o \tag{A.6}$$

or

$$s = s(T) - R_{mol} \ln p + s_o + 11.526 \tag{A.7}$$

where the pressure p is in N/m^2.

For a mixture of N species i the total enthalpy H is given by

$$H = \sum_{i=1}^{i=N} M_i h_i \tag{A.8}$$

where the enthalpy of a species i is h_i and the number of mols is M_i.

The enthalpy h_i is given by

$$h_i = h_i(T) + h_{oi} \tag{A.9}$$

where h_{oi} is the enthalpy of species i at the absolute zero.

We can then write (A.8) as

$$H = \sum_{i=1}^{i=N} M_i \, (h_i(T) + h_{oi}) \tag{A.10}$$

and using (A.1),

$$H = \sum_{i=1}^{i=N} M_i \left(R_{mol} \sum_{j=1}^{j=5} a_{ij} \, T^j + h_{oi}\right) \tag{A.11}$$

The coefficient a_{ij} is the coefficient a_j for species i.

Using the same notation we will have the total number of mols in the mixture

$$M = \sum_{i=1}^{i=N} M_i \tag{A.12}$$

and the internal energy E

$$E = H - R_{mol}T$$

or

$$E = R_{mol}\left(\sum_{i=1}^{i=N} M_i\left(\left(\left(\sum_{j=1}^{j=5} a_{ij}\ T^j\right)-T\right) + \frac{h_{oi}}{R_{mol}}\right)\right) \tag{A.13}$$

The specific heat at constant volume for the mixture, C_v, is

$$C_v = \frac{\partial e}{\partial T} = \frac{1}{M}\frac{\partial E}{\partial T}$$

or

$$C_v = \frac{R_{mol}}{M} \sum_{i=1}^{i=N} M_i \left(\left(\sum_{j=1}^{j=5} j\ a_{ij}\ T^{j-1}\right)-1\right) \tag{A.14}$$

By setting a matrix for the coefficients the properties H, E, C_v, can be readily obtained with simple computer programming. An example of the application of the method for cycle calculations is given by Benson and Baruah†.

The equilibrium constant K_p is calculated from equations (4.83) and (A.3). Since the constant a_6 is dependent on the reference pressure p_o the values given in Table A.4, are also dependent on p_o. These are expressed in the form:

$$K_p = \frac{\Pi\left(\left(\frac{p}{p_o}\right)^{\nu}\right)_{products}}{\Pi\left(\left(\frac{p}{p_o}\right)^{\nu}\right)_{reactants}}$$

when p is in N/m , p_o = 101325 N/m and
when p is in bars, p_o = 1.01325 bars.

---

†R.S. Benson and P.C. Baruah. A Generalized Calculation for an Ideal Otto Cycle with Hydrocarbon-Air Mixture Allowing for Dissociation and Variable Specific Heats. Int. J. Mech. Eng. Edcn. Vol.4, No.1, p.93.

TABLES

TABLE A.1 POLYNOMIAL

	Temperature Range Deg.K	e a_5	d a_4	c a_3
H_2	500-3000	0.00000	$-1.44392E^{-11}$	$9.66990E^{-8}$
	3000-6000	0.00000	$7.66560E^{-13}$	$-2.28839E^{-8}$
CO	500-3000	0.00000	$-2.19450E^{-12}$	$-3.22080E^{-8}$
	3000-6000	0.00000	$1.56492E^{-12}$	$-3.27672E^{-8}$
N_2	500-3000	0.00000	$-6.57470E^{-12}$	$1.95300E^{-9}$
	3000-6000	0.00000	$1.24154E^{-12}$	$-2.84107E^{-8}$
NO	500-3000	0.00000	$-4.90360E^{-12}$	$-9.58800E^{-9}$
	3000-6000	0.00000	$7.70280E^{-13}$	$-1.88284E^{-8}$
CO_2	500-3000	0.00000	$8.66002E^{-11}$	$-7.88542E^{-7}$
	3000-6000	0.00000	$2.04685E^{-12}$	$-5.61431E^{-8}$
O_2	500-3000	0.00000	$1.53897E^{-11}$	$-1.49524E^{-7}$
	3000-6000	0.00000	$1.00528E^{-12}$	$-2.87575E^{-8}$
H_2O	500-3000	0.00000	$-1.81802E^{-11}$	$4.95240E^{-8}$
	3000-6000	0.00000	$2.15227E^{-12}$	$0.21913E^{-8}$
CH_4	Up to 6000	$-8.58611E^{-15}$	$1.62497E^{-10}$	$-1.24402E^{-6}$

High Temperature Coefficients - Atoms

	Temperature Range Deg.K	e a_5	d a_4	c a_3
O	500-3000	0.00000	$-1.38670E^{-11}$	$1.00187E^{-7}$
	3000-6000	0.00000	$-3.20943E^{-13}$	$7.51428E^{-9}$
N	500-3000	0.00000	$6.15151E^{-13}$	$-2.44816E^{-9}$
	3000-6000	0.00000	$-1.17893E^{-12}$	$2.84705E^{-8}$
A	Up to 6000	0.00000	0.00000	0.00000
H	Up to 6000	0.00000	0.00000	0.00000
Kr		0.00000	0.00000	0.00000
Xe		0.00000	0.00000	0.00000

COEFFICIENTS

b a_2	a a_1	k a_6	h_o J/kg mol
$-8.18100E^{-6}$ $2.87156E^{-4}$	3.43328 3.21299	-3.84470 -2.78598	0.00000
$3.76970E^{-4}$ $2.73436E^{-4}$	3.31700 3.53114	4.63284 3.41176	$-1.13882E^{+08}$
$2.94260E^{-4}$ $2.58314E^{-4}$	3.34435 3.51443	3.75863 2.74113	0.00000
$2.99380E^{-4}$ $1.94975E^{-4}$	3.50174 3.74493	5.11346 3.71329	$8.99147E^{+07}$
$2.73114E^{-3}$ $5.90837E^{-4}$	3.09590 5.20816	6.58393 -4.32923	$-3.93405E^{+08}$
$6.52350E^{-4}$ $3.20312E^{-4}$	3.25304 3.55064	5.71243 4.20945	0.00000
$5.65590E^{-4}$ $6.58323E^{-4}$	3.74292 3.92119	$9.65140E^{-1}$ $-2.42340E^{-1}$	$-2.39082E^{+08}$
$4.96462E^{-3}$	1.93529	8.15300	$-6.69305E^{+07}$

b a_2	a a_1	k a_6	h_o J/kg mol
$-2.51427E^{-4}$ $-3.84318E^{-5}$	2.76403 2.59409	3.73309 4.57993	$2.46923E^{+08}$
$2.87441E^{-6}$ $-1.41798E^{-4}$	2.49906 2.70375	4.18504 3.06136	$4.71369E^{+08}$
0.00000	2.50000	0.00000	0.00000
0.00000	2.50000	-0.45931	$2.16110E^{+08}$
0.00000	2.50000	5.47600	0.00000
0.00000	2.50000	6.15100	0.00000

TABLE A.2 IDEAL GAS PROPERTIES 0-6000°K

Basic Relationships

Internal energy	u	$= u(T) + u_o$
Enthalpy	h	$= h(T) + h_o$
Entropy	s	$= s(T) - R_{mol} \ln \frac{p}{p_o} + s_o$
Gibbs function	g	$= g(T) + R_{mol}T \ln \frac{p}{p_o} + g_o$

Tabulated Quantities and Units

Temperatures T	°K	
Internal energy	$u(T)$	kJ/kg.mol
Enthalpy	$h(T)$	kJ/kg.mol
Entropy	$s(T)$	kJ/kg.mol°K
Gibbs function	$g(T)$	kJ/kg.mol

$$u_o = h_o = g_o$$

Universal Gas Constant R_{mol} = 8.314 kJ/kg.mol°K

Reference pressure p_o = 10135 N/m^2.

$h_o = 0$

MOLECULAR OXYGEN O_2

T	h(T)	u(T)	s(T)	g(T)
50	1365.7	950.0	153.8346	-6326.0
100	2757.6	1926.2	173.1098	-14553.4
150	4174.8	2927.7	184.5954	-23514.5
200	5616.4	3953.6	192.8865	-32960.9
250	7081.5	5003.0	199.4233	-42774.3
300	8569.3	6075.1	204.8473	-52884.9
350	10079.0	7169.1	209.5009	-63246.3
400	11609.8	8284.2	213.5884	-73825.6
450	13160.9	9419.6	217.2417	-84597.9
500	14731.4	10574.4	220.5508	-95544.0
550	16320.7	11748.0	223.5801	-106648.3
600	17928.0	12939.6	226.3770	-117898.1
650	19552.7	14148.6	228.9776	-129282.8
700	21193.9	15374.1	231.4101	-140793.1
750	22851.2	16615.7	233.6967	-152421.3
800	24523.7	17872.5	235.8554	-164160.6
850	26210.8	19143.9	237.9010	-176005.0
900	27912.0	20429.4	239.8457	-187949.1
950	29626.7	21728.4	241.6997	-199988.1
1000	31354.2	23040.2	243.4719	-212117.7
1050	33094.1	24364.4	245.1696	-224334.0
1100	34845.7	25700.3	246.7992	-236633.5
1150	36608.5	27047.4	248.3664	-249012.9
1200	38382.1	28405.3	249.8761	-261469.2
1250	40166.0	29773.5	251.3325	-273999.6
1300	41959.7	31151.5	252.7395	-286601.6
1350	43762.8	32538.9	254.1004	-299272.8
1400	45574.8	33935.2	255.4184	-312010.9
1450	47395.3	35340.0	256.6960	-324814.0
1500	49224.0	36753.0	257.9359	-337679.9
1550	51060.5	38173.8	259.1403	-350607.0
1600	52904.4	39602.0	260.3111	-363593.4
1650	54755.4	41037.3	261.4503	-376637.5
1700	56613.2	42479.4	262.5595	-389737.9
1750	58477.6	43928.1	263.6403	-402893.0
1800	60348.1	45382.9	264.6942	-416101.5
1850	62224.7	46843.8	265.7226	-429362.0
1900	64107.0	48310.4	266.7265	-442673.3
1950	65994.9	49782.6	267.7073	-456034.3
2000	67888.2	51260.2	268.6659	-469443.7

O_2 (cont.)

T	h(T)	u(T)	s(T)	g(T)
2050	69786.6	52742.9	269.6035	-482900.5
2100	71690.0	54230.6	270.5208	-496403.7
2150	73598.3	55723.2	271.4189	-509952.3
2200	75511.4	57220.6	272.2985	-523545.3
2250	77429.2	58722.7	273.1605	-537181.8
2300	79351.6	60229.4	274.0055	-550861.1
2350	81278.5	61740.6	274.8343	-564582.1
2400	83209.9	63256.3	275.6475	-578344.2
2450	85145.8	64776.5	276.4459	-592146.6
2500	87086.1	66301.1	277.2299	-605988.6
2550	89031.0	67830.3	278.0001	-619869.4
2600	90980.3	69363.9	278.7572	-633788.4
2650	92934.3	70902.2	279.5016	-647744.9
2700	94892.9	72445.1	280.2338	-661738.3
2750	96856.3	73992.8	280.9543	-675768.1
2800	98824.5	75545.3	281.6636	-689833.6
2850	100797.8	77102.9	282.3621	-703934.3
2900	102770.3	78659.7	283.0483	-718069.6
2950	104757.2	80230.9	283.7275	-732239.0
3000	106749.2	81807.2	284.3971	-746442.2
3050	108746.5	83388.8	285.0574	-760678.6
3100	110749.2	84975.8	285.7087	-774947.8
3150	112762.3	86573.2	286.3529	-789249.3
3200	114775.8	88171.0	286.9871	-803582.9
3250	116793.7	89773.2	287.6128	-817947.9
3300	118816.0	91379.8	288.2303	-832344.0
3350	120842.4	92990.5	288.8398	-846770.8
3400	122872.9	94605.3	289.4414	-861227.9
3450	124907.5	96224.2	290.0354	-875714.8
3500	126946.0	97847.0	290.6221	-890231.3
3550	128988.3	99473.6	291.2015	-904776.9
3600	131034.3	101103.9	291.7738	-919351.3
3650	133084.1	102738.0	292.3392	-933954.2
3700	135137.3	104375.5	292.8980	-948585.1
3750	137194.1	106016.6	293.4501	-963243.9
3800	139254.3	107661.1	293.9959	-977930.0
3850	141317.7	109308.8	294.5353	-992643.3
3900	143384.4	110959.8	295.0687	-1007383.5
3950	145454.2	112613.9	295.5960	-1022150.1
4000	147527.1	114271.1	296.1175	-1036943.0

O_2 (cont.)

T	h(T)	u(T)	s(T)	g(T)
4050	149603.0	115931.3	296.6333	-1051761.8
4100	151681.8	117594.4	297.1434	-1066606.2
4150	153763.4	119260.3	297.6481	-1081476.0
4200	155847.7	120928.9	298.1473	-1096370.9
4250	157934.8	122600.3	298.6413	-1111290.7
4300	160024.4	124274.2	299.1301	-1126235.0
4350	162116.6	125950.7	299.6138	-1141203.6
4400	164211.2	127629.6	300.0926	-1156196.3
4450	166308.2	129310.9	300.5665	-1171212.8
4500	168407.5	130994.5	301.0356	-1186252.8
4550	170509.1	132680.4	301.5001	-1201316.3
4600	172612.9	134368.5	301.9599	-1216402.8
4650	174718.8	136058.7	302.4153	-1231512.2
4700	176826.8	137751.0	302.8662	-1246644.2
4750	178936.7	139445.2	303.3127	-1261798.7
4800	181048.6	141141.4	303.7550	-1276975.4
4850	183162.4	142839.5	304.1931	-1292174.1
4900	185278.0	144539.4	304.6271	-1307394.7
4950	187395.3	146241.0	305.0570	-1322636.8
5000	189514.4	147944.4	305.4829	-1337900.3
5050	191635.1	149649.4	305.9050	-1353185.0
5100	193757.4	151356.0	306.3232	-1368490.7
5150	195881.3	153064.2	306.7376	-1383817.3
5200	198006.7	154773.9	307.1483	-1399164.4
5250	200133.5	156485.0	307.5553	-1414532.0
5300	202261.7	158197.5	307.9588	-1429919.9
5350	204391.2	159911.3	308.3587	-1445327.9
5400	206522.1	161626.5	308.7551	-1460755.7
5450	208654.2	163342.9	309.1482	-1476203.3
5500	210787.5	165060.5	309.5378	-1491670.5
5550	212922.0	166779.3	309.9242	-1507157.0
5600	215057.7	168499.3	310.3072	-1522662.8
5650	217194.4	170220.3	310.6871	-1538187.7
5700	219332.2	171942.4	311.0638	-1553731.5
5750	221471.0	173665.5	311.4374	-1569294.0
5800	223610.8	175389.6	311.8079	-1584875.2
5850	225751.6	177114.7	312.1754	-1600474.8
5900	227893.3	178840.7	312.5400	-1616092.7
5950	230035.8	180567.5	312.9016	-1631728.7
6000	232179.2	182295.2	313.2603	-1647382.8

$h_0 = 0$

MOLECULAR NITROGEN N_2

T	h(T)	u(T)	s(T)	g(T)
50	1396.4	980.7	140.2675	-5617.0
100	2805.0	1973.6	159.7851	-13173.5
150	4225.8	2978.7	171.3038	-21469.8
200	5658.9	3996.1	179.5476	-30250.6
250	7104.2	5025.7	185.9967	-39395.0
300	8561.7	6067.5	191.3106	-48831.5
350	10031.3	7121.4	195.8410	-58513.1
400	11513.0	8187.4	199.7979	-68406.1
450	13006.9	9265.6	203.3165	-78485.6
500	14512.7	10355.7	206.4894	-88732.0
550	16030.5	11457.8	209.3824	-99129.9
600	17560.1	12571.7	212.0442	-109666.4
650	19101.5	13697.4	214.5117	-120331.1
700	20654.7	14834.9	216.8136	-131114.8
750	22219.4	15983.9	218.9726	-142010.0
800	23795.6	17144.4	221.0070	-153010.0
850	25383.2	18316.3	222.9319	-164108.9
900	26982.1	19499.5	224.7596	-175301.6
950	28592.0	20693.7	226.5005	-186583.4
1000	30213.0	21899.0	228.1633	-197950.3
1050	31844.8	23115.1	229.7556	-209398.6
1100	33487.2	24341.8	231.2837	-220924.8
1150	35140.2	25579.1	232.7532	-232526.0
1200	36803.5	26826.7	234.1690	-244199.2
1250	38477.0	28084.5	235.5353	-255942.0
1300	40160.5	29352.3	236.8558	-267752.0
1350	41853.7	30629.8	238.1338	-279626.9
1400	43556.6	31917.0	239.3723	-291564.7
1450	45268.7	33213.4	240.5740	-303563.5
1500	46990.0	34519.0	241.7411	-315621.5
1550	48720.3	35833.6	242.8757	-327737.1
1600	50459.1	37156.7	243.9798	-339908.6
1650	52206.4	38488.3	245.0552	-352134.6
1700	53961.9	39828.1	246.1033	-364413.7
1750	55725.3	41175.8	247.1256	-376744.5
1800	57496.3	42531.1	248.1234	-389125.8
1850	59274.7	43893.8	249.0979	-401556.5
1900	61060.2	45263.6	250.0502	-414035.2
1950	62852.4	46640.1	250.9813	-426561.1
2000	64651.1	48023.1	251.8921	-439133.0

N_2 (cont.)

T	h(T)	u(T)	s(T)	g(T)
2050	66455.9	49412.2	252.7834	-451750.0
2100	68266.6	50807.2	253.6560	-464411.1
2150	70082.8	52207.7	254.5108	-477115.3
2200	71904.2	53613.4	255.3482	-489861.8
2250	73730.4	55023.9	256.1690	-502649.8
2300	75561.1	56438.9	256.9737	-515478.5
2350	77395.9	57858.0	257.7629	-528347.0
2400	79234.4	59280.8	258.5371	-541254.5
2450	81076.4	60707.1	259.2966	-554200.4
2500	82921.3	62136.3	260.0421	-567184.0
2550	84768.8	63568.1	260.7738	-580204.4
2600	86618.5	65002.1	261.4921	-593261.1
2650	88469.9	66437.8	262.1975	-606353.4
2700	90322.8	67875.0	262.8901	-619480.6
2750	92176.5	69313.0	263.5704	-632642.2
2800	94030.8	70751.6	264.2387	-645837.5
2850	95885.1	72190.2	264.8951	-659065.9
2900	97752.4	73641.8	265.5445	-672326.8
2950	99597.8	75071.5	266.1754	-685619.8
3000	101443.9	76501.9	266.7960	-698944.1
3050	103290.7	77933.0	267.4065	-712299.3
3100	105137.7	79364.3	268.0072	-725684.6
3150	106983.0	80793.9	268.5977	-739099.9
3200	108834.7	82229.9	269.1810	-752544.4
3250	110689.0	83668.5	269.7559	-766017.8
3300	112545.8	85109.6	270 3229	-779519.8
3350	114405.0	86553.1	270.8821	-793050.0
3400	116266.5	87998.9	271.4337	-806607.9
3450	118130.4	89447.1	271.9779	-820193.2
3500	119996.4	90897.4	272.5148	-833805.6
3550	121864.5	92349.8	273.0448	-847444.6
3600	123734.8	93804.4	273.5680	-861109.9
3650	125607.0	95260.9	274.0845	-874801.3
3700	127481.1	96719.3	274.5944	-888518.3
3750	129357.1	98179.6	275.0981	-902260.6
3800	131234.9	99641.7	275.5955	-916028.0
3850	133114.5	101105.6	276.0869	-929820.1
3900	134995.8	102571.2	276.5724	-943636.6
3950	136878.6	104038.3	277.0521	-957477.2
4000	138763.1	105507.1	277.5262	-971341.7

N_2 (cont.)

T	h(T)	u(T)	s(T)	g(T)
4050	140649.1	106977.4	277.9948	-985229.7
4100	142536.5	108449.1	278.4580	-999141.1
4150	144425.4	109922.3	278.9159	-1013075.5
4200	146315.6	111396.8	279.3686	-1027032.6
4250	148207.2	112872.7	279.8163	-1041012.2
4300	150100.0	114349.8	280.2591	-1055014.1
4350	151994.0	115828.1	280.6970	-1069038.1
4400	153889.3	117307.7	281.1302	-1083083.8
4450	155785.7	118788.4	281.5588	-1097151.0
4500	157683.1	120270.1	281.9828	-1111239.6
4550	159581.7	121753.0	282.4024	-1125349.2
4600	161481.3	123236.9	282.8176	-1139479.7
4650	163381.8	124721.7	283.2285	-1153630.9
4700	165283.4	126207.6	283.6353	-1167802.5
4750	167185.8	127694.3	284.0379	-1181994.4
4800	169089.2	129182.0	284.4365	-1196206.2
4850	170993.4	130670.5	284.8312	-1210437.9
4900	172898.4	132159.8	285.2220	-1224689.3
4950	174804.3	133650.0	285.6090	-1238960.1
5000	176711.0	135141.0	285.9922	-1253250.1
5050	178618.4	136632.7	286.3718	-1267559.2
5100	180526.5	138125.1	286.7478	-1281887.3
5150	182435.4	139618.3	287.1203	-1296234.0
5200	184345.0	141112.2	287.4893	-1310599.2
5250	186255.3	142606.8	287.8549	-1324982.8
5300	188166.2	144102.0	288.2171	-1339384.7
5350	190077.8	145597.9	288.5761	-1353804.5
5400	191990.1	147094.5	288.9319	-1368242.2
5450	193903.0	148591.7	289.2845	-1382697.6
5500	195816.5	150089.5	289.6340	-1397170.6
5550	197730.6	151587.9	289.9805	-1411661.0
5600	199645.4	153087.0	290.3239	-1426168.6
5650	201560.7	154586.6	290.6644	-1440693.3
5700	203476.7	156086.9	291.0020	-1455235.0
5750	205393.2	157587.7	291.3368	-1469793.5
5800	207310.4	159089.2	291.6688	-1484368.6
5850	209228.2	160591.3	291.9980	-1498960.3
5900	211146.5	162093.9	292.3246	-1513568.4
5950	213065.5	163597.2	292.6484	-1528192.7
6000	214985.1	165101.1	292.9697	-1542833.2

$h_o = 0$

MOLECULAR HYDROGEN H_2

T	h(T)	u(T)	s(T)	g(T)
50	1427.1	1011.4	79.6973	-2557.7
100	2854.5	2023.1	99.4848	-7093.9
150	4282.8	3035.7	111.0664	-12377.2
200	5712.4	4049.6	119.2916	-18145.9
250	7143.9	5065.4	125.6802	-24276.1
300	8577.9	6083.7	130.9090	-30694.8
350	10014.8	7104.9	135.3390	-37353.8
400	11455.2	8129.6	139.1856	-44219.0
450	12899.5	9158.2	142.5877	-51265.0
500	14348.1	10191.1	145.6402	-58472.0
550	15801.6	11228.9	148.4106	-65824.3
600	17260.2	12271.8	150.9489	-73309.2
650	18724.4	13320.3	153.2929	-80916.0
700	20194.6	14374.8	155.4719	-88635.7
750	21671.1	15435.6	157.5092	-96460.8
800	23154.4	16503.2	159.4237	-104384.6
850	24644.6	17577.7	161.2305	-112401.4
900	26142.1	18659.5	162.9424	-120506.1
950	27647.2	19748.9	164.5699	-128694.2
1000	29160.2	20846.2	166.1220	-136961.8
1050	30681.3	21951.6	167.6062	-145305.3
1100	32210.7	23065.3	169.0292	-153721.4
1150	33748.7	24187.6	170.3965	-162207.3
1200	35295.5	25318.7	171.7131	-170760.2
1250	36851.2	26458.7	172.9832	-179377.8
1300	38416.1	27607.9	174.2107	-188057.8
1350	39990.1	28766.2	175.3988	-196798.2
1400	41573.6	29934.0	176.5505	-205597.1
1450	43166.5	31111.2	177.6684	-214452.7
1500	44769.0	32298.0	178.7549	-223363.4
1550	46381.2	33494.5	179.8122	-232327.7
1600	48003.0	34700.6	180.8420	-241344.2
1650	49634.6	35916.5	181.8461	-250411.5
1700	51275.9	37142.1	182.8260	-259528.4
1750	52927.0	38377.5	183.7832	-268693.7
1800	54587.8	39622.6	184.7190	-277906.3
1850	56258.3	40877.4	185.6344	-287165.2
1900	57938.5	42141.9	186.5305	-296469.4
1950	59628.2	43415.9	187.4083	-305818.0
2000	61327.4	44699.4	188.2687	-315210.0

H_2 (cont.)

T	h(T)	u(T)	s(T)	g(T)
2050	63036.0	45992.3	189.1125	-324644.6
2100	64753.8	47294.4	189.9404	-334121.0
2150	66480.7	48605.6	190.7531	-343638.4
2200	68216.6	49925.8	191.5512	-353196.0
2250	69961.2	51254.7	192.3353	-362793.2
2300	71714.4	52592.2	193.1060	-372429.3
2350	73475.9	53938.0	193.8636	-382103.6
2400	75245.5	55291.9	194.6087	-391815.5
2450	77023.0	56653.7	195.3418	-401564.3
2500	78808.1	58023.1	196.0630	-411349.5
2550	80600.4	59399.7	196.7729	-421170.4
2600	82399.8	60783.4	197.4717	-431026.6
2650	84205.8	62173.7	198.1597	-440917.4
2700	86018.2	63570.4	198.8372	-450842.4
2750	87836.5	64973.0	199.5045	-460800.9
2800	89660.4	66381.2	200.1618	-470792.6
2850	91489.5	67794.6	200.8093	-480817.0
2900	93339.6	69229.0	201.4528	-490873.5
2950	95169.6	70643.3	202.0784	-500961.8
3000	97004.0	72062.0	202.6951	-511081.2
3050	98842.8	73485.1	203.3029	-521231.1
3100	100685.5	74912.1	203.9022	-531411.3
3150	102515.3	76326.2	204.4878	-541621.1
3200	104362.0	77757.2	205.0694	-551860.1
3250	106213.5	79193.0	205.6435	-562128.0
3300	108069.7	80633.5	206.2103	-572424.3
3350	109930.5	82078.6	206.7700	-582748.9
3400	111795.9	83528.3	207.3227	-593101.2
3450	113665.6	84982.3	207.8686	-603481.0
3500	115539.7	86440.7	208.4079	-613888.0
3550	117418.2	87903.5	208.9408	-624321.7
3600	119300.8	89370.4	209.4674	-634781.9
3650	121187.6	90841.5	209.9879	-645268.3
3700	123078.4	92316.6	210.5024	-655780.6
3750	124973.3	93795.8	211.0111	-666318.5
3800	126872.0	95278.8	211.5141	-676881.6
3850	128774.7	96765.8	212.0115	-687469.8
3900	130681.1	98256.5	212.5035	-698082.7
3950	132591.2	99750.9	212.9902	-708720.1
4000	134504.9	101248.9	213.4716	-719381.6

H_2 (cont.)

T	h(T)	u(T)	s(T)	g(T)
4050	136422.3	102750.6	213.9480	-730067.2
4100	138343.1	104255.7	214.4194	-740776.4
4150	140267.5	105764.4	214.8859	-751509.0
4200	142195.2	107276.4	215.3476	-762264.9
4250	144126.2	108791.7	215.8047	-773043.7
4300	146060.5	110310.3	216.2571	-783845.3
4350	147997.9	111832.0	216.7051	-794669.3
4400	149938.6	113357.0	217.1487	-805515.7
4450	151882.3	114885.0	217.5880	-816384.1
4500	153829.0	116416.0	218.0230	-827274.4
4550	155778.7	117950.0	218.4539	-838186.4
4600	157731.3	119486.9	218.8807	-849119.7
4650	159686.8	121026.7	219.3035	-860074.4
4700	161645.0	122569.2	219.7224	-871050.0
4750	163606.0	124114.5	220.1374	-882046.5
4800	165569.8	125662.6	220.5486	-893063.7
4850	167536.1	127213.2	220.9562	-904101.3
4900	169505.0	128766.4	221.3601	-915159.3
4950	171476.5	130322.2	221.7604	-926237.3
5000	173450.5	131880.5	222.1571	-937335.2
5050	175426.9	133441.2	222.5505	-948452.9
5100	177405.8	135004.4	222.9404	-959590.2
5150	179387.0	136569.9	223.3270	-970746.9
5200	181370.5	138137.7	223.7103	-981922.9
5250	183356.2	139707.7	224.0903	-993117.9
5300	185344.2	141280.0	224.4672	-1004331.8
5350	187334.4	142854.5	224.8409	-1015564.6
5400	189326.7	144431.1	225.2116	-1026815.9
5450	191321.1	146009.8	225.5792	-1038085.7
5500	193317.6	147590.6	225.9439	-1049373.8
5550	195316.1	149173.4	226.3056	-1060680.0
5600	197316.6	150758.2	226.6644	-1072004.3
5650	199319.0	152344.9	227.0204	-1083346.4
5700	201323.4	153933.6	227.3736	-1094706.3
5750	203329.6	155524.1	227.7241	-1106083.7
5800	205337.7	157116.5	228.0718	-1117478.6
5850	207347.6	158710.7	228.4168	-1128890.9
5900	209359.3	160306.7	228.7592	-1140320.3
5950	211372.7	161904.4	229.0991	-1151766.7
6000	213387.9	163503.9	229.4363	-1163230.1

$h_o = -239,081.7$

STEAM H_2O

T	h(T)	u(T)	s(T)	g(T)
50	1567.7	1152.0	130.2327	-4943.9
100	3159.3	2327.9	152.2772	-12068.4
150	4774.9	3527.8	165.3722	-20030.9
200	6414.9	4752.1	174.8046	-28546.0
250	8079.4	6000.9	182.2311	-37478.4
300	9768.7	7274.5	188.3896	-46748.2
350	11482.9	8573.0	193.6737	-56302.9
400	13222.3	9896.7	198.3182	-66105.0
450	14986.9	11245.6	202.4744	-76126.6
500	16776.9	12619.9	206.2458	-86346.0
550	18592.4	14019.7	209.7061	-96746.0
600	20433.4	15445.0	212.9095	-107312.3
650	22299.9	16895.8	215.8973	-118033.3
700	24192.1	18372.3	218.7016	-128899.0
750	26109.9	19874.4	221.3477	-139900.9
800	28053.3	21402.1	223.8560	-151031.5
850	30022.2	22955.3	226.2431	-162284.4
900	32016.6	24534.0	228.5229	-173654.0
950	34036.4	26138.1	230.7069	-185135.1
1000	36081.5	27767.5	232.8048	-196723.3
1050	38151.8	29422.1	234.8249	-208414.3
1100	40247.0	31101.6	236.7742	-220204.6
1150	42367.1	32806.0	238.6589	-232090.7
1200	44511.8	34535.0	240.4844	-244069.5
1250	46680.8	36288.3	242.2552	-256138.2
1300	48874.0	38065.8	243.9755	-268294.2
1350	51091.1	39867.2	245.6490	-280535.0
1400	53331.8	41692.2	247.2787	-292858.3
1450	55595.7	43540.4	248.8675	-305262.1
1500	57882.6	45411.6	250.4180	-317744.4
1550	60192.0	47305.3	251.9325	-330303.3
1600	62523.7	49221.3	253.4130	-342937.1
1650	64877.1	51159.0	254.8613	-355644.1
1700	67251.8	53118.0	256.2792	-368422.7
1750	69647.5	55098.0	257.6680	-381271.5
1800	72063.6	57098.4	259.0293	-394189.1
1850	74499.7	59118.8	260.3642	-407174.0
1900	76955.1	61158.5	261.6738	-420225.1
1950	79429.4	63217.1	262.9592	-433341.0
2000	81922.1	65294.1	264.2213	-446520.6

H_2O (cont.)

T	h(T)	u(T)	s(T)	g(T)
2050	84432.4	67388.7	265.4611	-459762.8
2100	86959.9	69500.5	266.6792	-473066.4
2150	89503.9	71628.8	267.8764	-486430.3
2200	92063.7	73772.9	269.0533	-499853.7
2250	94638.6	75932.1	270.2106	-513335.3
2300	97228.0	78105.8	271.3489	-526874.4
2350	99831.1	80293.2	272.4685	-540469.9
2400	102447.2	82493.6	273.5701	-554121.0
2450	105075.5	84706.2	274.6539	-567826.6
2500	107715.2	86930.2	275.7205	-581586.1
2550	110365.6	89164.9	276.7702	-595398.4
2600	113025.7	91409.3	277.8033	-609262.8
2650	115694.7	93662.6	278.8201	-623178.5
2700	118371.8	95924.0	279.8209	-637144.5
2750	121056.0	98192.5	280.8059	-651160.3
2800	123746.4	100467.2	281.7754	-665224.9
2850	126442.0	102747.1	282.7297	-679337.6
2900	135712.3	111601.7	287.0597	-696760.8
2950	138735.8	114209.5	288.0934	-711139.7
3000	141777.0	116835.0	289.1157	-725570.0
3050	144835.2	119477.5	290.1267	-740051.1
3100	147910.1	122136.7	291.1267	-754582.5
3150	159332.4	133143.3	296.0882	-773345.4
3200	162842.3	136237.5	297.1937	-788177.5
3250	166386.0	139365.5	298.2925	-803064.7
3300	169963.6	142527.4	299.3849	-818006.7
3350	173575.2	145723.3	300.4711	-833003.1
3400	177221.2	148953.6	301.5514	-848053.7
3450	180901.7	152218.4	302.6261	-863158.1
3500	184616.9	155517.9	303.6952	-878316.2
3550	188367.0	158852.3	304.7590	-893527.6
3600	192152.2	162221.8	305.8178	-908792.0
3650	195972.7	165626.6	306.8718	-924109.3
3700	199828.7	169066.9	307.9210	-939479.1
3750	203720.5	172543.0	308.9658	-954901.3
3800	207648.2	176055.0	310.0063	-970375.6
3850	211612.0	179603.1	311.0426	-985901.9
3900	215612.2	183187.6	312.0749	-1001479.8
3950	219649.1	186808.8	313.1034	-1017109.3
4000	223722.7	190466.7	314.1282	-1032790.1

H_2O (cont.)

T	h(T)	u(T)	s(T)	g(T)
4050	227833.4	194161.7	315.1495	-1048522.0
4100	231981.3	197893.9	316.1674	-1064305.0
4150	236166.8	201663.7	317.1821	-1080138.7
4200	240390.0	205471.2	318.1936	-1096023.1
4250	244651.3	209316.8	319.2022	-1111958.0
4300	248950.7	213200.5	320.2079	-1127943.3
4350	253288.6	217122.7	321.2109	-1143978.8
4400	257665.2	221083.6	322.2113	-1160064.4
4450	262080.8	225083.5	323.2091	-1176199.9
4500	266535.6	229122.6	324.2046	-1192385.2
4550	271029.8	233201.1	325.1978	-1208620.3
4600	275563.8	237319.4	326.1889	-1224905.0
4650	280137.8	241477.7	327.1778	-1241239.1
4700	284752.0	245676.2	328.1648	-1257622.7
4750	289406.7	249915.2	329.1500	-1274055.6
4800	294102.2	254195.0	330.1333	-1290537.7
4850	298838.7	258515.8	331.1150	-1307068.9
4900	303616.6	262878.0	332.0950	-1323649.2
4950	308436.0	267281.7	333.0736	-1340278.4
5000	313297.3	271727.3	334.0508	-1356956.5
5050	318200.8	276215.1	335.0266	-1373683.4
5100	323146.7	280745.3	336.0012	-1390459.1
5150	328135.4	285318.3	336.9745	-1407283.5
5200	333167.0	289934.2	337.9468	-1424156.6
5250	338242.0	294593.5	338.9181	-1441078.2
5300	343360.5	299296.3	339.8885	-1458048.4
5350	348523.0	304043.1	340.8579	-1475067.0
5400	353729.6	308834.0	341.8266	-1492134.1
5450	358980.8	313669.5	342.7946	-1509249.7
5500	364276.7	318549.7	343.7619	-1526413.6
5550	369617.7	323475.0	344.7286	-1543625.9
5600	375004.2	328445.8	345.6948	-1560886.4
5650	380436.4	333462.3	346.6605	-1578195.3
5700	385914.7	338524.9	347.6258	-1595552.5
5750	391439.3	343633.8	348.5908	-1612957.9
5800	397010.6	348789.4	349.5555	-1630411.6
5850	402628.9	353992.0	350.5201	-1647913.5
5900	408294.5	359241.9	351.4844	-1665463.6
5950	414007.8	364539.5	352.4487	-1683061.9
6000	419769.1	369885.1	353.4129	-1700708.4

h_o = -393,404.9

CARBON DIOXIDE CO_2

T	h(T)	u(T)	s(T)	g(T)
50	1342.9	927.2	157.6778	-6541.0
100	2794.5	1963.1	177.7167	-14977.2
150	4350.0	3102.9	190.3031	-24195.4
200	6004.8	4342.0	199.8108	-33957.3
250	7754.4	5675.9	207.6111	-44148.4
300	9594.2	7100.0	214.3151	-54700.3
350	11520.0	8610.1	220.2492	-65567.2
400	13527.6	10202.0	225.6083	-76715.7
450	15612.9	11871.6	230.5188	-88120.5
500	17771.8	13614.8	235.0668	-99761.5
550	20000.5	15427.8	239.3141	-111622.2
600	22295.2	17306.8	243.3066	-123688.7
650	24652.2	19248.1	247.0791	-135949.2
700	27068.0	21248.2	250.6591	-148393.4
750	29539.0	23303.5	254.0684	-161012.3
800	32062.0	25410.8	257.3246	-173797.7
850	34633.7	27566.8	260.4425	-186742.4
900	37250.9	29768.3	263.4342	-199839.9
950	39910.7	32012.4	266.3101	-213083.9
1000	42610.1	34296.1	269.0791	-226469.1
1050	45346.3	36616.6	271.7490	-239990.2
1100	48116.5	38971.1	274.3263	-253642.4
1150	50918.3	41357.2	276.8171	-267421.4
1200	53749.1	43772.3	279.2266	-281322.8
1250	56606.6	46214.1	281.5595	-295342.8
1300	59488.4	48680.2	283.8200	-309477.5
1350	62392.4	51168.5	286.0119	-323723.6
1400	65316.6	53677.0	288.1387	-338077.6
1450	68259.0	56203.7	290.2037	-352536.5
1500	71217.7	58746.7	292.2098	-367097.0
1550	74191.1	61304.4	294.1597	-381756.5
1600	77177.5	63875.1	296.0560	-396512.1
1650	80175.4	66457.3	297.9010	-411361.2
1700	83183.3	69049.5	299.6969	-426301.4
1750	86200.1	71650.6	301.4459	-441330.2
1800	89224.4	74259.2	303.1498	-456445.2
1850	92255.3	76874.4	304.8107	-471644.4
1900	95291.7	79495.1	306.4302	-486925.6
1950	98332.8	82120.5	308.0100	-502286.8
2000	101377.8	84749.8	309.5519	-517726.0

CO_2 (cont.)

T	h(T)	u(T)	s(T)	g(T)
2050	104426.1	87382.4	311.0573	-533241.4
2100	107477.1	90017.7	312.5277	-548831.1
2150	110530.3	92655.2	313.9646	-564493.6
2200	113585.6	95294.8	315.3694	-580227.1
2250	116642.5	97936.0	316.7433	-596030.0
2300	119701.1	100578.9	318.0878	-611900.9
2350	122761.4	103223.5	319.4041	-627838.3
2400	125823.3	105869.7	320.6934	-643840.9
2450	128887.2	108517.9	321.9569	-659907.2
2500	131953.4	111168.4	323.1958	-676036.1
2550	135022.2	113821.5	324.4112	-692226.4
2600	138094.3	116477.9	325.6043	-708476.9
2650	141170.3	119138.2	326.7761	-724786.5
2700	144250.9	121803.1	327.9278	-741154.2
2750	147337.0	124473.5	329.0603	-757578.9
2800	150429.5	127150.3	330.1748	-774059.9
2850	153529.6	129834.7	331.2722	-790596.1
2900	156670.7	132560.1	332.3646	-807186.5
2950	159774.1	135247.8	333.4256	-823831.3
3000	162887.4	137945.4	334.4721	-840528.8
3050	166011.2	140653.5	335.5048	-857278.3
3100	169146.3	143372.9	336.5243	-874079.0
3150	172224.6	146035.5	337.5091	-890929.1
3200	175352.3	148747.5	338.4943	-907829.3
3250	178487.5	151467.0	339.4664	-924778.4
3300	181629.9	154193.7	340.4259	-941775.7
3350	184779.3	156927.4	341.3731	-958820.7
3400	187935.5	159667.9	342.3083	-975912.8
3450	191098.3	162415.0	343.2318	-993051.4
3500	194267.7	165168.7	344.1439	-1010235.8
3550	197443.4	167928.7	345.0448	-1027465.6
3600	200625.2	170694.8	345.9348	-1044740.1
3650	203812.9	173466.8	346.8142	-1062058.9
3700	207006.5	176244.7	347.6832	-1079421.4
3750	210205.7	179028.2	348.5421	-1096827.0
3800	213410.5	181817.3	349.3910	-1114275.4
3850	216620.5	184611.6	350.2303	-1131766.0
3900	219835.7	187411.1	351.0600	-1149298.3
3950	223055.9	190215.6	351.8804	-1166871.8
4000	226281.0	193025.0	352.6918	-1184486.2

CO_2 (cont.)

T	h(T)	u(T)	s(T)	g(T)
4050	229510.9	195839.2	353.4942	-1202140.9
4100	232745.3	198657.9	354.2880	-1219835.4
4150	235984.1	201481.0	355.0732	-1237569.5
4200	239227.2	204308.4	355.8500	-1255342.6
4250	242474.5	207140.0	356.6186	-1273154.4
4300	245725.9	209975.7	357.3791	-1291004.3
4350	248981.1	212815.2	358.1318	-1308892.1
4400	252240.1	215658.5	358.8767	-1326817.4
4450	255502.7	218505.4	359.6140	-1344779.7
4500	258768.8	221355.8	360.3439	-1362778.7
4550	262038.3	224209.6	361.0664	-1380814.0
4600	265311.1	227066.7	361.7818	-1398885.2
4650	268587.1	229927.0	362.4901	-1416992.0
4700	271866.1	232790.3	363.1915	-1435134.1
4750	275148.0	235656.5	363.8861	-1453311.1
4800	278432.8	238525.6	364.5740	-1471522.6
4850	281720.2	241397.3	365.2554	-1489768.4
4900	285010.3	244271.7	365.9303	-1508048.0
4950	288302.9	247148.6	366.5988	-1526361.3
5000	291597.9	250027.9	367.2611	-1544707.8
5050	294895.2	252909.5	367.9173	-1563087.3
5100	298194.8	255793.4	368.5675	-1581499.4
5150	301496.4	258679.3	369.2117	-1599943.9
5200	304800.2	261567.4	369.8501	-1618420.5
5250	308105.8	264457.3	370.4828	-1636928.9
5300	311413.4	267349.2	371.1098	-1655468.7
5350	314722.7	270242.8	371.7313	-1674039.7
5400	318033.8	273138.2	372.3473	-1692641.7
5450	321346.5	276035.2	372.9580	-1711274.4
5500	324660.8	278933.8	373.5633	-1729937.4
5550	327976.5	281833.8	374.1635	-1748630.6
5600	331293.7	284735.3	374.7585	-1767353.7
5650	334612.3	287638.2	375.3484	-1786106.4
5700	337932.1	290542.3	375.9334	-1804888.5
5750	341253.2	293447.7	376.5135	-1823699.7
5800	344575.5	296354.3	377.0888	-1842539.7
5850	347898.9	299262.0	377.6594	-1861408.5
5900	351223.3	302170.7	378.2252	-1880305.6
5950	354548.8	305080.5	378.7865	-1899230.9
6000	357875.3	307991.3	379.3432	-1918184.2

$h_o = -113,882.3$

CARBON MONOXIDE CO

T	h(T)	u(T)	s(T)	g(T)
50	1386.7	971.0	146.7138	-5949.0
100	2788.8	1957.4	166.1395	-13825.1
150	4206.2	2959.1	177.6295	-22438.2
200	5638.7	3975.9	185.8694	-31535.2
250	7086.0	5007.5	192.3273	-40995.8
300	8548.0	6053.8	197.6574	-50749.3
350	10024.3	7114.4	202.2084	-60748.6
400	11514.9	8189.3	206.1888	-70960.6
450	13019.4	9278.1	209.7326	-81360.3
500	14537.7	10380.7	212.9317	-91928.2
550	16069.5	11496.8	215.8514	-102648.8
600	17614.6	12626.2	218.5401	-113509.5
650	19172.8	13768.7	221.0344	-124499.6
700	20743.8	14924.0	223.3627	-135610.1
750	22327.4	16091.9	225.5477	-146833.4
800	23923.3	17272.1	227.6076	-158162.8
850	25531.3	18464.4	229.5573	-169592.4
900	27151.2	19668.6	231.4091	-181116.9
950	28782.8	20884.5	233.1733	-192731.8
1000	30425.6	22111.6	234.8586	-204432.9
1050	32079.6	23349.9	236.4725	-216216.5
1100	33744.5	24599.1	238.0214	-228079.1
1150	35419.9	25858.8	239.5109	-240017.6
1200	37105.6	27128.8	240.9458	-252029.3
1250	38801.5	28409.0	242.3303	-264111.4
1300	40507.1	29698.9	243.6681	-276261.5
1350	42222.2	30998.3	244.9627	-288477.5
1400	43946.6	32307.0	246.2169	-300757.1
1450	45679.9	33624.6	247.4334	-313098.5
1500	47422.0	34951.0	248.6146	-325499.9
1550	49172.4	36285.7	249.7625	-337959.4
1600	50931.0	37628.6	250.8792	-350475.6
1650	52697.5	38979.4	251.9663	-363046.9
1700	54471.5	40337.7	253.0254	-375671.8
1750	56252.7	41703.2	254.0581	-388349.0
1800	58040.9	43075.7	255.0656	-401077.2
1850	59835.8	44454.9	256.0492	-413855.1
1900	61637.1	45840.5	257.0099	-426681.7
1950	63444.4	47232.1	257.9488	-439555.8
2000	65257.5	48629.5	258.8668	-452476.2

CO (cont.)

T	h(T)	u(T)	s(T)	g(T)
2050	67076.0	50032.3	259.7649	-465442.1
2100	68899.6	51440.2	260.6438	-478452.4
2150	70728.1	52853.0	261.5043	-491506.2
2200	72561.1	54270.3	262.3471	-504602.5
2250	74398.2	55691.7	263.1728	-517740.6
2300	76239.3	57117.1	263.9821	-530919.6
2350	78083.8	58545.9	264.7755	-544138.6
2400	79931.6	59978.0	265.5535	-557396.8
2450	81782.2	61412.9	266.3167	-570693.7
2500	83635.4	62850.4	267.0655	-584028.3
2550	85490.8	64290.1	267.8003	-597400.0
2600	87348.1	65731.7	268.5216	-610808.1
2650	89206.9	67174.8	269.2297	-624251.9
2700	91066.9	68619.1	269.9251	-637730.8
2750	92927.7	70064.2	270.6080	-651244.2
2800	94789.0	71509.8	271.2787	-664791.4
2850	96650.4	72955.5	271.9377	-678371.9
2900	98522.2	74411.6	272.5886	-691984.6
2950	100376.7	75850.4	273.2226	-705629.9
3000	102232.1	77290.1	273.8462	-719306.7
3050	104088.1	78730.4	274.4598	-733014.4
3100	105944.5	80171.1	275.0636	-746752.5
3150	107800.7	81611.6	275.6574	-760520.2
3200	109661.7	83056.9	276.2436	-774317.8
3250	111525.1	84504.6	276.8214	-788144.5
3300	113390.6	85954.4	277.3910	-801999.8
3350	115258.2	87406.3	277.9527	-815883.4
3400	117128.0	88860.4	278.5067	-829794.9
3450	118999.7	90316.4	279.0532	-843734.0
3500	120873.3	91774.3	279.5924	-857700.1
3550	122748.7	93234.0	280.1245	-871693.1
3600	124626.0	94695.6	280.6496	-885712.5
3650	126504.9	96158.8	281.1679	-899757.9
3700	128385.5	97623.7	281.6796	-913829.2
3750	130267.7	99090.2	282.1849	-927925.8
3800	132151.5	100558.3	282.6840	-942047.5
3850	134036.7	102027.8	283.1768	-956194.1
3900	135923.3	103498.7	283.6637	-970365.1
3950	137811.3	104971.0	284.1447	-984560.4
4000	139700.6	106444.6	284.6200	-998779.5

CO (cont.)

T	h(T)	u(T)	s(T)	g(T)
4050	141591.2	107919.5	285.0897	-1013022.3
4100	143483.0	109395.6	285.5540	-1027288.4
4150	145375.9	110872.8	286.0129	-1041577.6
4200	147270.0	112351.2	286.4666	-1055889.6
4250	149165.2	113830.7	286.9151	-1070224.2
4300	151061.4	115311.2	287.3587	-1084581.0
4350	152958.7	116792.8	287.7974	-1098959.9
4400	154856.9	118275.3	288.2313	-1113360.7
4450	156756.1	119758.8	288.6605	-1127783.0
4500	158656.2	121243.2	289.0851	-1142226.6
4550	160557.1	122728.4	289.5052	-1156691.4
4600	162458.9	124214.5	289.9209	-1171177.1
4650	164361.6	125701.5	290.3323	-1185683.4
4700	166265.0	127189.2	290.7394	-1200210.2
4750	168169.3	128677.8	291.1424	-1214757.3
4800	170074.3	130167.1	291.5414	-1229324.4
4850	171980.0	131657.1	291.9364	-1243911.4
4900	173886.5	133147.9	292.3274	-1258518.0
4950	175793.7	134639.4	292.7147	-1273144.1
5000	177701.6	136131.6	293.0982	-1287789.4
5050	179610.1	137624.4	293.4780	-1302453.8
5100	181519.4	139118.0	293.8542	-1317137.1
5150	183429.4	140612.3	294.2269	-1331839.2
5200	185340.0	142107.2	294.5961	-1346559.8
5250	187251.3	143602.8	294.9619	-1361298.7
5300	189163.2	145099.0	295.3244	-1376055.9
5350	191075.9	146596.0	295.6836	-1390831.1
5400	192989.2	148093.6	296.0395	-1405624.2
5450	194903.2	149591.9	296.3923	-1420435.0
5500	196817.9	151090.9	296.7420	-1435263.4
5550	198733.3	152590.6	297.0887	-1450109.2
5600	200649.3	154090.9	297.4324	-1464972.2
5650	202566.2	155592.1	297.7732	-1479852.4
5700	204483.7	157093.9	298.1111	-1494749.5
5750	206402.0	158596.5	298.4462	-1509663.4
5800	208321.1	160099.9	298.7785	-1524594.1
5850	210241.0	161604.1	299.1081	-1539541.2
5900	212161.7	163109.1	299.4350	-1554504.8
5950	214083.3	164615.0	299.7593	-1569484.7
6000	216005.7	166121.7	300.0811	-1584480.7

h_o = -66,930.5

METHANE CH_4

T	h(T)	u(T)	s(T)	g(T)
50	906.4	490.7	134.8175	-5834.5
100	2011.6	1180.2	149.9831	-12986.8
150	3308.0	2060.9	160.4449	-20758.8
200	4788.4	3125.6	168.9380	-28999.2
250	6445.8	4367.3	176.3204	-37634.3
300	8273.3	5779.1	182.9751	-46619.2
350	10264.2	7354.3	189.1067	-55923.1
400	12412.1	9086.5	194.8381	-65523.2
450	14710.5	10969.2	200.2490	-75401.6
500	17153.3	12996.3	205.3940	-85543.7
550	19734.7	15162.0	210.3126	-95937.2
600	22448.8	17460.4	215.0341	-106571.6
650	25290.0	19885.9	219.5812	-117437.7
700	28253.0	22433.2	223.9716	-128527.2
750	31332.3	25096.8	228.2198	-139832.5
800	34523.0	27871.8	232.3375	-151347.0
850	37820.1	30753.2	236.3345	-163064.2
900	41218.8	33736.2	240.2193	-174978.5
950	44714.5	36816.2	243.9988	-187084.4
1000	48302.7	39988.7	247.6795	-199376.8
1050	51979.1	43249.4	251.2666	-211850.8
1100	55739.6	46594.2	254.7650	-224502.0
1150	59580.1	50019.0	258.1791	-237325.9
1200	63496.7	53519.9	261.5127	-250318.5
1250	67485.8	57093.3	264.7693	-263475.9
1300	71543.6	60735.4	267.9522	-276794.2
1350	75666.9	64443.0	271.0643	-290269.9
1400	79852.2	68212.6	274.1083	-303899.5
1450	84096.3	72041.0	277.0869	-317679.7
1500	88396.1	75925.1	280.0022	-331607.2
1550	92748.8	79862.1	282.8566	-345678.9
1600	97151.5	83849.1	285.6521	-359891.8
1650	101601.5	87883.4	288.3907	-374243.1
1700	106096.2	91962.4	291.0743	-388730.0
1750	110633.2	96083.7	293.7045	-403349.7
1800	115210.0	100244.8	296.2831	-418099.6
1850	119824.5	104443.6	298.8117	-432977.2
1900	124474.5	108677.9	301.2918	-447979.9
1950	129157.9	112945.6	303.7248	-463105.6
2000	133872.8	117244.8	306.1122	-478351.7

CH_4 (cont.)

T	h(T)	u(T)	s(T)	g(T)
2050	138617.4	121573.7	308.4553	-493716.0
2100	143389.9	125930.5	310.7554	-509196.5
2150	148188.7	130313.6	313.0138	-524790.9
2200	153012.2	134721.4	315.2316	-540497.2
2250	157859.0	139152.5	317.4099	-556313.4
2300	162727.6	143605.4	319.5500	-572237.5
2350	167616.7	148078.8	321.6530	-588267.8
2400	172525.1	152571.5	323.7197	-604402.2
2450	177451.7	157082.4	325.7514	-620639.2
2500	182395.4	161610.4	327.7489	-636976.8
2550	187355.1	166154.4	329.7132	-653413.5
2600	192330.0	170713.6	331.6452	-669947.6
2650	197319.1	175287.0	333.5459	-686577.5
2700	202321.7	179873.9	335.4161	-703301.7
2750	207337.0	184473.5	337.2566	-720118.6
2800	212364.3	189085.1	339.0683	-737026.8
2850	217403.1	193708.2	340.8519	-754025.0
2900	222452.6	198342.0	342.6083	-771111.6
2950	227512.4	202986.1	344.3382	-788285.4
3000	232582.0	207640.0	346.0423	-805545.0
3050	237661.0	212303.3	347.7214	-822889.2
3100	242748.9	216975.5	349.3760	-840316.7
3150	247845.4	221656.3	351.0069	-857826.4
3200	252950.2	226345.4	352.6148	-875417.0
3250	258063.1	231042.6	354.2002	-893087.5
3300	263183.6	235747.4	355.7637	-910836.7
3350	268311.7	240459.8	357.3060	-928663.5
3400	273447.1	245179.5	358.8276	-946566.9
3450	278589.6	249906.3	360.3291	-964545.9
3500	283739.1	254640.1	361.8110	-982599.5
3550	288895.5	259380.8	363.2739	-1000726.7
3600	294058.6	264128.2	364.7181	-1018926.6
3650	299228.3	268882.2	366.1442	-1037198.2
3700	304404.5	273642.7	367.5528	-1055540.7
3750	309587.1	278409.6	368.9441	-1073953.2
3800	314776.1	283182.9	370.3187	-1092434.8
3850	319971.3	287962.4	371.6769	-1110984.8
3900	325172.6	292748.0	373.0192	-1129602.3
3950	330380.0	297539.7	374.3459	-1148286.5
4000	335593.4	302337.4	375.6575	-1167036.6

CH_4 (cont.)

T	h(T)	u(T)	s(T)	g(T)
4050	340812.6	307140.9	376.9542	-1185852.0
4100	346037.6	311950.2	378.2364	-1204731.8
4150	351268.2	316765.1	379.5045	-1223675.4
4200	356504.3	321585.5	380.7586	-1242682.0
4250	361745.6	326411.1	381.9992	-1261751.0
4300	366992.1	331241.9	383.2265	-1280881.7
4350	372243.4	336077.5	384.4407	-1300073.4
4400	377499.4	340917.8	385.6420	-1319325.6
4450	382759.7	345762.4	386.8308	-1338637.4
4500	388024.2	350611.2	388.0072	-1358008.4
4550	393292.3	355463.6	389.1715	-1377438.0
4600	398563.8	360319.4	390.3237	-1396925.4
4650	403838.2	365178.1	391.4642	-1416470.1
4700	409115.0	370039.2	392.5929	-1436071.6
4750	414393.9	374902.4	393.7101	-1455729.2
4800	419674.1	379766.9	394.8159	-1475442.4
4850	424955.0	384632.1	395.9104	-1495210.6
4900	430236.1	389497.5	396.9938	-1515033.3
4950	435516.6	394362.3	398.0659	-1534909.8
5000	440795.7	399225.7	399.1271	-1554839.7
5050	446072.5	404086.8	400.1772	-1574822.3
5100	451346.2	408944.8	401.2164	-1594857.2
5150	456615.7	413798.6	402.2446	-1614943.8
5200	461880.1	418647.3	403.2618	-1635081.5
5250	467138.2	423489.7	404.2682	-1655269.8
5300	472388.8	428324.6	405.2636	-1675508.1
5350	477630.6	433150.7	406.2480	-1695796.0
5400	482862.3	437966.7	407.2213	-1716132.8
5450	488082.4	442771.1	408.1835	-1736517.9
5500	493289.4	447562.4	409.1346	-1756950.9
5550	498481.8	452339.1	410.0744	-1777431.2
5600	503657.7	457099.3	411.0028	-1797958.2
5650	508815.4	461841.3	411.9198	-1818531.3
5700	513952.9	466563.1	412.8251	-1839150.0
5750	519068.4	471262.9	413.7186	-1859813.6
5800	524159.6	475938.4	414.6002	-1880521.6
5850	529224.4	480587.5	415.4697	-1901273.4
5900	534260.5	485207.9	416.3269	-1922068.4
5950	539265.4	489797.1	417.1716	-1942905.9
6000	544236.6	494352.6	418.0037	-1963785.3

$h_o = 89,914.7$

NITRIC OXIDE NO

T	h(T)	u(T)	s(T)	g(T)
50	1461.9	1046.2	156.6545	-6370.8
100	2936.2	2104.8	177.0823	-14772.1
150	4422.7	3175.6	189.1341	-23947.4
200	5921.6	4258.8	197.7561	-33629.7
250	7432.5	5354.0	204.4984	-43692.1
300	8955.6	6461.4	210.0514	-54059.8
350	10490.6	7580.7	214.7834	-64683.6
400	12037.5	8711.9	218.9142	-75528.2
450	13596.2	9854.9	222.5857	-86567.4
500	15166.5	11009.5	225.8944	-97780.7
550	16748.3	12175.6	228.9096	-109151.9
600	18341.6	13353.2	231.6822	-120667.7
650	19946.2	14542.1	234.2507	-132316.8
700	21561.9	15742.1	236.6454	-144089.8
750	23188.7	16953.2	238.8900	-155978.8
800	24826.2	18175.0	241.0036	-167976.7
850	26474.5	19407.6	243.0021	-180077.3
900	28133.4	20650.8	244.8984	-192275.2
950	29802.6	21904.3	246.7033	-204565.6
1000	31482.0	23168.0	248.4262	-216944.1
1050	33171.5	24441.8	250.0747	-229407.0
1100	34870.8	25725.4	251.6557	-241950.5
1150	36579.7	27018.6	253.1750	-254571.5
1200	38298.1	28321.3	254.6376	-267267.0
1250	40025.7	29633.2	256.0481	-280034.4
1300	41762.4	30954.2	257.4104	-292871.0
1350	43507.9	32284.0	258.7279	-305774.7
1400	45262.0	33622.4	260.0037	-318743.1
1450	47024.5	34969.2	261.2406	-331774.4
1500	48795.1	36324.1	262.4411	-344866.6
1550	50573.6	37686.9	263.6075	-358017.9
1600	52359.8	39057.4	264.7416	-371226.8
1650	54153.4	40435.3	265.8454	-384491.6
1700	55954.1	41820.3	266.9206	-397810.9
1750	57761.7	43212.2	267.9685	-411183.2
1800	59575.9	44610.7	268.9906	-424607.3
1850	61396.4	46015.5	269.9883	-438081.9
1900	63223.0	47426.4	270.9625	-451605.7
1950	65055.3	48843.0	271.9144	-465177.7
2000	66893.1	50265.1	272.8450	-478796.8

NO (cont.)

T	h(T)	u(T)	s(T)	g(T)
2050	68736.1	51692.4	273.7551	-492461.9
2100	70583.9	53124.5	274.6456	-506172.0
2150	72436.2	54561.1	275.5174	-519926.1
2200	74292.8	56002.0	276.3710	-533723.4
2250	76153.2	57446.7	277.2072	-547563.0
2300	78017.3	58895.1	278.0266	-561443.9
2350	79884.5	60346.6	278.8297	-575365.3
2400	81754.6	61801.0	279.6172	-589326.6
2450	83627.3	63258.0	280.3894	-603326.8
2500	85502.1	64717.1	281.1470	-617365.3
2550	87378.8	66178.1	281.8902	-631441.3
2600	89256.9	67640.5	282.6196	-645554.1
2650	91136.0	69103.9	283.3355	-659703.0
2700	93015.9	70568.1	284.0382	-673887.4
2750	94896.0	72032.5	284.7282	-688106.6
2800	96776.1	73496.9	285.4057	-702360.0
2850	98655.6	74960.7	286.0711	-716647.0
2900	100547.4	76436.8	286.7289	-730966.4
2950	102417.1	77890.8	287.3681	-745318.9
3000	104287.3	79345.3	287.9968	-759703.1
3050	106157.9	80800.2	288.6152	-774118.4
3100	108028.6	82255.2	289.2236	-788564.4
3150	109898.7	83709.6	289.8219	-803040.2
3200	111774.4	85169.6	290.4127	-817546.1
3250	113652.7	86632.2	290.9951	-832081.3
3300	115533.5	88097.3	291.5694	-846645.5
3350	117416.7	89564.8	292.1358	-861238.1
3400	119302.4	91034.8	292.6945	-875858.9
3450	121190.4	92507.1	293.2457	-890507.5
3500	123080.6	93981.6	293.7897	-905183.4
3550	124973.1	95458.4	294.3266	-919886.3
3600	126867.8	96937.4	294.8566	-934615.9
3650	128764.7	98418.6	295.3799	-949371.9
3700	130663.6	99901.8	295.8966	-964153.8
3750	132564.6	101387.1	296.4069	-978961.4
3800	134467.6	102874.4	296.9110	-993794.4
3850	136372.5	104363.6	297.4091	-1008652.4
3900	138279.4	105854.8	297.9012	-1023535.2
3950	140188.1	107347.8	298.3875	-1038442.4
4000	142098.7	108842.7	298.8681	-1053373.8

NO (cont.)

T	h(T)	u(T)	s(T)	g(T)
4050	144011.0	110339.3	299.3433	-1068329.2
4100	145925.1	111837.7	299.8130	-1083308.1
4150	147840.9	113337.8	300.2774	-1098310.4
4200	149758.4	114839.6	300.7367	-1113335.7
4250	151677.5	116343.0	301.1909	-1128384.0
4300	153598.2	117848.0	301.6402	-1143454.7
4350	155520.4	119354.5	302.0847	-1158547.9
4400	157444.2	120862.6	302.5244	-1173663.1
4450	159369.5	122372.2	302.9595	-1188800.3
4500	161296.2	123883.2	303.3900	-1203959.0
4550	163224.4	125395.7	303.8162	-1219139.2
4600	165153.9	126909.5	304.2379	-1234340.6
4650	167084.9	128424.8	304.6554	-1249562.9
4700	169017.1	129941.3	305.0688	-1264806.0
4750	170950.7	131459.2	305.4780	-1280069.7
4800	172885.6	132978.4	305.8832	-1295353.8
4850	174821.7	134498.8	306.2845	-1310658.0
4900	176759.1	136020.5	306.6819	-1325982.1
4950	178697.7	137543.4	307.0755	-1341326.1
5000	180637.4	139067.4	307.4654	-1356689.6
5050	182578.4	140592.7	307.8517	-1372072.6
5100	184520.5	142119.1	308.2344	-1387474.7
5150	186463.7	143646.6	308.6135	-1402896.0
5200	188408.0	145175.2	308.9892	-1418336.0
5250	190353.4	146704.9	309.3616	-1433794.8
5300	192299.9	148235.7	309.7306	-1449272.1
5350	194247.4	149767.5	310.0963	-1464767.8
5400	196196.0	151300.4	310.4588	-1480281.7
5450	198145.6	152834.3	310.8182	-1495813.7
5500	200096.3	154369.3	311.1745	-1511363.5
5550	202047.9	155905.2	311.5277	-1526931.1
5600	204000.5	157442.1	311.8780	-1542516.2
5650	205954.1	158980.0	312.2253	-1558118.8
5700	207908.7	160518.9	312.5697	-1573738.7
5750	209864.2	162058.7	312.9113	-1589375.7
5800	211820.7	163599.5	313.2501	-1605029.8
5850	213778.2	165141.3	313.5861	-1620700.7
5900	215736.5	166683.9	313.9195	-1636388.3
5950	217695.9	168227.6	314.2502	-1652092.6
6000	219656.1	169772.1	314.5782	-1667813.3

h_o = 246,923.1

ATOMIC OXYGEN O

T	h(T)	u(T)	s(T)	g(T)
50	1143.9	728.2	120.7298	-4892.6
100	2277.9	1446.5	136.4587	-11367.9
150	3402.7	2155.6	145.5825	-18434.6
200	4518.9	2856.1	152.0056	-25882.2
250	5627.0	3548.5	156.9514	-33610.9
300	6727.5	4233.3	160.9647	-41562.0
350	7821.0	4911.1	164.3363	-49696.7
400	8908.0	5582.4	167.2394	-57987.8
450	9988.9	6247.6	169.7860	-66414.8
500	11064.4	6907.4	172.0523	-74961.7
550	12134.8	7562.1	174.0927	-83616.2
600	13200.5	8212.1	175.9474	-92367.9
650	14262.1	8858.0	177.6469	-101208.4
700	15319.8	9500.0	179.2147	-110130.4
750	16374.2	10138.7	180.6696	-119128.0
800	17425.5	10774.3	182.0266	-128195.8
850	18474.2	11407.3	183.2981	-137329.2
900	19520.5	12037.9	184.4943	-146524.3
950	20564.8	12666.5	185.6235	-155777.5
1000	21607.4	13293.4	186.6931	-165085.7
1050	22648.6	13918.9	187.7091	-174445.9
1100	23688.7	14543.3	188.6768	-183855.8
1150	24727.8	15166.7	189.6006	-193312.9
1200	25766.3	15789.5	190.4846	-202815.2
1250	26804.4	16411.9	191.3321	-212360.7
1300	27842.2	17034.0	192.1462	-221947.8
1350	28880.0	17656.1	192.9295	-231574.8
1400	29917.8	18278.2	193.6844	-241240.3
1450	30955.9	18900.6	194.4130	-250942.8
1500	31994.5	19523.5	195.1171	-260681.2
1550	33033.5	20146.8	195.7985	-270454.2
1600	34073.1	20770.7	196.4586	-280260.7
1650	35113.4	21395.3	197.0989	-290099.7
1700	36154.5	22020.7	197.7204	-299970.3
1750	37196.3	22646.8	198.3244	-309871.5
1800	38239.0	23273.8	198.9119	-319802.4
1850	39282.5	23901.6	199.4837	-329762.4
1900	40326.8	24530.2	200.0407	-339750.6
1950	41371.9	25159.6	200.5837	-349766.2
2000	42417.8	25789.8	201.1133	-359808.7

O (cont.)

T	h(T)	u(T)	s(T)	g(T)
2050	43464.4	26420.7	201.6301	-369877.3
2100	44511.6	27052.2	202.1348	-379971.5
2150	45559.4	27684.3	202.6279	-390090.6
2200	46607.5	28316.7	203.1098	-400234.1
2250	47656.0	28949.5	203.5811	-410401.4
2300	48704.6	29582.4	204.0420	-420592.0
2350	49753.2	30215.3	204.4930	-430805.5
2400	50801.6	30848.0	204.9345	-441041.2
2450	51849.6	31480.3	205.3667	-451298.8
2500	52897.0	32112.0	205.7899	-461577.7
2550	53943.6	32742.9	206.2044	-471877.6
2600	54989.0	33372.6	206.6104	-482198.0
2650	56033.2	34001.1	207.0082	-492538.5
2700	57075.7	34627.9	207.3979	-502898.7
2750	58116.3	35252.8	207.7798	-513278.2
2800	59154.6	35875.4	208.1540	-523676.5
2850	60190.3	36495.4	208.5206	-534093.4
2900	61208.0	37097.4	208.8749	-544529.2
2950	62248.6	37722.3	209.2307	-554981.9
3000	63287.6	38345.6	209.5799	-565452.2
3050	64324.8	38967.1	209.9228	-575939.8
3100	65360.0	39586.6	210.2595	-586444.4
3150	66456.4	40267.3	210.6109	-596968.0
3200	67510.7	40905.9	210.9430	-607506.8
3250	68565.6	41545.1	211.2701	-618062.2
3300	69621.1	42184.9	211.5924	-628633.8
3350	70677.2	42825.3	211.9100	-639221.3
3400	71733.9	43466.3	212.2231	-649824.7
3450	72791.3	44108.0	212.5319	-660443.6
3500	73849.4	44750.4	212.8364	-671077.8
3550	74908.2	45393.5	213.1367	-681727.2
3600	75967.7	46037.3	213.4331	-692391.4
3650	77028.0	46681.9	213.7256	-703070.4
3700	78089.0	47327.2	214.0143	-713763.9
3750	79150.8	47973.3	214.2994	-724471.8
3800	80213.4	48620.2	214.5808	-735193.8
3850	81276.8	49267.9	214.8588	-745929.8
3900	82341.0	49916.4	215.1335	-756679.6
3950	83406.0	50565.7	215.4048	-767443.1
4000	84471.9	51215.9	215.6730	-778220.0

O (cont.)

T	h(T)	u(T)	s(T)	g(T)
4050	85538.7	51867.0	215.9380	-789010.3
4100	86606.4	52519.0	216.2000	-799813.8
4150	87674.9	53171.8	216.4591	-810630.3
4200	88744.4	53825.6	216.7153	-821459.7
4250	89814.8	54480.3	216.9686	-832301.8
4300	90886.1	55135.9	217.2192	-843156.5
4350	91958.4	55792.5	217.4671	-854023.6
4400	93031.7	56450.1	217.7125	-864903.1
4450	94105.9	57108.6	217.9552	-875794.8
4500	95181.1	57768.1	218.1955	-886698.6
4550	96257.3	58428.6	218.4333	-897614.4
4600	97334.5	59090.1	218.6688	-908541.9
4650	98412.8	59752.7	218.9019	-919481.2
4700	99492.1	60416.3	219.1328	-930432.1
4750	100572.4	61080.9	219.3614	-941394.4
4800	101653.7	61746.5	219.5879	-952368.2
4850	102736.1	62413.2	219.8122	-963353.2
4900	103819.6	63081.0	220.0345	-974349.4
4950	104904.2	63749.9	220.2547	-985356.6
5000	105989.8	64419.8	220.4729	-996374.8
5050	107076.5	65090.8	220.6892	-1007403.9
5100	108164.3	65762.9	220.9035	-1018443.7
5150	109253.2	66436.1	221.1160	-1029494.2
5200	110343.2	67110.4	221.3266	-1040555.3
5250	111434.4	67785.9	221.5355	-1051626.8
5300	112526.6	68462.4	221.7425	-1062708.8
5350	113620.0	69140.1	221.9479	-1073801.0
5400	114714.4	69818.8	222.1515	-1084903.5
5450	115810.1	70498.8	222.3534	-1096016.2
5500	116906.8	71179.8	222.5538	-1107138.9
5550	118004.7	71862.0	222.7525	-1118271.5
5600	119103.7	72545.3	222.9496	-1129414.1
5650	120203.9	73229.8	223.1452	-1140566.5
5700	121305.2	73915.4	223.3392	-1151728.6
5750	122407.6	74602.1	223.5318	-1162900.4
5800	123511.2	75290.0	223.7229	-1174081.7
5850	124615.9	75979.0	223.9126	-1185272.6
5900	125721.8	76669.2	224.1008	-1196473.0
5950	126828.8	77360.5	224.2876	-1207682.7
6000	127937.0	78053.0	224.4731	-1218901.7

$h_o = 471,368.6$

ATOMIC NITROGEN N

T	h(T)	u(T)	s(T)	g(T)
50	1038.9	623.2	116.0776	-4765.0
100	2077.9	1246.5	130.4814	-10970.2
150	3117.0	1869.9	138.9078	-17719.1
200	4156.2	2493.4	144.8869	-24821.1
250	5195.5	3117.0	149.5250	-32185.8
300	6234.8	3740.6	153.3147	-39759.6
350	7274.1	4364.2	156.5191	-47507.5
400	8313.5	4987.9	159.2949	-55404.4
450	9352.9	5611.6	161.7433	-63431.6
500	10392.3	6235.3	163.9336	-71574.5
550	11431.8	6859.1	165.9149	-79821.5
600	12471.2	7482.8	167.7238	-88163.1
650	13510.6	8106.5	169.3877	-96591.4
700	14550.0	8730.2	170.9283	-105099.8
750	15589.4	9353.9	172.3625	-113682.5
800	16628.7	9977.5	173.7040	-122334.5
850	17668.0	10601.1	174.9642	-131051.5
900	18707.3	11224.7	176.1523	-139829.7
950	19746.6	11848.3	177.2761	-148665.7
1000	20785.8	12471.8	178.3422	-157556.4
1050	21825.0	13095.3	179.3563	-166499.0
1100	22864.2	13718.8	180.3231	-175491.2
1150	23903.4	14342.3	181.2469	-184530.6
1200	24942.5	14965.7	182.1314	-193615.3
1250	25981.6	15589.1	182.9798	-202743.2
1300	27020.6	16212.4	183.7948	-211912.7
1350	28059.7	16835.8	184.5791	-221122.1
1400	29098.7	17459.1	185.3349	-230370.1
1450	30137.7	18082.4	186.0641	-239655.2
1500	31176.7	18705.7	186.7686	-248976.1
1550	32215.8	19329.1	187.4500	-258331.7
1600	33254.8	19952.4	188.1097	-267720.7
1650	34293.9	20575.8	188.7492	-277142.3
1700	35333.0	21199.2	189.3696	-286595.3
1750	36372.1	21822.6	189.9721	-296079.0
1800	37411.3	22446.1	190.5576	-305592.3
1850	38450.6	23069.7	191.1271	-315134.5
1900	39490.0	23693.4	191.6814	-324704.7
1950	40529.4	24317.1	192.2214	-334302.4
2000	41569.0	24941.0	192.7478	-343926.6

N (cont.)

T	h(T)	u(T)	s(T)	g(T)
2050	42608.6	25564.9	193.2612	-353576.9
2100	43648.4	26189.0	193.7624	-363252.6
2150	44688.4	26813.3	194.2518	-372953.0
2200	45728.5	27437.7	194.7300	-382677.6
2250	46768.9	28062.4	195.1976	-392425.8
2300	47809.4	28687.2	195.6550	-402197.1
2350	48850.2	29312.3	196.1027	-411991.1
2400	49891.2	29937.6	196.5410	-421807.3
2450	50932.5	30563.2	196.9704	-431645.1
2500	51974.1	31189.1	197.3913	-441504.2
2550	53016.0	31815.3	197.8039	-451384.1
2600	54058.2	32441.8	198.2087	-461284.4
2650	55100.8	33068.7	198.6059	-471204.8
2700	56143.8	33696.0	198.9958	-481144.9
2750	57187.2	34323.7	199.3787	-491104.3
2800	58231.0	34951.8	199.7549	-501082.6
2850	59275.3	35580.4	200.1246	-511079.7
2900	60337.1	36226.5	200.4913	-521087.7
2950	61376.7	36850.4	200.8467	-531121.1
3000	62417.6	37475.6	201.1966	-541172.2
3050	63459.8	38102.1	201.5412	-551240.7
3100	64503.4	38730.0	201.8805	-561326.3
3150	65544.4	39355.3	202.2113	-571421.2
3200	66589.2	39984.4	202.5404	-581540.0
3250	67636.5	40616.0	202.8651	-591675.1
3300	68686.3	41250.1	203.1857	-601826.4
3350	69738.8	41886.9	203.5022	-611993.6
3400	70793.9	42526.3	203.8149	-622176.6
3450	71851.8	43168.5	204.1237	-632375.1
3500	72912.6	43813.6	204.4290	-642588.9
3550	73976.3	44461.6	204.7308	-652817.9
3600	75043.0	45112.6	205.0291	-663061.9
3650	76112.8	45766.7	205.3243	-673320.8
3700	77185.7	46423.9	205.6162	-683594.3
3750	78261.9	47084.4	205.9051	-693882.3
3800	79341.3	47748.1	206.1911	-704184.7
3850	80424.1	48415.2	206.4742	-714501.4
3900	81510.3	49085.7	206.7545	-724832.1
3950	82600.0	49759.7	207.0321	-735176.8
4000	83693.2	50437.2	207.3071	-745535.3

N (cont.)

T	h(T)	u(T)	s(T)	g(T)
4050	84790.0	51118.3	207.5796	-755907.5
4100	85890.5	51803.1	207.8497	-766293.2
4150	86994.7	52491.6	208.1174	-776692.4
4200	88102.7	53183.9	208.3828	-787104.9
4250	89214.5	53880.0	208.6459	-797530.6
4300	90330.2	54580.0	208.9069	-807969.5
4350	91449.8	55283.9	209.1658	-818421.3
4400	92573.5	55991.9	209.4226	-828886.0
4450	93701.1	56703.8	209.6774	-839363.5
4500	94832.8	57419.8	209.9303	-849853.7
4550	95968.7	58140.0	210.1814	-860356.5
4600	97108.8	58864.4	210.4306	-870871.8
4650	98253.0	59592.9	210.6780	-881399.5
4700	99401.5	60325.7	210.9236	-891939.6
4750	100554.3	61062.8	211.1676	-902491.9
4800	101711.5	61804.3	211.4100	-913056.3
4850	102873.0	62550.1	211.6507	-923632.8
4900	104038.9	63300.3	211.8899	-934221.4
4950	105209.3	64055.0	212.1275	-944821.8
5000	106384.1	64814.1	212.3636	-955434.1
5050	107563.5	65577.8	212.5983	-966058.1
5100	108747.4	66346.0	212.8316	-976693.9
5150	109935.8	67118.7	213.0635	-987341.3
5200	111128.9	67896.1	213.2941	-998000.2
5250	112326.5	68678.0	213.5233	-1008670.7
5300	113528.8	69464.6	213.7512	-1019352.5
5350	114735.8	70255.9	213.9779	-1030045.8
5400	115947.4	71051.8	214.2033	-1040750.3
5450	117163.8	71852.5	214.4275	-1051466.1
5500	118384.9	72657.9	214.6505	-1062193.0
5550	119610.7	73468.0	214.8724	-1072931.1
5600	120841.3	74282.9	215.0931	-1083680.2
5650	122076.6	75102.5	215.3127	-1094440.4
5700	123316.7	75926.9	215.5313	-1105211.5
5750	124561.6	76756.1	215.7487	-1115993.5
5800	125811.3	77590.1	215.9651	-1126786.4
5850	127065.8	78428.9	216.1805	-1137590.0
5900	128325.2	79272.6	216.3948	-1148404.4
5950	129589.3	80121.0	216.6082	-1159229.5
6000	130858.3	80974.3	216.8206	-1170065.2

h_o = 216,110.1

ATOMIC HYDROGEN H

T	h(T)	u(T)	s(T)	g(T)
50	1039.3	623.6	77.4927	-2835.4
100	2078.5	1247.1	91.8998	-7111.5
150	3117.8	1870.7	100.3274	-11931.4
200	4157.0	2494.2	106.3068	-17104.4
250	5196.3	3117.8	110.9449	-22540.0
300	6235.5	3741.3	114.7344	-28184.8
350	7274.8	4364.9	117.9384	-34003.7
400	8314.0	4988.4	120.7139	-39971.6
450	9353.3	5612.0	123.1620	-46069.7
500	10392.5	6235.5	125.3519	-52283.5
550	11431.8	6859.1	127.3329	-58601.4
600	12471.0	7482.6	129.1415	-65013.9
650	13510.3	8106.2	130.8052	-71513.1
700	14549.5	8729.7	132.3455	-78092.4
750	15588.8	9353.3	133.7795	-84745.9
800	16628.0	9976.8	135.1210	-91468.8
850	17667.3	10600.4	136.3810	-98256.6
900	18706.5	11223.9	137.5691	-105105.7
950	19745.8	11847.5	138.6929	-112012.5
1000	20785.0	12471.0	139.7590	-118974.0
1050	21824.3	13094.6	140.7731	-125987.5
1100	22863.5	13718.1	141.7400	-133050.5
1150	23902.8	14341.7	142.6639	-140160.8
1200	24942.0	14965.2	143.5485	-147316.3
1250	25981.3	15588.8	144.3970	-154515.0
1300	27020.5	16212.3	145.2122	-161755.4
1350	28059.8	16835.9	145.9967	-169035.7
1400	29099.0	17459.4	146.7526	-176354.6
1450	30138.3	18083.0	147.4819	-183710.6
1500	31177.5	18706.5	148.1866	-191102.4
1550	32216.8	19330.1	148.8681	-198528.8
1600	33256.0	19953.6	149.5280	-205988.8
1650	34295.3	20577.2	150.1676	-213481.3
1700	35334.5	21200.7	150.7881	-221005.3
1750	36373.8	21824.3	151.3906	-228559.8
1800	37413.0	22447.8	151.9761	-236144.0
1850	38452.3	23071.4	152.5456	-243757.2
1900	39491.5	23694.9	153.0999	-251398.4
1950	40530.8	24318.5	153.6398	-259066.9
2000	41570.0	24942.0	154.1661	-266762.1

H (cont.)

T	h(T)	u(T)	s(T)	g(T)
2050	42609.3	25565.6	154.6793	-274483.3
2100	43648.5	26189.1	155.1802	-282229.8
2150	44687.8	26812.7	155.6692	-290001.1
2200	45727.0	27436.2	156.1471	-297796.6
2250	46766.3	28059.8	156.6142	-305615.6
2300	47805.5	28683.3	157.0710	-313457.8
2350	48844.8	29306.9	157.5180	-321322.6
2400	49884.0	29930.4	157.9556	-329209.5
2450	50923.3	30554.0	158.3842	-337118.0
2500	51962.5	31177.5	158.8041	-345047.7
2550	53001.8	31801.1	159.2157	-352998.3
2600	54041.0	32424.6	159.6193	-360969.2
2650	55080.3	33048.2	160.0152	-368960.1
2700	56119.5	33671.7	160.4037	-376970.6
2750	57158.8	34295.3	160.7851	-385000.3
2800	58198.0	34918.8	161.1596	-393049.0
2850	59237.3	35542.4	161.5275	-401116.2
2900	60276.5	36165.9	161.8890	-409201.6
2950	61315.8	36789.5	162.2443	-417305.0
3000	62355.0	37413.0	162.5936	-425425.9
3050	63394.3	38036.6	162.9372	-433564.2
3100	64433.5	38660.1	163.2752	-441719.6
3150	65472.8	39283.7	163.6078	-449891.7
3200	66512.0	39907.2	163.9351	-458080.3
3250	67551.3	40530.8	164.2573	-466285.1
3300	68590.5	41154.3	164.5747	-474505.9
3350	69629.8	41777.9	164.8872	-482742.5
3400	70669.0	42401.4	165.1952	-490994.6
3450	71708.3	43025.0	165.4986	-499261.9
3500	72747.5	43648.5	165.7977	-507544.3
3550	73786.8	44272.1	166.0925	-515841.6
3600	74826.0	44895.6	166.3832	-524153.5
3650	75865.2	45519.2	166.6699	-532479.9
3700	76904.5	46142.7	166.9527	-540820.4
3750	77943.8	46766.3	167.2317	-549175.1
3800	78983.0	47389.8	167.5070	-557543.6
3850	80022.3	48013.4	167.7787	-565925.7
3900	81061.5	48636.9	168.0469	-574321.4
3950	82100.8	49260.5	168.3117	-582730.3
4000	83140.0	49884.0	168.5731	-591152.5

H (cont.)

T	h(T)	u(T)	s(T)	g(T)
4050	84179.3	50507.5	168.8313	-599587.6
4100	85218.5	51131.1	169.0864	-608035.6
4150	86257.7	51754.7	169.3383	-616496.2
4200	87297.0	52378.2	169.5872	-624969.3
4250	88336.3	53001.8	169.8332	-633454.9
4300	89375.5	53625.3	170.0763	-641952.6
4350	90414.7	54248.9	170.3166	-650462.4
4400	91454.0	54872.4	170.5541	-658984.2
4450	92493.3	55496.0	170.7890	-667517.8
4500	93532.5	56119.5	171.0212	-676063.1
4550	94571.7	56743.0	171.2509	-684619.9
4600	95611.0	57366.6	171.4781	-693188.1
4650	96650.3	57990.2	171.7028	-701767.7
4700	97689.5	58613.7	171.9251	-710358.4
4750	98728.8	59237.3	172.1450	-718960.1
4800	99768.0	59860.8	172.3627	-727572.8
4850	100807.3	60484.4	172.5781	-736196.4
4900	101846.5	61107.9	172.7912	-744830.6
4950	102885.8	61731.5	173.0023	-753475.4
5000	103925.0	62355.0	173.2112	-762130.8
5050	104964.3	62978.5	173.4180	-770796.5
5100	106003.5	63602.1	173.6228	-779472.5
5150	107042.8	64225.7	173.8255	-788158.8
5200	108082.0	64849.2	174.0264	-796855.1
5250	109121.3	65472.8	174.2253	-805561.4
5300	110160.5	66096.3	174.4223	-814277.6
5350	111199.8	66719.9	174.6174	-823003.6
5400	112239.0	67343.4	174.8108	-831739.3
5450	113278.3	67967.0	175.0024	-840484.6
5500	114317.5	68590.5	175.1922	-849239.5
5550	115356.8	69214.0	175.3803	-858003.8
5600	116396.0	69837.6	175.5667	-866777.5
5650	117435.3	70461.2	175.7515	-875560.4
5700	118474.5	71084.7	175.9346	-884352.6
5750	119513.8	71708.3	176.1161	-893153.9
5800	120553.0	72331.8	176.2961	-901964.2
5850	121592.3	72955.4	176.4745	-910783.5
5900	122631.5	73578.9	176.6514	-919611.6
5950	123670.8	74202.5	176.8268	-928448.6
6000	124710.0	74826.0	177.0007	-937294.3

$h_o = 0$

ATOMIC ARGON A

T	h(T)	u(T)	s(T)	g(T)
50	1039.3	623.6	81.3114	-3026.3
100	2078.5	1247.1	95.7185	-7493.3
150	3117.8	1870.7	104.1461	-12504.2
200	4157.0	2494.2	110.1255	-17868.1
250	5196.3	3117.8	114.7636	-23494.6
300	6235.5	3741.3	118.5531	-29330.4
350	7274.8	4364.9	121.7571	-35340.2
400	8314.0	4988.4	124.5326	-41499.0
450	9353.3	5612.0	126.9807	-47788.1
500	10392.5	6235.5	129.1706	-54192.8
550	11431.8	6859.1	131.1517	-60701.7
600	12471.0	7482.6	132.9602	-67305.1
650	13510.3	8106.2	134.6239	-73995.3
700	14549.5	8729.7	136.1642	-80765.4
750	15588.8	9353.3	137.5982	-87609.9
800	16628.0	9976.8	138.9397	-94523.7
850	17667.3	10600.4	140.1997	-101502.5
900	18706.5	11223.9	141.3878	-108542.5
950	19745.8	11847.5	142.5116	-115640.2
1000	20785.0	12471.0	143.5777	-122792.7
1050	21824.3	13094.6	144.5918	-129997.1
1100	22863.5	13718.1	145.5587	-137251.1
1150	23902.8	14341.7	146.4826	-144552.3
1200	24942.0	14965.2	147.3672	-151898.7
1250	25981.3	15588.8	148.2157	-159288.4
1300	27020.5	16212.3	149.0309	-166719.7
1350	28059.8	16835.9	149.8154	-174191.0
1400	29099.0	17459.4	150.5713	-181700.8
1450	30138.3	18083.0	151.3006	-189247.7
1500	31177.5	18706.5	152.0053	-196830.4
1550	32216.8	19330.1	152.6868	-204447.8
1600	33256.0	19953.6	153.3467	-212098.8
1650	34295.3	20577.2	153.9863	-219782.2
1700	35334.5	21200.7	154.6068	-227497.1
1750	36373.8	21824.3	155.2093	-235242.5
1800	37413.0	22447.8	155.7948	-243017.7
1850	38452.3	23071.4	156.3643	-250821.8
1900	39491.5	23694.9	156.9186	-258653.9
1950	40530.8	24318.5	157.4585	-266513.4
2000	41570.0	24942.0	157.9848	-274399.5

A (cont.)

T	h(T)	u(T)	s(T)	g(T)
2050	42609.3	25565.6	158.4980	-282311.6
2100	43648.5	26189.1	158.9989	-290249.1
2150	44687.8	26812.7	159.4879	-298211.3
2200	45727.0	27436.2	159.9658	-306197.7
2250	46766.3	28059.8	160.4329	-314207.7
2300	47805.5	28683.3	160.8897	-322240.8
2350	48844.8	29306.9	161.3367	-330296.5
2400	49884.0	29930.4	161.7743	-338374.3
2450	50923.3	30554.0	162.2029	-346473.8
2500	51962.5	31177.5	162.6228	-354594.5
2550	53001.8	31801.1	163.0344	-362736.0
2600	54041.0	32424.6	163.4380	-370897.8
2650	55080.3	33048.2	163.8339	-379079.6
2700	56119.5	33671.7	164.2224	-387281.1
2750	57158.8	34295.3	164.6038	-395501.8
2800	58198.0	34918.8	164.9783	-403741.3
2850	59237.3	35542.4	165.3462	-411999.5
2900	60276.5	36165.9	165.7077	-420275.8
2950	61315.8	36789.5	166.0630	-428570.1
3000	62355.0	37413.0	166.4123	-436882.0
3050	63394.3	38036.6	166.7559	-445211.3
3100	64433.5	38660.1	167.0939	-453557.5
3150	65472.8	39283.7	167.4265	-461920.6
3200	66512.0	39907.2	167.7538	-470300.1
3250	67551.3	40530.8	168.0760	-478695.9
3300	68590.5	41154.3	168.3934	-487107.6
3350	69629.8	41777.9	168.7059	-495535.1
3400	70669.0	42401.4	169.0139	-503978.1
3450	71708.3	43025.0	169.3173	-512436.4
3500	72747.5	43648.5	169.6164	-520909.8
3550	73786.8	44272.1	169.9112	-529398.0
3600	74826.0	44895.6	170.2019	-537900.9
3650	75865.2	45519.2	170.4886	-546418.1
3700	76904.5	46142.7	170.7714	-554949.6
3750	77943.8	46766.3	171.0504	-563495.2
3800	78983.0	47389.8	171.3257	-572054.6
3850	80022.3	48013.4	171.5974	-580627.7
3900	81061.5	48636.9	171.8656	-589214.3
3950	82100.8	49260.5	172.1304	-597814.2
4000	83140.0	49884.0	172.3918	-606427.3

A (cont.)

T	h(T)	u(T)	s(T)	g(T)
4050	84179.3	50507.5	172.6500	-615053.3
4100	85218.5	51131.1	172.9051	-623692.2
4150	86257.7	51754.7	173.1570	-632343.8
4200	87297.0	52378.2	173.4059	-641007.9
4250	88336.3	53001.8	173.6519	-649684.3
4300	89375.5	53625.3	173.8950	-658373.0
4350	90414.7	54248.9	174.1353	-667073.8
4400	91454.0	54872.4	174.3728	-675786.5
4450	92493.3	55496.0	174.6077	-684511.0
4500	93532.5	56119.5	174.8399	-693247.2
4550	94571.7	56743.0	175.0696	-701995.0
4600	95611.0	57366.6	175.2968	-710754.2
4650	96650.3	57990.2	175.5215	-719524.6
4700	97689.5	58613.7	175.7438	-728306.3
4750	98728.8	59237.3	175.9637	-737099.0
4800	99768.0	59860.8	176.1814	-745902.6
4850	100807.3	60484.4	176.3968	-754717.1
4900	101846.5	61107.9	176.6099	-763542.2
4950	102885.8	61731.5	176.8210	-772378.0
5000	103925.0	62355.0	177.0299	-781224.3
5050	104964.3	62978.5	177.2367	-790081.0
5100	106003.5	63602.1	177.4415	-798947.9
5150	107042.8	64225.7	177.6442	-807825.1
5200	108082.0	64849.2	177.8451	-816712.3
5250	109121.3	65472.8	178.0440	-825609.6
5300	110160.5	66096.3	178.2410	-834516.7
5350	111199.8	66719.9	178.4361	-843433.6
5400	112239.0	67343.4	178.6295	-852360.3
5450	113278.3	67967.0	178.8211	-861296.5
5500	114317.5	68590.5	179.0109	-870242.4
5550	115356.8	69214.0	179.1990	-879197.6
5600	116396.0	69837.6	179.3854	-888162.2
5650	117435.3	70461.2	179.5702	-897136.1
5700	118474.5	71084.7	179.7533	-906119.2
5750	119513.8	71708.3	179.9348	-915111.4
5800	120553.0	72331.8	180.1148	-924112.7
5850	121592.3	72955.4	180.2932	-933122.9
5900	122631.5	73578.9	180.4701	-942142.0
5950	123670.8	74202.5	180.6455	-951169.9
6000	124710.0	74826.0	180.8194	-960206.5

TABLE A.3 IDEAL GAS HEATS OF REACTION 0-6000°K

Basic Relationships

Heat of Reaction at Constant Pressure

$$Q_p = \sum Mh(\text{products}) - \sum Mh(\text{reactants})$$

$$= \sum Mh_{(T)(\text{products})} - \sum Mh_{(T)(\text{reactants})} + \Delta H_o$$

where $\Delta H_o = \sum Mh_{o(\text{products})} - \sum Mh_{o(\text{reactants})}$.

Heat of Reaction at Constant Volume

$$Q_v = \sum Mu(\text{products}) - \sum Mu(\text{reactants})$$

$$= \sum Mu_{(T)(\text{products})} - \sum Mu_{(T)(\text{reactants})} + \Delta E_o$$

where $\Delta E_o = \sum Mu_{o(\text{products})} - \sum Mu_{o(\text{reactants})} + \Delta H_o$

$$Q_p - Q_v = R_{mol} T \left[\sum M(\text{products}) - \sum M(\text{reactants}) \right].$$

Q_p and Q_v are tabulated in units of kJ/kg mol for the following reactions 1 to 7.

Reaction 1 $CO + \frac{1}{2}O_2 \rightarrow CO_2$

Reaction 2 $H_2 + \frac{1}{2}O_2 \rightarrow H_2O$

Reaction 3 $CO + H_2O \rightarrow CO_2 + H_2$

Reaction 4 $CH_4 + 2O_2 \rightarrow CO_2 + 2H_2O$

Reaction 5 $N_2 \rightarrow 2N$

Reaction 6 $O_2 \rightarrow 2O$

Reaction 7 $H_2 \rightarrow 2H$

Notes: 1. The tabulated Q_p and Q_v values relate to reactions involving 1 mol of the reactant occurring first in the stoichiometric equations 1 to 7. Thus, in a constant pressure reaction at 1500°K in which one kg mol of CO reacts with 0.5 kg mol of O_2 to form one kg mol of CO_2 the heat of reaction is, from the table, -2.8034×10^5 kJ.

2. In reactions involving water, this is assumed to be in the gaseous phase. The tabulated values for reactions 2 and 4 therefore represent the <u>lower</u> heat of reaction.

HEAT OF REACTION AT

Q_p

T	$CO+\frac{1}{2}O_2 = CO_2$	$H_2+\frac{1}{2}O_2 = H_2O$	$CO+H_2O = CO_2+H_2$
0	-2.79523, 5	-2.39082, 5	-4.04409, 4
50	-2.80249, 5	-2.39624, 5	-4.06252, 4
100	-2.80896, 5	-2.40156, 5	-4.07399, 4
150	-2.81466, 5	-2.40677, 5	-4.07892, 4
200	-2.81965, 5	-2.41187, 5	-4.07772, 4
250	-2.82395, 5	-2.41687, 5	-4.07080, 4
300	-2.82761, 5	-2.42176, 5	-4.05854, 4
350	-2.83066, 5	-2.42653, 5	-4.04133, 4
400	-2.83315, 5	-2.43120, 5	-4.01952, 4
450	-2.83510, 5	-2.43575, 5	-3.99348, 4
500	-2.83654, 5	-2.44019, 5	-3.96355, 4
550	-2.83752, 5	-2.44451, 5	-3.93007, 4
600	-2.83806, 5	-2.44873, 5	-3.89334, 4
650	-2.83819, 5	-2.45283, 5	-3.85370, 4
700	-2.83795, 5	-2.45681, 5	-3.81142, 4
750	-2.83737, 5	-2.46069, 5	-3.76680, 4
800	-2.83646, 5	-2.46445, 5	-3.72011, 4
850	-2.83526, 5	-2.46809, 5	-3.67162, 4
900	-2.83379, 5	-2.47163, 5	-3.62158, 4
950	-2.83208, 5	-2.47506, 5	-3.57022, 4
1000	-2.83015, 5	-2.47837, 5	-3.51778, 4
1050	-2.82803, 5	-2.48158, 5	-3.46448, 4
1100	-2.82573, 5	-2.48468, 5	-3.41051, 4
1150	-2.82328, 5	-2.48768, 5	-3.35608, 4
1200	-2.82070, 5	-2.49057, 5	-3.30136, 4
1250	-2.81800, 5	-2.49335, 5	-3.24654, 4
1300	-2.81521, 5	-2.49604, 5	-3.19175, 4
1350	-2.81234, 5	-2.49862, 5	-3.13717, 4
1400	-2.80940, 5	-2.50111, 5	-3.08291, 4
1450	-2.80641, 5	-2.50350, 5	-3.02911, 4
1500	-2.80339, 5	-2.50580, 5	-2.97587, 4
1550	-2.80034, 5	-2.50801, 5	-2.92331, 4
1600	-2.79728, 5	-2.51013, 5	-2.87151, 4
1650	-2.79422, 5	-2.51217, 5	-2.82055, 4
1700	-2.79117, 5	-2.51412, 5	-2.77050, 4
1750	-2.78814, 5	-2.51600, 5	-2.72140, 4
1800	-2.78513, 5	-2.51780, 5	-2.67332, 4
1850	-2.78215, 5	-2.51953, 5	-2.62627, 4
1900	-2.77921, 5	-2.52119, 5	-2.58029, 4
1950	-2.77632, 5	-2.52278, 5	-2.53537, 4
2000	-2.77346, 5	-2.52431, 5	-2.49152, 4

CONSTANT PRESSURE

T	$CH_4+2O_2 = CO_2+2H_2O$		$N_2 = 2N$		$O_2 = 2O$		$H_2 = 2H$	
0	-8.04638,	5	9.42737,	5	4.93846,	5	4.32220,	5
50	-8.03797,	5	9.43419,	5	4.94768,	5	4.32872,	5
100	-8.03051,	5	9.44088,	5	4.95645,	5	4.33523,	5
150	-8.02396,	5	9.44746,	5	4.96477,	5	4.34173,	5
200	-8.01824,	5	9.45391,	5	4.97268,	5	4.34822,	5
250	-8.01334,	5	9.46024,	5	4.98019,	5	4.35469,	5
300	-8.00918,	5	9.46645,	5	4.98732,	5	4.36113,	5
350	-8.00574,	5	9.47254,	5	4.99409,	5	4.36755,	5
400	-8.00297,	5	9.47851,	5	5.00052,	5	4.37393,	5
450	-8.00083,	5	9.48436,	5	5.00663,	5	4.38027,	5
500	-7.99928,	5	9.49009,	5	5.01244,	5	4.38657,	5
550	-7.99829,	5	9.49570,	5	5.01795,	5	4.39282,	5
600	-7.99781,	5	9.50120,	5	5.02319,	5	4.39902,	5
650	-7.99781,	5	9.50657,	5	5.02818,	5	4.40516,	5
700	-7.99827,	5	9.51183,	5	5.03292,	5	4.41125,	5
750	-7.99914,	5	9.51697,	5	5.03744,	5	4.41727,	5
800	-8.00040,	5	9.52199,	5	5.04174,	5	4.42322,	5
850	-8.00201,	5	9.52690,	5	5.04584,	5	4.42910,	5
900	-8.00397,	5	9.53170,	5	5.04975,	5	4.43491,	5
950	-8.00622,	5	9.53638,	5	5.05349,	5	4.44064,	5
1000	-8.00876,	5	9.54096,	5	5.05707,	5	4.44630,	5
1050	-8.01155,	5	9.54543,	5	5.06049,	5	4.45187,	5
1100	-8.01458,	5	9.54978,	5	5.06378,	5	4.45736,	5
1150	-8.01782,	5	9.55404,	5	5.06693,	5	4.46277,	5
1200	-8.02126,	5	9.55819,	5	5.06997,	5	4.46809,	5
1250	-8.02487,	5	9.56223,	5	5.07289,	5	4.47331,	5
1300	-8.02864,	5	9.56618,	5	5.07571,	5	4.47845,	5
1350	-8.03256,	5	9.57003,	5	5.07843,	5	4.48350,	5
1400	-8.03659,	5	9.57378,	5	5.08107,	5	4.48845,	5
1450	-8.04074,	5	9.57744,	5	5.08363,	5	4.49330,	5
1500	-8.04499,	5	9.58101,	5	5.08611,	5	4.49806,	5
1550	-8.04932,	5	9.58449,	5	5.08853,	5	4.50273,	5
1600	-8.05373,	5	9.58788,	5	5.09088,	5	4.50729,	5
1650	-8.05821,	5	9.59119,	5	5.09318,	5	4.51176,	5
1700	-8.06274,	5	9.59441,	5	5.09542,	5	4.51613,	5
1750	-8.06731,	5	9.59756,	5	5.09761,	5	4.52041,	5
1800	-8.07193,	5	9.60064,	5	5.09976,	5	4.52458,	5
1850	-8.07657,	5	9.60364,	5	5.10187,	5	4.52866,	5
1900	-8.08124,	5	9.60657,	5	5.10393,	5	4.53265,	5
1950	-8.08594,	5	9.60944,	5	5.10595,	5	4.53653,	5
2000	-8.09065,	5	9.61224,	5	5.10794,	5	4.54033,	5

HEAT OF REACTION AT

Q_p

T	$CO+\frac{1}{2}O_2 = CO_2$	$H_2+\frac{1}{2}O_2 = H_2O$	$CO+H_2O = CO_2+H_2$
2050	-2.77066, 5	-2.52579, 5	-2.44872, 4
2100	-2.76790, 5	-2.52721, 5	-2.40696, 4
2150	-2.76520, 5	-2.52858, 5	-2.36618, 4
2200	-2.76254, 5	-2.52990, 5	-2.32635, 4
2250	-2.75993, 5	-2.53119, 5	-2.28740, 4
2300	-2.75737, 5	-2.53244, 5	-2.24926, 4
2350	-2.75484, 5	-2.53366, 5	-2.21186, 4
2400	-2.75236, 5	-2.53485, 5	-2.17508, 4
2450	-2.74991, 5	-2.53602, 5	-2.13884, 4
2500	-2.74748, 5	-2.53718, 5	-2.10301, 4
2550	-2.74507, 5	-2.53832, 5	-2.06746, 4
2600	-2.74267, 5	-2.53946, 5	-2.03206, 4
2650	-2.74026, 5	-2.54060, 5	-1.99664, 4
2700	-2.73785, 5	-2.54175, 5	-1.96105, 4
2750	-2.73541, 5	-2.54290, 5	-1.92511, 4
2800	-2.73294, 5	-2.54408, 5	-1.88863, 4
2850	-2.73042, 5	-2.54528, 5	-1.85142, 4
2900	-2.72759, 5	-2.48094, 5	-2.46650, 4
2950	-2.72504, 5	-2.47894, 5	-2.46097, 4
3000	-2.72242, 5	-2.47683, 5	-2.45584, 4
3050	-2.71973, 5	-2.47463, 5	-2.45102, 4
3100	-2.71695, 5	-2.47232, 5	-2.44638, 4
3150	-2.71480, 5	-2.38646, 5	-3.28342, 4
3200	-2.71220, 5	-2.37989, 5	-3.32306, 4
3250	-2.70957, 5	-2.37306, 5	-3.36509, 4
3300	-2.70691, 5	-2.36596, 5	-3.40954, 4
3350	-2.70423, 5	-2.35858, 5	-3.45646, 4
3400	-2.70152, 5	-2.35093, 5	-3.50588, 4
3450	-2.69878, 5	-2.34299, 5	-3.55783, 4
3500	-2.69601, 5	-2.33478, 5	-3.61237, 4
3550	-2.69322, 5	-2.32627, 5	-3.66951, 4
3600	-2.69041, 5	-2.31747, 5	-3.72931, 4
3650	-2.68757, 5	-2.30839, 5	-3.79180, 4
3700	-2.68470, 5	-2.29900, 5	-3.85702, 4
3750	-2.68182, 5	-2.28932, 5	-3.92501, 4
3800	-2.67891, 5	-2.27933, 5	-3.99580, 4
3850	-2.67598, 5	-2.26903, 5	-4.06944, 4
3900	-2.67302, 5	-2.25843, 5	-4.14597, 4
3950	-2.67005, 5	-2.24751, 5	-4.22541, 4
4000	-2.66706, 5	-2.23628, 5	-4.30782, 4
4050	-2.66404, 5	-2.22472, 5	-4.39323, 4

CONSTANT PRESSURE

(cont.)

T	$CH_4+2O_2 = CO_2+2H_2O$		$N_2 = 2N$		$O_2 = 2O$		$H_2 = 2H$	
2050	-8.09537,	5	9.61499,	5	5.10989,	5	4.54403,	5
2100	-8.10011,	5	9.61768,	5	5.11179,	5	4.54763,	5
2150	-8.10485,	5	9.62031,	5	5.11367,	5	4.55115,	5
2200	-8.10960,	5	9.62290,	5	5.11550,	5	4.55458,	5
2250	-8.11436,	5	9.62545,	5	5.11729,	5	4.55791,	5
2300	-8.11911,	5	9.62795,	5	5.11904,	5	4.56117,	5
2350	-8.12388,	5	9.63042,	5	5.12074,	5	4.56434,	5
2400	-8.12865,	5	9.63285,	5	5.12240,	5	4.56743,	5
2450	-8.13343,	5	9.63526,	5	5.12400,	5	4.57044,	5
2500	-8.13822,	5	9.63764,	5	5.12554,	5	4.57337,	5
2550	-8.14301,	5	9.64000,	5	5.12702,	5	4.57623,	5
2600	-8.14783,	5	9.64235,	5	5.12844,	5	4.57902,	5
2650	-8.15266,	5	9.64469,	5	5.12978,	5	4.58175,	5
2700	-8.15751,	5	9.64702,	5	5.13105,	5	4.58441,	5
2750	-8.16239,	5	9.64935,	5	5.13223,	5	4.58701,	5
2800	-8.16729,	5	9.65169,	5	5.13331,	5	4.58956,	5
2850	-8.17223,	5	9.65403,	5	5.13429,	5	4.59205,	5
2900	-8.04536,	5	9.65659,	5	5.13492,	5	4.59434,	5
2950	-8.04419,	5	9.65893,	5	5.13586,	5	4.59682,	5
3000	-8.04277,	5	9.66129,	5	5.13672,	5	4.59926,	5
3050	-8.04110,	5	9.66366,	5	5.13749,	5	4.60166,	5
3100	-8.03919,	5	9.66606,	5	5.1381 ,	5	4.60402,	5
3150	-7.87119,	5	9.66843,	5	5.13997,	5	4.60650,	5
3200	-7.86103,	5	9.67081,	5	5.14092,	5	4.60882,	5
3250	-7.85029,	5	9.67321,	5	5.14184,	5	4.61109,	5
3300	-7.83896,	5	9.67564,	5	5.14272,	5	4.61331,	5
3350	-7.82705,	5	9.67810,	5	5.14358,	5	4.61549,	5
3400	-7.81453,	5	9.68059,	5	5.14441,	5	4.61762,	5
3450	-7.80141,	5	9.68311,	5	5.14521,	5	4.61971,	5
3500	-7.78767,	5	9.68566,	5	5.14599,	5	4.62175,	5
3550	-7.77332,	5	9.68825,	5	5.14674,	5	4.62375,	5
3600	-7.75836,	5	9.69089,	5	5.14747,	5	4.62571,	5
3650	-7.74276,	5	9.69356,	5	5.14818,	5	4.62763,	5
3700	-7.72653,	5	9.69628,	5	5.14887,	5	4.62951,	5
3750	-7.70966,	5	9.69904,	5	5.14954,	5	4.63134,	5
3800	-7.69216,	5	9.70185,	5	5.15019,	5	4.63314,	5
3850	-7.67400,	5	9.70471,	5	5.15082,	5	4.63490,	5
3900	-7.65519,	5	9.70762,	5	5.15144,	5	4.63662,	5
3950	-7.63572,	5	9.71059,	5	5.15204,	5	4.63830,	5
4000	-7.61559,	5	9.71361,	5	5.15263,	5	4.63995,	5
4050	-7.59479,	5	9.71668,	5	5.15321,	5	4.64156,	5

HEAT OF REACTION AT

Q_p

T	$CO+\frac{1}{2}O_2 = CO_2$	$H_2+\frac{1}{2}O_2 = H_2O$	$CO+H_2O = CO_2+H_2$
4100	-2.66101, 5	-2.21284, 5	-4.48168, 4
4150	-2.65796, 5	-2.20064, 5	-4.57321, 4
4200	-2.65489, 5	-2.18811, 5	-4.66786, 4
4250	-2.65181, 5	-2.17524, 5	-4.76566, 4
4300	-2.64870, 5	-2.16204, 5	-4.86667, 4
4350	-2.64559, 5	-2.14849, 5	-4.97091, 4
4400	-2.64245, 5	-2.13461, 5	-5.07844, 4
4450	-2.63930, 5	-2.12037, 5	-5.18928, 4
4500	-2.63614, 5	-2.10579, 5	-5.30348, 4
4550	-2.63296, 5	-2.09085, 5	-5.42108, 4
4600	-2.62977, 5	-2.07556, 5	-5.54212, 4
4650	-2.62657, 5	-2.05990, 5	-5.66664, 4
4700	-2.62335, 5	-2.04388, 5	-5.79468, 4
4750	-2.62012, 5	-2.02749, 5	-5.92628, 4
4800	-2.61688, 5	-2.01074, 5	-6.06148, 4
4850	-2.61364, 5	-1.99360, 5	-6.20033, 4
4900	-2.61038, 5	-1.97609, 5	-6.34286, 4
4950	-2.60711, 5	-1.95820, 5	-6.48911, 4
5000	-2.60383, 5	-1.93992, 5	-6.63914, 4
5050	-2.60055, 5	-1.92125, 5	-6.79297, 4
5100	-2.59726, 5	-1.90219, 5	-6.95065, 4
5150	-2.59396, 5	-1.88274, 5	-7.11222, 4
5200	-2.59066, 5	-1.86288, 5	-7.27773, 4
5250	-2.58735, 5	-1.84263, 5	-7.44721, 4
5300	-2.58403, 5	-1.82196, 5	-7.62071, 4
5350	-2.58071, 5	-1.80089, 5	-7.79826, 4
5400	-2.57739, 5	-1.77940, 5	-7.97992, 4
5450	-2.57406, 5	-1.75749, 5	-8.16572, 4
5500	-2.57073, 5	-1.73516, 5	-8.35571, 4
5550	-2.56740, 5	-1.71241, 5	-8.54993, 4
5600	-2.56407, 5	-1.68923, 5	-8.74842, 4
5650	-2.56074, 5	-1.66562, 5	-8.95122, 4
5700	-2.55740, 5	-1.64157, 5	-9.15838, 4
5750	-2.55407, 5	-1.61708, 5	-9.36994, 4
5800	-2.55074, 5	-1.59214, 5	-9.58594, 4
5850	-2.54741, 5	-1.56676, 5	-9.80643, 4
5900	-2.54408, 5	-1.54093, 5	-1.00314, 5
5950	-2.54075, 5	-1.51465, 5	-1.02610, 5
6000	-2.53743, 5	-1.48790, 5	-1.04952, 5

CONSTANT PRESSURE

(cont.)

T	$CH_4+2O_2 = CO_2+2H_2O$		N_2 = 2N		O_2 = 2O		H_2 = 2H	
4100	-7.57331,	5	9.71982,	5	5.15377,	5	4.64314,	5
4150	-7.55115,	5	9.72301,	5	5.15433,	5	4.64468,	5
4200	-7.52830,	5	9.72627,	5	5.15487,	5	4.64619,	5
4250	-7.50476,	5	9.72959,	5	5.15541,	5	4.64766,	5
4300	-7.48052,	5	9.73298,	5	5.15594,	5	4.64911,	5
4350	-7.45556,	5	9.73643,	5	5.15647,	5	4.65052,	5
4400	-7.42989,	5	9.73995,	5	5.15698,	5	4.65190,	5
4450	-7.40350,	5	9.74354,	5	5.15750,	5	4.65324,	5
4500	-7.37637,	5	9.74720,	5	5.15801,	5	4.65456,	5
4550	-7.34850,	5	9.75093,	5	5.15852,	5	4.65585,	5
4600	-7.31989,	5	9.75474,	5	5.15902,	5	4.65711,	5
4650	-7.29051,	5	9.75861,	5	5.15953,	5	4.65834,	5
4700	-7.26036,	5	9.76257,	5	5.16004,	5	4.65954,	5
4750	-7.22944,	5	9.76660,	5	5.16054,	5	4.66072,	5
4800	-7.19772,	5	9.77071,	5	5.16105,	5	4.66186,	5
4850	-7.16520,	5	9.77490,	5	5.16156,	5	4.66299,	5
4900	-7.13186,	5	9.77917,	5	5.16208,	5	4.66408,	5
4950	-7.09770,	5	9.78352,	5	5.16259,	5	4.66515,	5
5000	-7.06270,	5	9.78795,	5	5.16311,	5	4.66620,	5
5050	-7.02684,	5	9.79246,	5	5.16364,	5	4.66722,	5
5100	-6.99011,	5	9.79705,	5	5.16417,	5	4.66821,	5
5150	-6.95249,	5	9.80174,	5	5.16471,	5	4.66919,	5
5200	-6.91397,	5	9.80650,	5	5.16526,	5	4.67014,	5
5250	-6.87453,	5	9.81135,	5	5.16582,	5	4.67106,	5
5300	-6.83415,	5	9.81629,	5	5.16638,	5	4.67197,	5
5350	-6.79282,	5	9.82131,	5	5.16695,	5	4.67285,	5
5400	-6.75051,	5	9.82642,	5	5.16753,	5	4.67371,	5
5450	-6.70721,	5	9.83162,	5	5.16812,	5	4.67456,	5
5500	-6.66288,	5	9.83691,	5	5.16872,	5	4.67538,	5
5550	-6.61752,	5	9.84228,	5	5.16934,	5	4.67618,	5
5600	-6.57109,	5	9.84774,	5	5.16996,	5	4.67696,	5
5650	-6.52357,	5	9.85330,	5	5.17060,	5	4.67772,	5
5700	-6.47494,	5	9.85894,	5	5.17124,	5	4.67846,	5
5750	-6.42517,	5	9.86467,	5	5.17190,	5	4.67918,	5
5800	-6.37423,	5	9.87050,	5	5.17258,	5	4.67988,	5
5850	-6.32209,	5	9.87641,	5	5.17327,	5	4.68057,	5
5900	-6.26873,	5	9.88241,	5	5.17397,	5	4.68124,	5
5950	-6.21410,	5	9.88850,	5	5.17468,	5	4.68189,	5
6000	-6.15819,	5	9.89469,	5	5.17541,	5	4.68252,	5

HEAT OF REACTION AT

Q_V

T	$CO+\frac{1}{2}O_2 = CO_2$	$H+\frac{1}{2}O_2 = H_2O$	$CO+H_2O = CO_2+H_2$
0	-2.79523, 5	-2.39082, 5	-4.04409, 4
50	-2.80041, 5	-2.39416, 5	-4.06252, 4
100	-2.80480, 5	-2.39740, 5	-4.07399, 4
150	-2.80843, 5	-2.40053, 5	-4.07892, 4
200	-2.81133, 5	-2.40356, 5	-4.07772, 4
250	-2.81356, 5	-2.40648, 5	-4.07080, 4
300	-2.81514, 5	-2.40928, 5	-4.05854, 4
350	-2.81611, 5	-2.41198, 5	-4.04133, 4
400	-2.81652, 5	-2.41457, 5	-4.01952, 4
450	-2.81639, 5	-2.41704, 5	-3.99348, 4
500	-2.81576, 5	-2.41940, 5	-3.96355, 4
550	-2.81466, 5	-2.42165, 5	-3.93007, 4
600	-2.81312, 5	-2.42378, 5	-3.89334, 4
650	-2.81117, 5	-2.42580, 5	-3.85370, 4
700	-2.80885, 5	-2.42771, 5	-3.81142, 4
750	-2.80619, 5	-2.42951, 5	-3.76680, 4
800	-2.80320, 5	-2.43119, 5	-3.72011, 4
850	-2.79992, 5	-2.43276, 5	-3.67162, 4
900	-2.79638, 5	-2.43422, 5	-3.62158, 4
950	-2.79259, 5	-2.43557, 5	-3.57022, 4
1000	-2.78858, 5	-2.43680, 5	-3.51778, 4
1050	-2.78438, 5	-2.43793, 5	-3.46448, 4
1100	-2.78001, 5	-2.43896, 5	-3.41051, 4
1150	-2.77548, 5	-2.43987, 5	-3.35608, 4
1200	-2.77082, 5	-2.44068, 5	-3.30136, 4
1250	-2.76604, 5	-2.44139, 5	-3.24654, 4
1300	-2.76117, 5	-2.44199, 5	-3.19175, 4
1350	-2.75622, 5	-2.44250, 5	-3.13717, 4
1400	-2.75120, 5	-2.44291, 5	-3.08291, 4
1450	-2.74614, 5	-2.44322, 5	-3.02911, 4
1500	-2.74103, 5	-2.44345, 5	-2.97587, 4
1550	-2.73591, 5	-2.44358, 5	-2.92331, 4
1600	-2.73077, 5	-2.44362, 5	-2.87151, 4
1650	-2.72563, 5	-2.44358, 5	-2.82055, 4
1700	-2.72050, 5	-2.44345, 5	-2.77050, 4
1750	-2.71539, 5	-2.44325, 5	-2.72140, 4
1800	-2.71031, 5	-2.44297, 5	-2.67332, 4
1850	-2.70525, 5	-2.44262, 5	-2.62627, 4
1900	-2.70023, 5	-2.44220, 5	-2.58029, 4
1950	-2.69525, 5	-2.44172, 5	-2.53537, 4
2000	-2.69032, 5	-2.44117, 5	-2.49152, 4

CONSTANT VOLUME

T	$CH_4+2O_2 = CO_2+2H_2O$		$N_2 = 2N$		$O_2 = 2O$		$H_2 = 2H$	
0	-8.04638,	5	9.42737,	5	4.93846,	5	4.32220,	5
50	-8.03797,	5	9.43003,	5	4.94353,	5	4.32456,	5
100	-8.03051,	5	9.43257,	5	4.94813,	5	4.32691,	5
150	-8.02396,	5	9.43498,	5	4.95230,	5	4.32926,	5
200	-8.01824,	5	9.43728,	5	4.95605,	5	4.33159,	5
250	-8.01334,	5	9.43946,	5	4.95940,	5	4.33390,	5
300	-8.00918,	5	9.44151,	5	4.96238,	5	4.33619,	5
350	-8.00574,	5	9.44344,	5	4.96499,	5	4.33845,	5
400	-8.00297,	5	9.44526,	5	4.96727,	5	4.34067,	5
450	-8.00083,	5	9.44695,	5	4.96922,	5	4.34286,	5
500	-7.99928,	5	9.44852,	5	4.97087,	5	4.34500,	5
550	-7.99829,	5	9.44998,	5	4.97222,	5	4.34709,	5
600	-7.99781,	5	9.45131,	5	4.97331,	5	4.34914,	5
650	-7.99781,	5	9.45253,	5	4.97414,	5	4.35112,	5
700	-7.99827,	5	9.45363,	5	4.97472,	5	4.35305,	5
750	-7.99914,	5	9.45461,	5	4.97508,	5	4.35491,	5
800	-8.00040,	5	9.45548,	5	4.97522,	5	4.35671,	5
850	-8.00201,	5	9.45623,	5	4.97517,	5	4.35843,	5
900	-8.00397,	5	9.45687,	5	4.97493,	5	4.36008,	5
950	-8.00622,	5	9.45740,	5	4.97451,	5	4.36166,	5
1000	-8.00876,	5	9.45782,	5	4.97393,	5	4.36316,	5
1050	-8.01155,	5	9.45813,	5	4.97320,	5	4.36458,	5
1100	-8.01458,	5	9.45833,	5	4.97233,	5	4.36591,	5
1150	-8.01782,	5	9.45843,	5	4.97132,	5	4.36716,	5
1200	-8.02126,	5	9.45842,	5	4.97020,	5	4.36832,	5
1250	-8.02487,	5	9.45831,	5	4.96897,	5	4.36939,	5
1300	-8.02864,	5	9.45810,	5	4.96763,	5	4.37037,	5
1350	-8.03256,	5	9.45779,	5	4.96620,	5	4.37126,	5
1400	-8.03659,	5	9.45739,	5	4.96468,	5	4.37205,	5
1450	-8.04074,	5	9.45689,	5	4.96308,	5	4.37275,	5
1500	-8.04499,	5	9.45630,	5	4.96140,	5	4.37335,	5
1550	-8.04932,	5	9.45562,	5	4.95966,	5	4.37386,	5
1600	-8.05373,	5	9.45485,	5	4.95786,	5	4.37427,	5
1650	-8.05821,	5	9.45401,	5	4.95600,	5	4.37458,	5
1700	-8.06274,	5	9.45308,	5	4.95408,	5	4.37479,	5
1750	-8.06731,	5	9.45207,	5	4.95212,	5	4.37491,	5
1800	-8.07193,	5	9.45098,	5	4.95011,	5	4.37493,	5
1850	-8.07657,	5	9.44983,	5	4.94806,	5	4.37485,	5
1900	-8.08124,	5	9.44860,	5	4.94596,	5	4.37468,	5
1950	-8.08594,	5	9.44731,	5	4.94383,	5	4.37441,	5
2000	-8.09065,	5	9.44596,	5	4.94166,	5	4.37405,	5

HEAT OF REACTION AT

Q_V

T	$CO+\frac{1}{2}O_2 = CO_2$	$H_2+\frac{1}{2}O_2 = H_2O$	$CO+H_2O = CO_2+H_2$
2050	-2.68544, 5	-2.44057, 5	-2.44872, 4
2100	-2.68060, 5	-2.43991, 5	-2.40696, 4
2150	-2.67582, 5	-2.43920, 5	-2.36618, 4
2200	-2.67108, 5	-2.43845, 5	-2.32635, 4
2250	-2.66640, 5	-2.43766, 5	-2.28740, 4
2300	-2.66175, 5	-2.43683, 5	-2.24926, 4
2350	-2.65715, 5	-2.43597, 5	-2.21186, 4
2400	-2.65259, 5	-2.43508, 5	-2.17508, 4
2450	-2.64806, 5	-2.43417, 5	-2.13884, 4
2500	-2.64355, 5	-2.43325, 5	-2.10301, 4
2550	-2.63906, 5	-2.43232, 5	-2.06746, 4
2600	-2.63458, 5	-2.43138, 5	-2.03206, 4
2650	-2.63010, 5	-2.43044, 5	-1.99664, 4
2700	-2.62561, 5	-2.42951, 5	-1.96105, 4
2750	-2.62110, 5	-2.42859, 5	-1.92511, 4
2800	-2.61655, 5	-2.42768, 5	-1.88863, 4
2850	-2.61195, 5	-2.42681, 5	-1.85142, 4
2900	-2.60704, 5	-2.36039, 5	-2.46650, 4
2950	-2.60241, 5	-2.35631, 5	-2.46097, 4
3000	-2.59771, 5	-2.35212, 5	-2.45584, 4
3050	-2.59294, 5	-2.34784, 5	-2.45102, 4
3100	-2.58809, 5	-2.34345, 5	-2.44638, 4
3150	-2.58385, 5	-2.25551, 5	-3.28342, 4
3200	-2.57918, 5	-2.24687, 5	-3.32306, 4
3250	-2.57447, 5	-2.23796, 5	-3.36509, 4
3300	-2.56973, 5	-2.22878, 5	-3.40954, 4
3350	-2.56497, 5	-2.21932, 5	-3.45646, 4
3400	-2.56018, 5	-2.20959, 5	-3.50588, 4
3450	-2.55536, 5	-2.19958, 5	-3.55783, 4
3500	-2.55052, 5	-2.18928, 5	-3.61237, 4
3550	-2.54565, 5	-2.17870, 5	-3.66951, 4
3600	-2.54075, 5	-2.16782, 5	-3.72931, 4
3650	-2.53584, 5	-2.15666, 5	-3.79180, 4
3700	-2.53089, 5	-2.14519, 5	-3.85702, 4
3750	-2.52593, 5	-2.13343, 5	-3.92501, 4
3800	-2.52094, 5	-2.12136, 5	-3.99580, 4
3850	-2.51593, 5	-2.10899, 5	-4.06944, 4
3900	-2.51090, 5	-2.09630, 5	-4.14597, 4
3950	-2.50585, 5	-2.08331, 5	-4.22541, 4
4000	-2.50078, 5	-2.07000, 5	-4.30782, 4
4050	-2.49569, 5	-2.05636, 5	-4.39323, 4

CONSTANT VOLUME

(cont.)

T	$CH_4+2O = CO_2+2H_2O$		$N_2 = 2N$		$O_2 = 2O$		$H_2 = 2H$	
2050	-8.09537,	5	9.44455,	5	4.93945,	5	4.37359,	5
2100	-8.10011,	5	9.44308,	5	4.93720,	5	4.37304,	5
2150	-8.10485,	5	9.44156,	5	4.93492,	5	4.37240,	5
2200	-8.10960,	5	9.43999,	5	4.93259,	5	4.37167,	5
2250	-8.11436,	5	9.43838,	5	4.93023,	5	4.37085,	5
2300	-8.11911,	5	9.43673,	5	4.92782,	5	4.36995,	5
2350	-8.12388,	5	9.43504,	5	4.92536,	5	4.36896,	5
2400	-8.12865,	5	9.43332,	5	4.92286,	5	4.36789,	5
2450	-8.13343,	5	9.43157,	5	4.92030,	5	4.36674,	5
2500	-8.13822,	5	9.42979,	5	4.91769,	5	4.36552,	5
2550	-8.14301,	5	9.42800,	5	4.91502,	5	4.36423,	5
2600	-8.14783,	5	9.42619,	5	4.91228,	5	4.36286,	5
2650	-8.15266,	5	9.42437,	5	4.90946,	5	4.36143,	5
2700	-8.15751,	5	9.42254,	5	4.90657,	5	4.35993,	5
2750	-8.16239,	5	9.42072,	5	4.90359,	5	4.35838,	5
2800	-8.16729,	5	9.41889,	5	4.90052,	5	4.35677,	5
2850	-8.17223,	5	9.41708,	5	4.89734,	5	4.35510,	5
2900	-8.04536,	5	9.41549,	5	4.89381,	5	4.35323,	5
2950	-8.04419,	5	9.41367,	5	4.89060,	5	4.35156,	5
3000	-8.04277,	5	9.41187,	5	4.88730,	5	4.34984,	5
3050	-8.04110,	5	9.41009,	5	4.88392,	5	4.34808,	5
3100	-8.03919,	5	9.40833,	5	4.88044,	5	4.34628,	5
3150	-7.87119,	5	9.40654,	5	4.87808,	5	4.34461,	5
3200	-7.86103,	5	9.40476,	5	4.87487,	5	4.34277,	5
3250	-7.85029,	5	9.40301,	5	4.87163,	5	4.34089,	5
3300	-7.83896,	5	9.40128,	5	4.86836,	5	4.33895,	5
3350	-7.82705,	5	9.39958,	5	4.86506,	5	4.33697,	5
3400	-7.81453,	5	9.39791,	5	4.86174,	5	4.33495,	5
3450	-7.80141,	5	9.39627,	5	4.85838,	5	4.33288,	5
3500	-7.78767,	5	9.39467,	5	4.85500,	5	4.33076,	5
3550	-7.77332,	5	9.39311,	5	4.85160,	5	4.32861,	5
3600	-7.75836,	5	9.39158,	5	4.84817,	5	4.32641,	5
3650	-7.74276,	5	9.39010,	5	4.84472,	5	4.32417,	5
3700	-7.72653,	5	9.38866,	5	4.84125,	5	4.32189,	5
3750	-7.70966,	5	9.38726,	5	4.83776,	5	4.31957,	5
3800	-7.69216,	5	9.38592,	5	4.83426,	5	4.31721,	5
3850	-7.67400,	5	9.38462,	5	4.83073,	5	4.31481,	5
3900	-7.65519,	5	9.38338,	5	4.82719,	5	4.31237,	5
3950	-7.63572,	5	9.38218,	5	4.82364,	5	4.30990,	5
4000	-7.61559,	5	9.38105,	5	4.82007,	5	4.30739,	5
4050	-7.59479,	5	9.37997,	5	4.81649,	5	4.30485,	5

HEAT   OF   REACTION   AT

Q_v

T	$CO + \frac{1}{2}O_2 = CO_2$		$H_2 + \frac{1}{2}O_2 = H_2O$		$\begin{array}{c} CO + H_2O \\ = CO_2 + H_2 \end{array}$	
4100	-2.49057,	5	-2.04241,	5	-4.48168,	4
4150	-2.48545,	5	-2.02812,	5	-4.57321,	4
4200	-2.48030,	5	-2.01351,	5	-4.66786,	4
4250	-2.47513,	5	-1.99857,	5	-4.76566,	4
4300	-2.46995,	5	-1.98329,	5	-4.86667,	4
4350	-2.46476,	5	-1.96766,	5	-4.97091,	4
4400	-2.45954,	5	-1.95170,	5	-5.07844,	4
4450	-2.45431,	5	-1.93539,	5	-5.18928,	4
4500	-2.44907,	5	-1.91872,	5	-5.30348,	4
4550	-2.44382,	5	-1.90171,	5	-5.42108,	4
4600	-2.43855,	5	-1.88433,	5	-5.54212,	4
4650	-2.43326,	5	-1.86660,	5	-5.66664,	4
4700	-2.42797,	5	-1.84850,	5	-5.79468,	4
4750	-2.42266,	5	-1.83004,	5	-5.92628,	4
4800	-2.41735,	5	-1.81120,	5	-6.06148,	4
4850	-2.41202,	5	-1.79199,	5	-6.20033,	4
4900	-2.40668,	5	-1.77240,	5	-6.34286,	4
4950	-2.40134,	5	-1.75243,	5	-6.48911,	4
5000	-2.39598,	5	-1.73207,	5	-6.63914,	4
5050	-2.39062,	5	-1.71133,	5	-6.79297,	4
5100	-2.38525,	5	-1.69019,	5	-6.95065,	4
5150	-2.37988,	5	-1.66865,	5	-7.11222,	4
5200	-2.37449,	5	-1.64672,	5	-7.27773,	4
5250	-2.36911,	5	-1.62438,	5	-7.44721,	4
5300	-2.36371,	5	-1.60164,	5	-7.62071,	4
5350	-2.35831,	5	-1.57849,	5	-7.79826,	4
5400	-2.35291,	5	-1.55492,	5	-7.97992,	4
5450	-2.34751,	5	-1.53094,	5	-8.16572,	4
5500	-2.34210,	5	-1.50653,	5	-8.35571,	4
5550	-2.33669,	5	-1.48170,	5	-8.54993,	4
5600	-2.33128,	5	-1.45644,	5	-8.74842,	4
5650	-2.32587,	5	-1.43074,	5	-8.95122,	4
5700	-2.32045,	5	-1.40462,	5	-9.15838,	4
5750	-2.31504,	5	-1.37805,	5	-9.36994,	4
5800	-2.30963,	5	-1.35104,	5	-9.58594,	4
5850	-2.30422,	5	-1.32358,	5	-9.80643,	4
5900	-2.29881,	5	-1.29567,	5	-1.00314,	5
5950	-2.29341,	5	-1.26730,	5	-1.02610,	5
6000	-2.28801,	5	-1.23848,	5	-1.04952,	5

CONSTANT VOLUME

(cont.)

T	CH_4+2O_2 $= CO_2+2H_2O$	$N_2 = 2N$	$O_2 = 2O$	$H_2 = 2H$
4100	-7.57331, 5	9.37894, 5	4.812 0, 5	4.30227, 5
4150	-7.55115, 5	9.37798, 5	4.80930, 5	4.29965, 5
4200	-7.52830, 5	9.37708, 5	4.80569, 5	4.29700, 5
4250	-7.50476, 5	9.37625, 5	4.80207, 5	4.29432, 5
4300	-7.48052, 5	9.37548, 5	4.79844, 5	4.29160, 5
4350	-7.45556, 5	9.37477, 5	4.79481, 5	4.28886, 5
4400	-7.42989, 5	9.37413, 5	4.79117, 5	4.28608, 5
4450	-7.40350, 5	9.37357, 5	4.78753, 5	4.28327, 5
4500	-7.37637, 5	9.37307, 5	4.78388, 5	4.28043, 5
4550	-7.34850, 5	9.37264, 5	4.78023, 5	4.27756, 5
4600	-7.31989, 5	9.37229, 5	4.77658, 5	4.27466, 5
4650	-7.29051, 5	9.37201, 5	4.77293, 5	4.27174, 5
4700	-7.26036, 5	9.37181, 5	4.76928, 5	4.26878, 5
4750	-7.22944, 5	9.37169, 5	4.76563, 5	4.26580, 5
4800	-7.19772, 5	9.37164, 5	4.76198, 5	4.26279, 5
4850	-7.16520, 5	9.37167, 5	4.75833, 5	4.25976, 5
4900	-7.13186, 5	9.37178, 5	4.75469, 5	4.25670, 5
4950	-7.09770, 5	9.37197, 5	4.75105, 5	4.25361, 5
5000	-7.06270, 5	9.37225, 5	4.74741, 5	4.25050, 5
5050	-7.02684, 5	9.37260, 5	4.74378, 5	4.24736, 5
5100	-6.99011, 5	9.37304, 5	4.74016, 5	4.24420, 5
5150	-6.95249, 5	9.37356, 5	4.73654, 5	4.24102, 5
5200	-6.91397, 5	9.37417, 5	4.73293, 5	4.23781, 5
5250	-6.87453, 5	9.37487, 5	4.72933, 5	4.23458, 5
5300	-6.83415, 5	9.37565, 5	4.72574, 5	4.23133, 5
5350	-6.79282, 5	9.37651, 5	4.72215, 5	4.22805, 5
5400	-6.75051, 5	9.37747, 5	4.71857, 5	4.22476, 5
5450	-6.70721, 5	9.37851, 5	4.71501, 5	4.22144, 5
5500	-6.66288, 5	9.37964, 5	4.71145, 5	4.21811, 5
5550	-6.61752, 5	9.38085, 5	4.70791, 5	4.21475, 5
5600	-6.57109, 5	9.38216, 5	4.70438, 5	4.21137, 5
5650	-6.52357, 5	9.38356, 5	4.70085, 5	4.20798, 5
5700	-6.47494, 5	9.38504, 5	4.69735, 5	4.20456, 5
5750	-6.42517, 5	9.38662, 5	4.69385, 5	4.20113, 5
5800	-6.37423, 5	9.38828, 5	4.69037, 5	4.19767, 5
5850	-6.32209, 5	9.39004, 5	4.68690, 5	4.19420, 5
5900	-6.26873, 5	9.39188, 5	4.68344, 5	4.19071, 5
5950	-6.21410, 5	9.39382, 5	4.68000, 5	4.18721, 5
6000	-6.15819, 5	9.39585, 5	4.67657, 5	4.18368, 5

TABLE A.4 IDEAL GAS REACTION EQUILIBRIUM CONSTANTS 500-6000°K

Basic Relationships

$$K_p = \frac{\Pi(p/p_o)^{\nu}_{products}}{\Pi(p/p_o)^{\nu}_{reactants}}$$

$$\text{Ln } K_p = -\frac{1}{R_{mol}T}\left[\sum(\nu g^{\circ})_{products} - \sum(\nu g^{\circ})_{reactants}\right]$$

where v denotes a stoichiometric coefficient

and

$$g^{\circ} = g - R_{mol}T \ln \frac{p}{p_o}$$

$$= g(T) - g_o.$$

K_p is tabulated for the reactions 1 to 7 below

Reaction 1	$CO+\frac{1}{2}O_2 \rightarrow CO_2$
Reaction 2	$H_2+\frac{1}{2}O_2 \rightarrow H_2O$
Reaction 3	$CO+H_2O \rightarrow CO_2+H_2$
Reaction 4	$CH_4+2O_2 \rightarrow CO_2+2H_2O$
Reaction 5	$N_2 \rightarrow 2N$
Reaction 6	$O_2 \rightarrow 2O$
Reaction 7	$H_2 \rightarrow 2H$

†Reference pressure p_o = 1.01325 bars
or 101325 N/m^2.

---

†See footnote to p.82 Chapter 3.

EQUILIBRIUM

K_p / T	$\frac{(p/p_o)_{CO_2}}{(p/p_o)_{CO}(p/p_o)^{\frac{1}{2}}_{O_2}}$	$\frac{(p/p_o)_{H_2O}}{(p/p_o)_{H_2}(p/p_o)^{\frac{1}{2}}_{O_2}}$	$\frac{(p/p_o)_{CO_2}(p/p_o)_{H_2}}{(p/p_o)_{CO}(p/p_o)_{H_2O}}$
500	1.07169, 25	7.92127, 22	1.35293, 2
550	2.16553, 22	3.79505, 20	5.70620, 1
600	1.22896, 20	4.39391, 18	2.79697, 1
650	1.54464, 18	1.00369, 17	1.53896, 1
700	3.62806, 16	3.91281, 15	9.27228
750	1.40588, 15	2.34046, 14	6.00686
800	8.18490, 13	1.98312, 13	4.12728
850	6.66453, 12	2.23916, 12	2.97636
900	7.17808, 11	3.21251, 11	2.23442
950	9.78592, 10	5.64029, 10	1.73500
1000	1.63018, 10	1.17594, 10	1.38628
1050	3.22481, 9	2.84123, 9	1.13501
1100	7.40029, 8	7.79819, 8	9.48976, -1
1150	1.93228, 8	2.39155, 8	8.07963, -1
1200	5.64894, 7	8.08320, 7	6.98850, -1
1250	1.82411, 7	2.97633, 7	6.12873, -1
1300	6.43191, 6	1.18227, 7	5.44032, -1
1350	2.45236, 6	5.02411, 6	4.88119, -1
1400	1.00263, 6	2.26777, 6	4.42123, -1
1450	4.36386, 5	1.08056, 6	4.03851, -1
1500	2.00930, 5	5.40608, 5	3.71674, -1
1550	9.73399, 4	2.82661, 5	3.44370, -1
1600	4.93777, 4	1.53823, 5	3.21003, -1
1650	2.61178, 4	8.68134, 4	3.00850, -1
1700	1.43515, 4	5.06500, 4	2.83347, -1
1750	8.16557, 3	3.04633, 4	2.68046, -1
1800	4.79658, 3	1.88403, 4	2.54591, -1
1850	2.90135, 3	1.19548, 4	2.42694, -1
1900	1.80294, 3	7.76722, 3	2.32121, -1
1950	1.14858, 3	5.15798, 3	2.22681, -1
2000	7.48734, 2	3.49525, 3	2.14215, -1
2050	4.98584, 2	2.41337, 3	2.06592, -1
2100	3.38631, 2	1.69568, 3	1.99703, -1
2150	2.34254, 2	1.21091, 3	1.93454, -1
2200	1.64843, 2	8.77910, 2	1.87768, -1
2250	1.17863, 2	6.45547, 2	1.82578, -1
2300	8.55357, 1	4.81003, 2	1.77828, -1
2350	6.29456, 1	3.62864, 2	1.73469, -1
2400	4.69298, 1	2.76940, 2	1.69458, -1
2450	3.54198, 1	2.13681, 2	1.65761, -1
2500	2.70419, 1	1.66571, 2	1.62344, -1

CONSTANT K_p

K_p / T	$\frac{(p/p_o)_{CO_2}(p/p_o)^2_{H_2O}}{(p/p_o)_{CH_4}(p/p_o)^2_{O_2}}$	$\frac{(p/p_o)^2_{N}}{(p/p_o)_{N_2}}$	$\frac{(p/p_o)^2_{O}}{(p/p_o)_{O_2}}$	$\frac{(p/p_o)^2_{H}}{(p/p_o)_{H_2}}$
500	4.23125, 83	1.56479,-93	1.22339,-46	4.57427,-41
550	1.07056, 76	1.62296,-84	7.09101,-42	6.75098,-37
600	5.00591, 69	5.34466,-77	6.67206,-38	2.03473,-33
650	2.20487, 64	1.23771,-70	1.54855,-34	1.80522,-30
700	5.65342, 59	3.55739,-65	1.19558,-31	6.12240,-28
750	5.92998, 55	1.92500,-60	3.82398,-29	9.61482,-26
800	1.95317, 52	2.68091,-56	5.97389,-27	8.07358,-24
850	1.65028, 49	1.22048,-52	5.17043,-25	4.04668,-22
900	3.05654, 46	2.18760,-49	2.73472,-23	1.31897,-20
950	1.09616, 44	1.78807,-46	9.55125,-22	2.99130,-19
1000	6.89277, 41	7.49611,-44	2.34368,-20	4.98297,-18
1050	7.01318, 39	1.77278,-41	4.24858,-19	6.37043,-17
1100	1.08122, 38	2.55664,-39	5.92861,-18	6.47878,-16
1150	2.39237, 36	2.39791,-37	6.58850,-17	5.39945,-15
1200	7.26164, 34	1.54328,-35	5.99865,-16	3.77992,-14
1250	2.91106, 33	7.12989,-34	4.58245,-15	2.26947,-13
1300	1.49257, 32	2.45648,-32	2.99693,-14	1.18937,-12
1350	9.52376, 30	6.52012,-31	1.70709,-13	5.52310,-12
1400	7.38755, 29	1.37092,-29	8.59478,-13	2.30199,-11
1450	6.82713, 28	2.33896,-28	3.87379,-12	8.70770,-11
1500	7.38623, 27	3.30623,-27	1.58033,-11	3.01829,-10
1550	9.21374, 26	3.94298,-26	5.89166,-11	9.66758,-10
1600	1.30764, 26	4.03086,-25	2.02424,-10	2.88246, -9
1650	2.08685, 25	3.58192,-24	6.45700,-10	8.05197, -9
1700	3.70644, 24	2.80112,-23	1.92476, -9	2.11938, -8
1750	7.25956, 23	1.94889,-22	5.39277, -9	5.28297, -8
1800	1.55532, 23	1.21814,-21	1.42748, -8	1.25274, -7
1850	3.61852, 22	6.89958,-21	3.58626, -8	2.83725, -7
1900	9.08299, 21	3.56887,-20	8.58645, -8	6.15953, -7
1950	2.44557, 21	1.69763,-19	1.96646, -7	1.28591, -6
2000	7.02602, 20	7.47274,-19	4.32210, -7	2.58910, -6
2050	2.14368, 20	3.06123,-18	9.14426, -7	5.04080, -6
2100	6.91632, 19	1.17305,-17	1.86733, -6	9.51245, -6
2150	2.35053, 19	4.22427,-17	3.68961, -6	1.74367, -5
2200	8.38493, 18	1.43562,-16	7.06967, -6	3.11075, -5
2250	3.12953, 18	4.62223,-16	1.31632, -5	5.41093, -5
2300	1.21851, 18	1.41485,-15	2.38603, -5	9.19161, -5
2350	4.93603, 17	4.13063,-15	4.21775, -5	1.52713, -4
2400	2.07520, 17	1.15358,-14	7.28207, -5	2.48494, -4
2450	9.03420, 16	3.09016,-14	1.22976, -4	3.96517, -4
2500	4.06408, 16	7.95973,-14	2.03400, -4	6.21180, -4

EQUILIBRIUM

K_p / T	$\frac{(p/p_o)_{CO_2}}{(p/p_o)_{CO}(p/p_o)^{\frac{1}{2}}_{O_2}}$	$\frac{(p/p_o)_{H_2O}}{(p/p_o)_{H_2}(p/p_o)^{\frac{1}{2}}_{O_2}}$	$\frac{(p/p_o)_{CO_2}(p/p_o)_{H_2}}{(p/p_o)_{CO}(p/p_o)_{H_2O}}$
2550	2.08700, 1	1.31108, 2	1.59182, -1
2600	1.62716, 1	1.04138, 2	1.56249, -1
2650	1.28088, 1	8.34306, 1	1.53526, -1
2700	1.01747, 1	6.73851, 1	1.50993, -1
2750	8.15187	5.48447, 1	1.48636, -1
2800	6.58437	4.49635, 1	1.46438, -1
2850	5.35928	3.71171, 1	1.44389, -1
2900	4.39404	3.53102, 1	1.24441, -1
2950	3.62768	2.96611, 1	1.22304, -1
3000	3.01471	2.50645, 1	1.20278, -1
3050	2.52100	2.13005, 1	1.18354, -1
3100	2.12072	1.81996, 1	1.16525, -1
3150	1.79405	1.83365, 1	9.78400, -2
3200	1.52589	1.59062, 1	9.59309, -2
3250	1.30450	1.38638, 1	9.40938, -2
3300	1.12070	1.21389, 1	9.23236, -2
3350	9.67317, -1	1.06750, 1	9.06155, -2
3400	8.38667, -1	9.42690	8.89653, -2
3450	7.30243, -1	8.35814	8.73690, -2
3500	6.38443, -1	7.43907	8.58230, -2
3550	5.60376, -1	6.64551	8.43239, -2
3600	4.93705, -1	5.95766	8.28688, -2
3650	4.36534, -1	5.35921	8.14549, -2
3700	3.87319, -1	4.83668	8.00795, -2
3750	3.44794, -1	4.37886	7.87404, -2
3800	3.07916, -1	3.97642	7.74354, -2
3850	2.75824, -1	3.62152	7.61625, -2
3900	2.47805, -1	3.30760	7.49198, -2
3950	2.23262, -1	3.02911	7.37056, -2
4000	2.01699, -1	2.78134	7.25184, -2
4050	1.82695, -1	2.56031	7.13567, -2
4100	1.65901, -1	2.36261	7.02192, -2
4150	1.51017, -1	2.18533	6.91046, -2
4200	1.37790, -1	2.02598	6.80117, -2
4250	1.26007, -1	1.88240	6.69396, -2
4300	1.15483, -1	1.75274	6.58871, -2
4350	1.06061, -1	1.63539	6.48534, -2
4400	9.76056, -2	1.52897	6.38377, -2
4450	9.00010, -2	1.43225	6.28390, -2
4500	8.31465, -2	1.34418	6.18568, -2
4550	7.69550, -2	1.26383	6.08902, -2

CONSTANT K_p (cont.)

K_p / T	$\frac{(p/p_o)_{CO_2}(p/p_o)^2_{H_2O}}{(p/p_o)_{CH_4}(p/p_o)_{O_2}}$	$\frac{(p/p_o)^2_N}{(p/p_o)_{N_2}}$	$\frac{(p/p_o)^2_O}{(p/p_o)_{O_2}}$	$\frac{(p/p_o)^2_H}{(p/p_o)_{H_2}}$
2550	1.88557, 16	1.97605,-13	3.29894, -4	9.56414, -4
2600	9.00663, 15	4.73810,-13	5.25263, -4	1.44870, -3
2650	4.42189, 15	1.09942,-12	8.21881, -4	2.16077, -3
2700	2.22802, 15	2.47328,-12	1.26499, -3	3.17619, -3
2750	1.15049, 15	5.40329,-12	1.91690, -3	4.60483, -3
2800	6.08045, 14	1.14816,-11	2.86221, -3	6.58943, -3
2850	3.28505, 14	2.37649,-11	4.21433, -3	9.31329, -3
2900	2.37556, 14	4.79498,-11	6.12377, -3	1.30093, -2
2950	1.34946, 14	9.45445,-11	8.78621, -3	1.79703, -2
3000	7.81230, 13	1.82275,-10	1.24562, -2	2.45615, -2
3050	4.60495, 13	3.43983,-10	1.74591, -2	3.32332, -2
3100	2.76137, 13	6.36088,-10	2.42074, -2	4.45369, -2
3150	2.31635, 13	1.15302, -9	3.32248, -2	5.91419, -2
3200	1.44872, 13	2.05298, -9	4.51499, -2	7.78545, -2
3250	9.19806, 12	3.59157, -9	6.07821, -2	1.01638, -1
3300	5.92457, 12	6.17849, -9	8.10969, -2	1.31637, -1
3350	3.86896, 12	1.04594, -8	1.07279, -1	1.69198, -1
3400	2.56009, 12	1.74366, -8	1.40758, -1	2.15902, -1
3450	1.71553, 12	2.86444, -8	1.83243, -1	2.73588, -1
3500	1.16359, 12	4.63994, -8	2.36770, -1	3.44385, -1
3550	7.98450, 11	7.41548, -8	3.03744, -1	4.30742, -1
3600	5.54035, 11	1.16994, -7	3.86988, -1	5.35466, -1
3650	3.88578, 11	1.82312, -7	4.89802, -1	6.61753, -1
3700	2.75353, 11	2.80744, -7	6.16015, -1	8.13226, -1
3750	1.97062, 11	4.27424, -7	7.70052, -1	9.93972, -1
3800	1.42382, 11	6.43657, -7	9.56995, -1	1.20858
3850	1.03822, 11	9.59141, -7	1.18266	1.46220
3900	7.63774, 10	1.41488, -6	1.45366	1.76053
3950	5.66682, 10	2.06695, -6	1.77749	2.10994
4000	4.23918, 10	2.99142, -6	2.16261	2.51742
4050	3.19644, 10	4.29050, -6	2.61851	2.99073
4100	2.42870, 10	6.10053, -6	3.15584	3.53832
4150	1.85905, 10	8.60187, -6	3.78643	4.16948
4200	1.43321, 10	1.20313, -5	4.52346	4.89431
4250	1.11256, 10	1.66976, -5	5.38147	5.72382
4300	8.69441, 9	2.30001, -5	6.37653	6.66991
4350	6.83850, 9	3.14527, -5	7.52631	7.74544
4400	5.41248, 9	4.27115, -5	8.85014	8.96430
4450	4.30985, 9	5.76092, -5	1.03692, 1	1.03414, 1
4500	3.45203, 9	7.71967, -5	1.21064, 1	1.18926, 1
4550	2.78072, 9	1.02792, -4	1.40869, 1	1.36352, 1

K_p / T	$\dfrac{(p/p_o)_{CO_2}}{(p/p_o)_{CO}(p/p_o)^{\frac{1}{2}}_{O_2}}$	$\dfrac{(p/p_o)_{H_2O}}{(p/p_o)_{H_2}(p/p_o)^{\frac{1}{2}}_{O_2}}$	$\dfrac{(p/p_o)_{CO_2}(p/p_o)_{H_2}}{(p/p_o)_{CO}(p/p_o)_{H_2O}}$
4600	7.13511, -2	1.19040	5.99388, -2
4650	6.62688, -2	1.12317	5.90018, -2
4700	6.16508, -2	1.06150	5.80788, -2
4750	5.74469, -2	1.00486	5.71692, -2
4800	5.36131, -2	9.52741, -1	5.62725, -2
4850	5.01106, -2	9.04714, -1	5.53884, -2
4900	4.69054, -2	8.60393, -1	5.45163, -2
4950	4.39675, -2	8.19433, -1	5.36560, -2
5000	4.12702, -2	7.81528, -1	5.28070, -2
5050	3.87900, -2	7.46405, -1	5.19691, -2
5100	3.65060, -2	7.13817, -1	5.11419, -2
5150	3.43996, -2	6.83545, -1	5.03252, -2
5200	3.24541, -2	6.55392, -1	4.95187, -2
5250	3.06550, -2	6.29180, -1	4.87221, -2
5300	2.89888, -2	6.04749, -1	4.79352, -2
5350	2.74437, -2	5.81955, -1	4.71578, -2
5400	2.60092, -2	5.60667, -1	4.63898, -2
5450	2.46756, -2	5.40766, -1	4.56308, -2
5500	2.34344, -2	5.22147, -1	4.48808, -2
5550	2.22778, -2	5.04711, -1	4.41397, -2
5600	2.11988, -2	4.88370, -1	4.34071, -2
5650	2.01910, -2	4.73044, -1	4.26831, -2
5700	1.92488, -2	4.58660, -1	4.19675, -2
5750	1.83670, -2	4.45150, -1	4.12602, -2
5800	1.75408, -2	4.32453, -1	4.05611, -2
5850	1.67659, -2	4.20514, -1	3.98700, -2
5900	1.60384, -2	4.09281, -1	3.91869, -2
5950	1.53549, -2	3.98707, -1	3.85117, -2
6000	1.47120, -2	3.88750, -1	3.78443, -2

CONSTANT K_p (cont.)

K_p / T	$\dfrac{(p/p_o)_{CO_2}(p/p_o)^2_{H_2O}}{(p/p_o)_{CH_4}(p/p_o)^2_{O_2}}$	$\dfrac{(p/p_o)^2_N}{(p/p_o)_{N_2}}$	$\dfrac{(p/p_o)^2_O}{(p/p_o)_{O_2}}$	$\dfrac{(p/p_o)^2_H}{(p/p_o)_{H_2}}$
4600	2.25233, 9	1.36039, -4	1.63377, 1	1.55872, 1
4650	1.83413, 9	1.78977, -4	1.88881, 1	1.77681, 1
4700	1.50135, 9	2.34122, -4	2.17696, 1	2.01985, 1
4750	1.23516, 9	3.04565, -4	2.50162, 1	2.29002, 1
4800	1.02114, 9	3.94079, -4	2.86643, 1	2.58962, 1
4850	8.48234, 8	5.07255, -4	3.27527, 1	2.92108, 1
4900	7.07865, 8	6.49647, -4	3.73230, 1	3.28698, 1
4950	5.93384, 8	8.27952, -4	4.24195, 1	3.69000, 1
5000	4.99595, 8	1.05020, -3	4.80892, 1	4.13297, 1
5050	4.22422, 8	1.32599, -3	5.43823, 1	4.61886, 1
5100	3.58651, 8	1.66675, -3	6.13514, 1	5.15074, 1
5150	3.05736, 8	2.08601, -3	6.90527, 1	5.73187, 1
5200	2.61654, 8	2.59976, -3	7.75451, 1	6.36559, 1
5250	2.24785, 8	3.22684, -3	8.68907, 1	7.05543, 1
5300	1.93833, 8	3.98929, -3	9.71551, 1	7.80500, 1
5350	1.67751, [illegible]	4.91290, -3	1.08407, 2	8.61810, 1
5400	1.45695, 8	6.02770, -3	1.20718, 2	9.49862, 1
5450	1.26977, 8	7.36853, -3	1.34164, 2	1.04506, 2
5500	1.11039, 8	8.97574, -3	1.48823, 2	1.14783, 2
5550	9.74223, 7	1.08959, -2	1.64777, 2	1.25859, 2
5600	8.57514, 7	1.31824, -2	1.82113, 2	1.37779, 2
5650	7.57169, 7	1.58967, -2	2.00918, 2	1.50590, 2
5700	6.70630, 7	1.91091, -2	2.21287, 2	1.64337, 2
5750	5.95774, 7	2.28997, -2	2.43315, 2	1.79069, 2
5800	5.30836, 7	2.73595, -2	2.67101, 2	1.94835, 2
5850	4.74343, 7	3.25920, -2	2.92748, 2	2.11687, 2
5900	4.25062, 7	3.87142, -2	3.20364, 2	2.29676, 2
5950	3.81957, 7	4.58583, -2	3.50059, 2	2.48855, 2
6000	3.44156, 7	5.41733, -2	3.81946, 2	2.69279, 2

ANSWERS

Chapter 1

5.

$T\,^{\circ}R$	941	976	1030	1080	1164
v_1 ft^3/mol	0.75	0.80	0.84	0.92	1.50
v_g ft^3/mol	6.00	4.80	3.75	3.00	1.50

Chapter 2

1. (1) $a = R_{mol}T_c/p_c e^2$; $b = 4R^2_{mol}T^2_c/p_c e^2$.

   (2) $2e^{-2} \sim 0.271$.

   (3) $p_R = \dfrac{T_R}{2v_R - 1} \exp\left[2\left(1 - \dfrac{1}{T_R v_R}\right)\right]$.

2. (1) $\dfrac{pa^2}{R^2_{mol}v^2T^3}$. (2) $-\dfrac{2a}{v^2T^2}$. (3) 0.

3. $R_{mol}\left(1 - \dfrac{2a(v-b)^2}{R_{mol}Tv^3}\right)^{-1}$.

4. $p_R = \dfrac{8T_R}{3v_R - 1} - \dfrac{3}{T_R\, v_R^2}$.

9. 1600°R, 178°R, 333 atm.

10. $p_i = \dfrac{a}{b^2}\left[1 - \sqrt{\dfrac{bR_{mol}T_i}{2a}}\right]\left[3\sqrt{\dfrac{bR_{mol}T_i}{2a}} - 1\right]$; 46.6 atm.

11. 0.357 lb/ft^3; 0.224 Btu/lb°R.

12.

	Ice	Acetic acid	Tin
Change in specific volume cm^3/g	-0.0907	0.159	0.0039
Change in entropy, cal/g/°K	0.292	0.154	0.0278

13. $k = \frac{1}{3}$.

15. 0.495 ft lbf; -1.81 Btu.

16. $B_1 = 1$; $B_2 = b - \frac{a}{R_{mol}T}$; $T_c = \frac{8a}{27bR_{mol}}$; $v_c = 3b$.

18. (a) 3.05×10^{-7} m^3°K/J. (b) 285°K.

21. $6T_c$, $0.5T_c$.

22. 439 kJ

23. $M = 4aT$ + constant, $N = \frac{1}{3}(T-T_o)^3 - 4ap_oT$ + constant

Chapter 3

1. (a) CH_4 4.02 lbf/in², 0.0455 lb,
N_2 36.33 lbf/in², 0.717 lb,
O_2 9.65 lbf/in², 0.218 lb.

(b) CH_4 4.02 lbf/in²,
N_2 36.37 lbf/in²,
O_2 1.608 lbf/in².

2. 8.48:1. 37.4 per cent; 144°F. 4.08 per cent.

3. (a) -0.0367 Btu/°R.
(b) 28 Btu.
(c) 382 Btu.

4. (a) $C_v = 7.71 + 0.000806\ T$. CHU/lb mol/°C.
(b) -1.27 CHU/°C.

5. 1.895, 1038.4 CHU/lb mol.

6. 111.5 Btu/lb.

7. $C_v = 0.175 + 0.0001\ T$ Volume ratio = 17.
$C_p = 0.24 + 0.0001\ T$.

8. 13.58 Btu/lb mol/°R.

9. $\frac{\gamma}{\gamma-1} R_{mol} \ln \frac{32}{37}$.

10. 2,198,000 Btu/lb mol; 247 atm abs; 4870°R.

11. $3002^\circ F$ rise; 23.8 atm abs; $610.4^\circ R$; 3.92 atm abs; 20,711 Btu/lb mol.

12. -756.6 kcal/g mol; 52.2 atm abs; $2586^\circ K$.

13. $Q_p$higher = $Q_v$higher = -385,000 Btu/lb mol.

$Q_p$lower = $Q_v$lower = -347,000 Btu/lb mol. $2.365^\circ F$ rise.

14. (a) Q_v = 233,000 Btu/lb mol; Q_p = 233,348 Btu/lb mol.

(b) Q_v = 232,140 Btu/lb mol; Q_p = 232,885 Btu/lb mol.

(c) 456 Btu/ft^3; 212 Btu/ft^3.

15. $5240^\circ R$; 1220 p.s.i.a.

16. $5370^\circ R$; 485 p.s.i.a.

17. $C_p \ln \left[\frac{1}{n}\left(\frac{M+n}{M+1}\right)^{M+1}\right]$, (a) $-R_{mol} \ln \left[\frac{1}{M}\left(\frac{M}{M+1}\right)^{M+1}\right]$; (b) 0.

18. 0.956

19. (a) $600^\circ K$, (b) -1874 kJ/kg mol, (c) -4.46 kJ/kg mol$^\circ$K.

20. $C_v = 19.7 \times 10^3 + 4T$ J/kg mol$^\circ$K.

(a) 1.14, (b) 2.49 kJ/kg mol$^\circ$K.

21. 1.444

22. (a) $28.7 \times 10^3 + 6.04T$ J/kg mol$^\circ$K, (b) 28.7,
(c) $902^\circ K$, (d) 1.535 Atm.

23. 1.766, (a) unaffected, (b) increased.

Chapter 4

1. 0.214.

2. (a) H_2 = 2.34 per cent.
O_2 = 1.17 per cent.
H_2O = 96.39 per cent.
pressure = 5.06 atm.

(b) H_2 = 14.2 per cent.
CO_2 = 14.2 per cent.
CO = 35.4 per cent.
H_2O = 35.4 per cent.
pressure = 7.5 atm.

3. H_2 = 19.7 per cent. CO_2 = 19.7 per cent.
CO = 13.7 per cent. H_2O = 47.0 per cent.

5. 85.6 ppm.

6. CO_2 = 30.8 per cent. CO = 3.9 per cent. H_2 = 0.9 per cent.
H_2O = 51.1 per cent. O_2 = 13.3 per cent.

7. H_2 = 5.72 per cent. CO = 27.62 per cent.
CO_2 = 39.04 per cent. H_2O = 27.62 per cent.
Final pressure = 2.5 atm.

8. (a) 3540°R.

(b) 4.681.

CO = 34.25 per cent. CO_2 = 15.75 per cent.
H_2 = 15.75 per cent. H_2O = 34.25 per cent.

9. CO_2 = 92.3 per cent. CO = 5.1 per cent.
O_2 = 2.6 per cent. Temperature = 4390°R.

10. Yes.

11. $pV = (1+\alpha)R_{mol}T.$

$$K_p = \left(\frac{4\alpha^2}{1-\alpha^2}\right)p$$

$$\mu C_p = 2R_{mol}T^2(4p+K_p)^{3/2}K_p^{-1/2}\frac{dK_p}{dT}.$$

12. $$\left(\frac{\partial\alpha}{\partial T}\right)_p = \left(\frac{(1-\alpha^2)^2}{8p}\right)\frac{dK_p}{dT}$$

$\alpha = 0.447.$

$$\left(\frac{\partial\alpha}{\partial T}\right)_p = 7.33 \times 10^{-4}\ {}^{\circ}C^{-1}.$$

13. 5460°R.

14. 3850°F.

15. 5306°R. 36.2 atm.

17. $$\left(\frac{5}{2}R_{mol} - \frac{\Delta H_o}{T}\right)(\alpha_2-\alpha_1) + 2R_{mol}\left[\ln\frac{\alpha_2}{\alpha_1} - \ln\frac{1-\alpha_2}{1-\alpha_1}\right].$$

18. $$C_p = \frac{5}{2}R_{mol}(1+\alpha) + \frac{R_{mol}\alpha}{2}(1-\alpha^2)\left(\frac{5}{2}+\frac{\theta_1}{T}\right)^2.$$

$$C_v = \frac{3}{2}R_{mol}(1+\alpha) + R_{mol}\alpha\left(\frac{1-\alpha}{2-\alpha}\right)\left(\frac{3}{2}+\frac{\theta_1}{T}\right)^2.$$

19. $\alpha = 0.598$; 37.7 ft^3.

20. $770^\circ K$, 33.4%.

21. 0.138, 3.03 $atm^{-\frac{1}{2}}$.

22. 26.2 bar (a) pressure increased, (b) pressure greater than (a).

23. 0.847.

24. 12.93%.

25. 0.176%.

26. 0.264, 0.159, Kinetic energy with frozen flow is less by 1.05×10^5 kJ/kg.mol.

27. 0.932 atm., 0.564.

28. 367 ppm, $985^\circ K$.

29. 0.0368% by volume, $973^\circ K$.

Chapter 5

5. F increases 10 per cent. G remains the same.

7. 0.0297 volts.

8. 1.185 volts.
   1.127 volts.
   1.151 volts.

17. 0.536 volts, 1.43×10^{-4} volt/$^\circ K$, 153 kJ/kg.

18. 1.233 volts, -53.8 MJ/kg of H_2.

19. 80.8%, 1.197 volt, 0.048 kg/h, 0.908 kW.

20. $350^\circ K$ to $274^\circ K$.

21. 45.9 gm/h, 115.7 gm/h.

22. $a = 5.67 \times 10^{-6}$, $b = 0.755 \times 10^{-8}$, 642.5 μV, 1463 μV.

AUTHOR INDEX

SUBJECT INDEX